# Bluetooth Demystified

Nathan J. Muller

**McGraw-Hill**
New York  San Francisco  Washington, D.C.  Auckland  Bogotá
Caracas  Lisbon  London  Madrid  Mexico City  Milan
Montreal  New Delhi  San Juan  Singapore
Sydney  Tokyo  Toronto

**Library of Congress Cataloging-in-Publication Data**

Muller, Nathan J.
    Bluetooth demystified / Nathan J. Muller
      p.   cm.
    Includes index.
    ISBN 0-07-136323-8
    1. Bluetooth technology.   2. Telecommunication—Equipment and supplies.
3. Computer network protocols.   I. Title.

TK5103.3 .M85    2000
004.6′2—dc21

                                                                                            00-063829

# McGraw-Hill
*A Division of The McGraw-Hill Companies*

Copyright © 2001 by The McGraw-Hill Companies, Inc. All rights reserved. Printed in the United States of America. Except as permitted under the United States Copyright Act of 1976, no part of this publication may be reproduced or distributed in any form or by any means, or stored in a data base or retrieval system, without the prior written permission of the publisher.

2 3 4 5 6 7 8 9 0   DOC/DOC   0 6 5 4 3 2 1 0

ISBN 0-07-136323-8

*The sponsoring editor for this book was Steven S. Chapman and the production supervisor was Pamela A. Pelton. It was set in Vendome by Patricia Wallenburg.*

*Printed and bound by R. R. Donnelley & Sons Company.*

 This book is printed on recycled, acid-free paper containing a minimum of 50% recycled, de-inked fiber.

Bluetooth is a trademark owned by Telefonaktiebolaget L M Ericsson, Sweden.

McGraw-Hill books are available at special quantity discounts to use as premiums and sales promotions, or for use in corporate training programs. For more information, please write to the Director of Special Sales, McGraw-Hill, Two Penn Plaza, New York, NY 10121-2298. Or contact your local bookstore.

---

Information contained in this work has been obtained by The McGraw-Hill Companies, Inc. ("McGraw-Hill") from sources believed to be reliable. However, neither McGraw-Hill nor its authors guarantee the accuracy or completeness of any information published herein and neither McGraw-Hill nor its authors shall be responsible for any errors, omissions, or damages arising out of use of this information. This work is published with the understanding that McGraw-Hill and its authors are supplying information but are not attempting to render engineering or other professional services. If such services are required, the assistance of an appropriate professional should be sought.

*To a new friend, who seems like an old friend...*
                                                    Genie Williams

# CONTENTS

| | |
|---|---|
| **Preface** | xv |
| **1 The Case for Bluetooth** | 1 |
| What About Infrared? | 2 |
| Infrared and Bluetooth? | 3 |
| Speed Differential | 4 |
| Wireless to Wireline | 4 |
| Dialup to the Internet | 5 |
| How About Wireless LANs? | 6 |
| HomeRF Networks | 8 |
| Bluetooth Advantage | 13 |
| Origin of Bluetooth | 14 |
|     What's With the Name? | 15 |
| Bluetooth Technology | 16 |
|     Types of Links | 17 |
|     Ad Hoc Networking | 17 |
|     Voice over Bluetooth | 18 |
|     Video over Bluetooth | 19 |
|     Radio Link | 21 |
|     Interference | 21 |
|     Safety | 22 |
| Personal Area Networks | 22 |
| Bluetooth Topology | 23 |
| Security | 25 |
| What Can You Do With Bluetooth? | 26 |
|     Presentations | 26 |
|     Card Scanning | 27 |
|     Collaboration | 27 |
|     Synchronizing Data | 27 |
|     Remote Synchronization | 28 |

|  |  |
|---|---|
| Printing | 28 |
| In-Car Systems | 29 |
| Communicator Platforms | 29 |
| Electronic Books | 30 |
| Travel | 31 |
| Home Entertainment | 32 |
| Payment Systems | 32 |
| Scanners | 33 |
| Behavior Enforcement | 34 |
| Mobile E-commerce | 34 |
| Java and Bluetooth | 37 |
| Jini and Bluetooth | 38 |
| Other Connectivity Solutions | 40 |
| JetSend | 40 |
| HAVi | 41 |
| Global 3G Wireless Framework | 42 |
| Problems with Bluetooth | 44 |
| Bluetooth Qualification Program | 45 |
| Market for Bluetooth | 45 |
| Summary | 46 |

## 2 Basic Concepts — 49

| | |
|---|---|
| Serial versus Parallel | 50 |
| Serial Transmission | 51 |
| Parallel Transmission | 51 |
| Asynchronous versus Synchronous | 53 |
| Asynchronous | 54 |
| Synchronous | 57 |
| Spread Spectrum | 60 |
| Spreading | 62 |
| Direct Sequence | 64 |
| Frequency Hopping | 65 |
| Circuit and Packet Switching | 66 |
| Time Division Duplexing | 68 |
| Physical Links | 74 |
| SCO Links | 74 |
| ACL Links | 75 |
| Peeking into Packets | 75 |

## Contents

| | |
|---|---|
| Bluetooth Packets | 77 |
|     Access Code | 77 |
|     Header | 79 |
|     Payload | 81 |
| Logical Channels | 82 |
| Client-Server Architecture | 83 |
|     Architectural Model | 83 |
| Service Discovery | 86 |
| Summary | 88 |

### 3 Bluetooth Protocol Architecture — 89

| | |
|---|---|
| What Are Protocols? | 90 |
| Open Systems Interconnection | 91 |
|     Application Layer | 92 |
|     Presentation Layer | 94 |
|     Session Layer | 95 |
|     Transport Layer | 96 |
|     Network Layer | 97 |
|     Data-Link Layer | 98 |
|     Physical Layer | 99 |
| Bluetooth Protocol Stack | 101 |
| Bluetooth Core Protocols | 103 |
|     Baseband | 104 |
|     Link Manager Protocol (LMP) | 104 |
|     Logical Link Control and Adaptation Protocol | 105 |
|     Service Discovery Protocol (SDP) | 105 |
| Cable Replacement Protocols | 105 |
|     RFCOMM | 105 |
|     Telephony Control Protocols | 107 |
| Adopted Protocols | 107 |
|     PPP | 107 |
|     TCP/UDP/IP | 108 |
|     OBEX Protocol | 110 |
|     Wireless Application Protocol (WAP) | 111 |
|     WAP Applications Environment (WAE) | 113 |
|     Content Formats | 116 |
| Usage Models and Profiles | 118 |
| Summary | 120 |

## 4 Link Management — 121

Types of PDUs — 123
General Response Messages — 127
Authentication — 127
Pairing — 128
Changing the Link Key — 129
Changing the Current Link Key — 130
    Changing a Temporary Link Key — 130
Encryption — 131
Clock Offset Request — 133
Slot Offset Information — 133
Timing Accuracy Information Request — 134
LMP Version — 134
Supported Features — 136
Switching of Master-Slave Role — 136
Name Request — 137
Detach — 137
Hold Mode — 137
Sniff Mode — 138
Park Mode — 139
Power Control — 140
Channel Quality-Driven Change of Data Rate — 141
Quality of Service (QoS) — 142
SCO Links — 143
Control of Multi-Slot Packets — 144
Paging Scheme — 145
Link Supervision — 146
Connection Establishment — 146
Test Modes — 147
Error Handling — 147
Summary — 148

## 5 Logical Link Control — 149

L2CAP Functions — 151
Basic Operation — 152
    Channel Identifiers — 153
    Segmentation and Reassembly — 154

|  |  |
|---|---|
| State Machine | 155 |
|     Events | 157 |
|     Actions | 161 |
|     Channel Operational States | 163 |
|     Mapping Events to Actions | 164 |
| Data Packet Format | 168 |
|     Connection-Oriented Channel | 168 |
|     Connectionless Data Channel | 169 |
| Signaling | 170 |
|     Packet Structure | 170 |
|     Signaling Commands | 171 |
| Configuration Parameter Options | 173 |
|     Packet Structure | 173 |
|     Options | 174 |
|     Configuration Process | 175 |
| Service Primitives | 176 |
|     Event Indication | 176 |
|     Connect | 176 |
|     Connect Response | 177 |
|     Configure | 177 |
|     Configuration Response | 177 |
|     Disconnect | 177 |
|     Write | 177 |
|     Read | 178 |
|     Group Create | 178 |
|     Group Close | 178 |
|     Group Add Member | 178 |
|     Group Remove Member | 178 |
|     Get Group Membership | 179 |
|     Ping | 179 |
|     Get Info | 179 |
|     Disable Connectionless Traffic | 179 |
|     Enable Connectionless Traffic | 179 |
| Summary | 179 |

## 6 Bluetooth General Profiles      181

|  |  |
|---|---|
| Generic Access Profile | 182 |
|     Common Parameters | 184 |

|  |  |
|---|---|
| Idle Mode Procedures | 186 |
| Bonding | 187 |
| Establishment Procedures | 188 |
| Serial Port Profile | 190 |
| Application-Level Procedures | 191 |
| Power Mode and Link Loss Handling | 193 |
| RS-232 Control Signals | 193 |
| L2CAP Interoperability Requirements | 194 |
| SDP Interoperability Requirements | 195 |
| Link Manager Interoperability Requirements | 195 |
| Service Discovery Application Profile | 196 |
| Client and Server Roles | 197 |
| Pairing | 199 |
| Service Discovery Application | 200 |
| Message Sequence | 202 |
| Service Discovery | 202 |
| Signaling | 203 |
| Configuration Options | 204 |
| SDP Transactions and L2CAP Connections | 204 |
| Link Manager | 206 |
| Link Control | 208 |
| Generic Object Exchange Profile (GOEP) | 210 |
| Profile Stack | 211 |
| Server and Client | 211 |
| Profile Basics | 212 |
| Features | 213 |
| OBEX Operations | 213 |
| Summary | 214 |

# 7 Bluetooth Profiles for Usage Models — 217

|  |  |
|---|---|
| Intercom Profile | 218 |
| Call Procedures | 221 |
| Message Summary | 223 |
| Call Failure | 223 |
| Cordless Telephony Profile | 225 |
| Device Roles | 226 |
| Typical Call Scenarios | 227 |
| Features | 229 |

## Contents

| | |
|---|---|
| Terminal-to-Gateway Connection | 230 |
| Terminal-to-Terminal Connection | 231 |
| Call Control | 232 |
| Group Management | 234 |
| Periodic Key Update | 235 |
| Inter-Piconet Capability | 236 |
| Service Discovery Procedures | 236 |
| LMP Procedures | 237 |
| Link Control Features | 238 |
| GAP Compliance | 239 |
| Headset Profile | 241 |
|    Profile Restrictions | 243 |
|    Basic Operation | 243 |
|    Features | 244 |
|    Link Control Features | 246 |
|    GAP Compliance | 247 |
| Dialup Networking Profile | 248 |
|    Profile Restrictions | 250 |
|    Basic Operation | 250 |
|    Services | 251 |
|    Gateway Commands | 251 |
|    Audio Feedback | 253 |
|    Service Discovery Procedures | 254 |
|    Link Control Features | 254 |
|    GAP Compliance | 254 |
| Fax Profile | 256 |
|    Profile Restrictions | 257 |
|    Basic Operation | 258 |
|    Services | 259 |
|    Gateway Commands | 259 |
|    Audio Feedback | 260 |
|    Service Discovery Procedures | 260 |
|    Link Control Features | 260 |
|    GAP Compliance | 260 |
| LAN Access Profile | 261 |
|    Profile Restrictions | 263 |
|    Basic Operation | 264 |
|    Security | 265 |

|  |  |
|---|---|
| GAP Compliance | 265 |
| Service Discovery Procedures | 266 |
| Link Control | 267 |
| Management Entity Procedures | 267 |
| File Transfer Profile | 268 |
|     Basic Operation | 270 |
|     Functions | 270 |
|     Features | 271 |
|     OBEX Operations | 272 |
|     Service Discovery Procedures | 273 |
| Object Push Profile | 273 |
|     Functions | 275 |
|     Basic Operation | 276 |
|     Features | 277 |
|     Content Formats | 277 |
|     OBEX Operations | 278 |
|     Service Discovery Procedures | 278 |
| Synchronization Profile | 279 |
|     Basic Operation | 281 |
|     Features | 283 |
|     OBEX Operations | 284 |
|     Service Discovery Procedures | 284 |
| Summary | 286 |

## 8 Bluetooth Security — 289

|  |  |
|---|---|
| Security Modes | 290 |
| Link-level Security | 291 |
| A Matter of Trust | 292 |
| Flexible Access | 293 |
| Implementation | 293 |
| Architecture Overview | 294 |
| Security Level of Services | 296 |
| Connection Setup | 296 |
|     Authentication on Baseband Link Setup | 297 |
|     Protocol Stack Handling | 298 |
|     Registration Procedures | 299 |
|     External Key Management | 301 |
|     Access Control Procedures | 301 |

## Contents

|   |   |
|---|---:|
| Connectionless L2CAP | 301 |
| Security Manager | 301 |
| Interface to L2CAP | 305 |
| Interface to Other Multiplexing Protocols | 306 |
| Interface to ESCE | 306 |
| Registration Procedures | 306 |
| Interface to HCI/Link Manager | 307 |
| Summary | 308 |
| **9 Bluetooth in the Global Scheme of 3G Wireless** | **309** |
| The IMT-2000 Vision | 311 |
| Spanning the Generations | 311 |
| Current 2G Networks | 314 |
|     Time Division Multiple Access | 314 |
|     Code Division Multiple Access | 316 |
|     CDMA versus TDMA | 317 |
|     GSM | 318 |
| Global 3G Initiative | 326 |
|     Standards Development | 326 |
|     Goals of IMT-2000 | 328 |
|     Universal Mobile Telecommunications System | 329 |
| U.S. Participation in 3G | 334 |
|     CDMA Proposals | 334 |
|     TDMA Proposal | 338 |
| Role of Bluetooth | 339 |
| Summary | 341 |
| **Appendix A** | **343** |
| Contributors to the Bluetooth Specification | 343 |
| **Appendix B** | **347** |
| Terms and Definitions | 347 |
| **Appendix C** | **371** |
| Acronyms | 371 |
| **Index** | **381** |

# PREFACE

Bluetooth™ wireless technology has become a global technology specification for "always on" wireless communication between portable devices and desktop machines and peripherals. Among the many things Bluetooth wireless technology enables users to do is swap data and synchronize files without having to cable devices together. Data exchange can be as simple as transferring business card and calendar information from a mobile phone to a hand-held computer, or synchronizing personal information between a palmtop and desktop computer, merely by having the devices come within range of each other. Even photos from a digital camera can be dropped off at a PC for editing or to a color printer for output—all without having to connect cables, load files, and open applications. And since the wireless link has a range of 30 feet (10 meters), users have more mobility than ever before. Without cables, the work environment also looks and feels more comfortable and inviting.

Bluetooth technology can also be used to make wireless data connections to conventional local area networks (LANs) via an access point equipped with a Bluetooth radio transceiver that is wired to the LAN. Once a wireless connection is established with one of these access points, a mobile device can access any of the resources on that LAN, including printers, database servers, and the Internet. If desired, a user might tap out an e-mail reply on a palmtop, tell it to make an Internet connection through a mobile phone, print a copy on a printer nearby, and archive the original on the desktop PC—all while walking down the hall to a meeting.

The Bluetooth baseband protocol is a combination of circuit- and packet-switching, making it suitable for voice as well as data. For example, instead of fumbling with a cell phone while driving, the user can wear a lightweight headset to answer a call and engage in a conversation without even taking the phone out of a briefcase or purse. Or the cell phone can be used to communicate through a base station in the home or office as if it were a cordless phone connected to the Public Switched Telephone Network (PSTN). Users can also connect to each other directly

---

Bluetooth is a trademark owned by Telefonaktiebolaget L M Ericsson, Sweden.

over a limited range through their telephones using Bluetooth wireless technology, without incurring usage charges from a service provider.

No matter what the application, data or voice, Bluetooth wireless technology is intended to replace cable connections between computers, peripherals, and other electronic devices. With Bluetooth technology, making connections is as easy as powering up the device. There is no need for the user to open an application or press a button to initiate a process. In fact, one of the main advantages of Bluetooth wireless technology is that it does not need to be set up—it is always on, running in the background. The Bluetooth protocols scan for other Bluetooth devices and when they come within range of each other, start to exchange messages so they can become aware of each other's capabilities, establish connections and, if needed, arrange for security to protect sensitive data during transmission. The devices do not even require a line of sight to communicate with each other.

Bluetooth technology can be used for a variety of purposes, eliminating the need for multiple types of cable connections. With a radio link, users can think about what they are working on, rather than how to cable everything together to achieve connectivity between the various devices. Within a few years, about 80 percent of new mobile phones are expected to carry a Bluetooth chip that can provide a wireless connection to similarly equipped notebook computers, printers and, potentially, any other digital device.

The Bluetooth radio transceivers operate in the globally available unlicensed ISM radio band of 2.4 GHz. The ISM (industrial, scientific, and medical) bands include the frequency ranges at 902 MHz to 928 MHz and 2.4 GHz to 2.484 GHz, which do not require an operator's license from a regulatory agency, such as the Federal Communications Commission (FCC) in the U.S. The use of a generally available frequency band means that devices using Bluetooth wireless technology can operate virtually anywhere in the world and link up with other such devices for ad hoc networking when they come within range of each other.

There will be points of convergence between Bluetooth and other wireless technologies. Infrared and Bluetooth technologies, for example, provide complementary implementations for data exchange and voice applications. Bluetooth complements infrared's point-and-shoot ease of use with omni-directional signaling, longer distance communications, and capacity to penetrate through walls. For some devices, having both Bluetooth and infrared will provide the optimal short-range wireless solution. For other devices, the choice of adding Bluetooth or infrared wireless technology will be based on the applications and intended usage models.

# Preface

As the number and types of computer and communications devices continue to proliferate, establishing connectivity between them becomes the critical issue. What everyone needs is an economical wireless solution that is convenient, reliable, easy to use, and operates over a longer distance than infrared without requiring a clear line of sight. Of the many emerging wireless solutions that attempt to address one or more of these needs, there is only one promising enough to elicit the support of a broad base of vendors representing all segments of the computer and communications markets—Bluetooth.

Since its initial development in 1994 by the Swedish telecommunications firm, Ericsson, over 2000 companies worldwide have signed on as members of the Bluetooth Special Interest Group (SIG) to build products to the wireless specification and promote the new technology in the marketplace. The sheer number and diversity of companies involved in the Bluetooth SIG is indicative of where the computer and communications industries are headed: Bluetooth wireless technology enables seamless voice and data transmission via short-range radio, allowing users to connect a wide range of devices easily and quickly, without the need for cables, thus expanding communications capabilities for mobile phones and handheld computing devices.

Bluetooth wireless technology has been greeted with unparalleled enthusiasm throughout the computer and communications industries. It is widely anticipated that soon people all over the world will enjoy the convenience, speed and security of instant wireless connections. To meet these expectations, Bluetooth wireless technology is expected to be embedded in hundreds of millions of mobile phones, PCs, laptops and a wide range of other electronic devices in the next few years. This book explains the advantages that Bluetooth wireless technology provides to users for a variety of applications and discusses the key operational details of the technology and its relationship to the emerging third-generation global wireless infrastructure.

The information contained in this book, especially as it relates to specific vendors and products, is believed to be accurate at the time it was written and is, of course, subject to change with continued advancements in technology and shifts in market forces. Mention of specific products and services is for illustration purposes only and does not constitute an endorsement of any kind by either the author or the publisher.

*Nathan J. Muller*

CHAPTER 1

# The Case for Bluetooth

Despite our best efforts at hiding them, cables for our computers and peripherals run rampant in our homes and offices. At the least, there are cables connecting the computer to a printer, scanner, palm-size device, and, perhaps, a business-card scanner. If we have a multimedia computer, there are more cables for microphone, speakers, and subwoofer. There may also be another cable for a video camera for conferencing over the Internet. If we prefer an external modem and ZIP drive, connecting these devices adds to the tangle of cables hanging off our desks. If we want to move data quickly and reliably between a palmtop, laptop, and desktop, this may entail even more cables. When we travel, our briefcases are stuffed with all kinds of devices that must be set up and cabled for use at various locations. Let's face it, in an increasingly plugged-in society, plugging in things has become a big hassle.

## What About Infrared?

Oh, sure—some devices can communicate through optical connections like infrared. This method of communication uses 850-nanometer (nm) infrared light between devices for voice as well as data. But this type of signal must have a clear, straight path from one device to another. Even with a line of sight, the devices must be positioned close to each other because the connection only works over very short distances of three feet (1 meter) or less. Whereas infrared is intended for point-to-point links between two devices for simple data transfers and file synchronization, Bluetooth™ wireless technology was designed from the start to support both data and multiple voice channels over a range of 30 feet (10 meters).

With both infrared and Bluetooth products, data exchange is considered to be a fundamental function. Data exchange can be as simple as transferring business card information from a mobile phone to a palmtop, or as sophisticated as synchronizing personal information between your palmtop and desktop PC. In fact, both technologies can support many of the same applications.

---

Bluetooth is a trademark owned by Telefonaktiebolaget L M Ericsson, Sweden.

## Infrared *and* Bluetooth?

This begs the question: if infrared and Bluetooth devices can support many of the same applications, why do we need both technologies? The answer lies in the fact that each technology has its advantages and disadvantages. Fortunately, the very scenarios that leave infrared devices falling short are the ones where Bluetooth devices excels and vice versa.

Take the electronic exchange of business card information between two devices. This application usually will take place in a conference room or exhibit floor where a number of other devices may be attempting to do the same thing. This is the situation where infrared excels. The short range and narrow angle of infrared—30 degrees or less—allows each user to aim his/her device at the intended recipient with point-and-shoot ease. Close proximity to another person is natural in a business card exchange situation, as is pointing one device at another. The limited range and angle of infrared allows other users to perform a similar activity with ample security and no interference.

In the same situation, a Bluetooth device would not perform as well as an infrared device. With its omnidirectional capability, the Bluetooth device must first discover the intended recipient. The user cannot simply point at the intended recipient—a Bluetooth device must perform a discovery operation that will probably reveal several other Bluetooth devices within range. Close proximity offers no advantage here. The user will be forced to select from a list of discovered devices and apply a security mechanism to prevent unauthorized access. All this makes the use of Bluetooth devices for business card exchange an awkward and needlessly time-consuming process.

However, in other data exchange situations Bluetooth products might be the preferred choice. The ability of Bluetooth wireless technology to penetrate solid objects and its capability to communicate with other devices in a "piconet" allows for data exchange opportunities that are very difficult or impossible with infrared.

For example, using Bluetooth wireless technology you could synchronize your mobile phone with a notebook computer without taking the phone out of your jacket pocket or purse. This would allow you to type a new address at the computer and move it to your mobile phone's directory without unpacking the phone and setting up a cable connection between the two devices. The omnidirectional capability of Bluetooth allows synchronization to occur instantly, assuming that the phone and computer are within 30 feet of each other.

Using products with Bluetooth wireless technology for synchronization does not require that the phone remain in a fixed location. If you carry the phone in your pocket, the synchronization can occur while you move around. This is not possible with infrared because the signal is not able to penetrate solid objects—even jacket pockets or leather purses—and the devices must be within a few feet of each other. Furthermore, the use of infrared requires that both devices remain stationary while the synchronization occurs.

## Speed Differential

When it comes to data transfers, infrared does offer a big speed advantage over the Bluetooth specification. While Bluetooth wireless technology moves data between devices at only 721 Kbps, infrared offers 4 Mbps of data throughput. A higher-speed version of infrared is now available that can transmit data between devices at up to 16 Mbps—a quadruple improvement over the previous version. The higher speed is achieved with the Very Fast Infrared (VFIR) protocol, which is designed to address the new demands of transferring large image files between digital cameras, scanners, and PCs. Even when the Bluetooth specification is enhanced for higher data rates in the future, infrared is likely to maintain its speed advantage for many years to come.

## Wireless to Wireline

An important feature of both Bluetooth and infrared technologies is their ability to forge wireless connections between portable devices and a wired network. Because there is no line-of-sight requirement for Bluetooth devices, however, it is better suited for this type of application because with it users have more flexibility in placing a LAN access point within the premises than users of infrared possess, since infrared requires close proximity as well as line of sight. An access point is a transceiver that accepts wireless signals from other devices and provides a wired connection to the LAN (Figure 1.1).

Also, once an infrared device is connected to the LAN, it must remain relatively stationary for the duration of the data session. With

# The Case for Bluetooth

Bluetooth wireless technology, the portable device can be in motion while connected to the LAN access point as long as the user stays within the 30-foot range.

**Figure 1.1**
An access point provides the bridge between wireless devices and a Local Area Network (LAN).

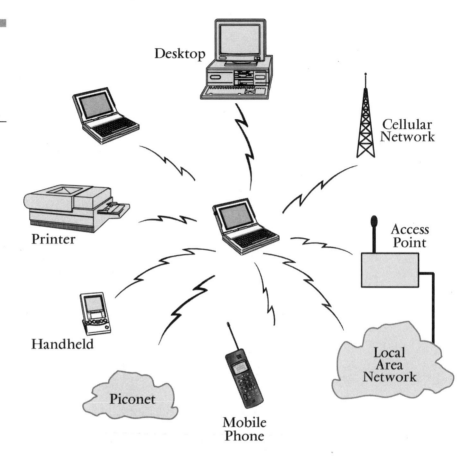

## Dialup to the Internet

Emulating an EIA/TIA 232 connection between a portable computer and a mobile phone for establishing a dialup connection to the Internet is another application targeted by both infrared and Bluetooth wireless technologies. While an infrared connection works well for this type of application, the primary advantage of a Bluetooth connection is that the user can leave the mobile phone clipped to his or her belt or in a pocket and walk around the room during the connection. Bluetooth technology does not require that the phone be positioned near any other device, as infrared does. Table 1.1 summarizes the performance characteristics of infrared products.

**TABLE 1.1**

Performance Characteristics of Infrared

| Feature/Function | Performance |
|---|---|
| Connection Type | Infrared, narrow beam (30 degree angle or less) |
| Spectrum | Optical, 850 nanometers (nm) |
| Transmission Power | 100 milliwatts (mW) |
| Data Rate | Up to 16 Mbps using Very Fast Infrared (VFIR) |
| Range | Up to 3 feet (1 meter) |
| Supported Devices | Two (2) |
| Voice Channels | One (1) |
| Data Security | The short range and narrow angle of the infrared beam provides a simple form of security; otherwise, there are no security capabilities at the link level. |
| Addressing | Each device has a 32-bit physical ID that is used to establish a connection with another device. |

## How About Wireless LANs?

Another wireless connectivity option is the local area network (LAN), as described by the 802.11 standard issued by the Institute of Electrical and Electronic Engineers (IEEE). Wirelss LANs using the 802.11 or Bluetooth specification are intended for completely different applications. Bluetooth devices require little power and arc intended for transmitting small amounts of data at almost 1 Mbps over short distances of up to 30 feet, whereas 802.11 connections can range from 1 Mbps or 2 Mbps and 11 Mbps over distances of several hundred feet, making them suited for corporate offices and campuses where it may not be practical or economical to install cable, or where configuration flexibility is required.

In a typical wireless LAN configuration, one or more access points connect to an Ethernet hub, making the connection to the wired network. The access points are essentially bridges equipped with transceivers that provide the interface between the wired and wireless net-

# The Case for Bluetooth

works (Figure 1.2). At a minimum, the access point receives, buffers, and transmits data between the wireless LAN and the wired network infrastructure. A single access point can support a small group of users who connect to it through wireless LAN adapters in their PCs or notebook computers, which include a built-in antenna (Figure 1.3).

**Figure 1.2**
A wireless LAN access point, the Cisco Aeronet 340.

**Figure 1.3**
A wireless LAN adapter for a notebook computer from Cisco Systems.

Wireless LANs have been around for years, but market acceptance has been slow. Growth has been hampered by the lack of interoperability, which has only recently been addressed with the IEEE 802.11 standard. Even with the interoperability problem solved, the components required to implement a wireless LAN are still expensive for most people—$200 for an adapter card and close to $1,000 for an access point.

Wireless LANs can provide a data rate of up to 11 Mbps using direct sequence spread spectrum and 1 or 2 Mbps using frequency-hopping spread spectrum. With direct sequence spread spectrum, the base digital bit stream is modulated with a higher-rate chipping code to produce a very high bit rate data stream that, when transmitted, is spread across a broad portion of the frequency spectrum. With frequency hopping, the bandwidth is divided into 1 MHz channels. The FCC requires that the transmitter visit at least 79 of the channels at least once every 30 seconds, which produces a minimum rate of 2.5 hops per second. The hop sequence itself is a pseudorandom pattern, so that to conventional radios the frequency-hopping transmission appears to be nothing more than low-level background noise.

Although direct sequence offers the higher data rate, frequency-hopping spread spectrum is more resistant to interference and is preferable in environments with electromechanical noise and more stringent security requirements. In addition, direct sequence uses more power than frequency-hopping spread spectrum and is also more expensive to implement.

While useful in minimizing the need for cables, wireless LANs are not intended for interconnecting the range of mobile devices we carry around every day between home and office. For this, Bluetooth wireless technology is needed. Table 1.2 summarizes the performance characteristics of 802.11 wireless LANs.

## HomeRF Networks

Another wireless technology that shares the unlicensed 2.4-GHz ISM band with the Bluetooth specification is called Home Radio Frequency (HomeRF); it is supported by more than 100 member companies belonging to the HomeRF Consortium. Many of these also belong to the Bluetooth Special Interest Group (SIG). HomeRF provides the foundation for a broad range of interoperable consumer devices by establishing an open industry specification for wireless digital communication between PCs and consumer electronic devices anywhere in and around the home. HomeRF devices run at 2 Mbps, and will eventually allow for speeds of 10 Mbps.

Like the Bluetooth specification, HomeRF uses frequency-hopping spread spectrum radio for reliability and security. The differences in the hop rate minimize the chance of interference between products

# The Case for Bluetooth

**TABLE 1.2**

Performance Characteristics of 802.11 Wireless LANs Using Frequency Hopping

| Feature/Function | Performance |
|---|---|
| Connection Type | Spread spectrum (direct sequence or frequency hopping) |
| Spectrum | 2.4 GHz ISM band |
| Transmission Power | 100 milliwatts (mW) |
| Data Rate | 1 Mbps or 2 Mbps using frequency hopping; 11 Mbps using direct sequence |
| Range | Up to 300 feet between access point and client |
| Supported Stations | Multiple devices per access point; multiple access points per network |
| Voice Channels | Voice over IP |
| Data Security | Authentication: challenge-response between access point and client via Wired Equivalent Privacy (WEP)  Encryption: 40-bit standard; 128-bit optional |
| Addressing | Each device has a 48-bit MAC address that is used to establish a connection with another device. |

that use both technologies in the same vicinity. Like the Bluetooth specification, the HomeRF specification defines support for voice as well as data, enabling the delivery of a broad range of affordable, interoperable consumer devices. The HomeRF specification is based on the Shared Wireless Access Protocol (SWAP), which defines a common interface that supports wireless voice and data networking in the home.

SWAP enables different consumer electronic devices, available from a large number of manufacturers to interoperate, while providing users with a compelling and complete home network solution that supports both voice and data traffic and provides interoperability with the Public Switched Telephone Network (PSTN) and the Internet. For example, a SWAP-compliant residential gateway for Digital Subscriber Line (DSL) can provide a connection point for wireless devices in the home (Figure 1.4). DSL provides high-speed data, telephony, and digital video services over standard telephone lines. In supporting SWAP, the gateway enables the whole family to access the Internet simultaneously. With the next generation of gateways, users will be able to dial in from a remote location to access and control any device within range of the gateway.

**Figure 1.4**
This pair of first-generation wireless gateways from Intel Corp., AnyPoint Wireless Home Network, acts as a connection point for all wireless devices in the home, enabling family members to share printers and files, as well as a DSL connection for Internet access.

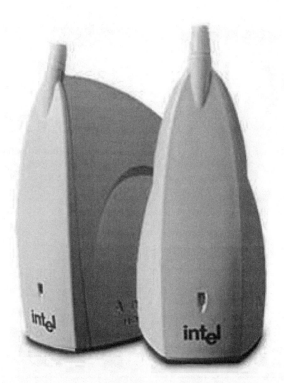

Some other examples of what users can do with the availability of products that adhere to the SWAP specification include:

- Set up a wireless home network to share voice and data between PCs, peripherals, PC-enhanced cordless phones, and new devices such as portable, remote display pads.
- Share files/modems/printers in multi-PC homes.
- Intelligently forward incoming telephone calls to multiple cordless handsets, fax machines, and voice mailboxes.
- Review incoming voice, fax, and e-mail messages from a small PC-enhanced cordless telephone handset.
- Activate other home electronic systems by speaking a command into a PC-enhanced cordless handset.
- Play multi-player games run from PC- or Internet-based resources.

SWAP was derived from extensions of existing cordless telephone (i.e., Digital Enhanced Cordless Telephone, or DECT) and wireless LAN technologies to enable a new class of home cordless services. It supports both a TDMA (Time Division Multiple Access) service to provide

# The Case for Bluetooth

delivery of interactive voice and other time-critical services, and CSMA/CA (Carrier Sense Multiple Access/Collision Avoidance) service for delivery of high-speed packet data. Table 1.3 summarizes the performance characteristics of HomeRF products.

**TABLE 1.3** Performance Characteristics of HomeRF

| Feature/Function | Performance |
| --- | --- |
| Connection Type | Spread spectrum (frequency hopping) |
| Spectrum | 2.4 GHz ISM band |
| Transmission Power | 100 milliwatts (mW) |
| Data Rate | 1 Mbps or 2 Mbps, depending on modulation scheme, using frequency hopping |
| Range | Covers typical home and yard |
| Supported Stations | Up to 127 devices per network |
| Voice Channels | Up to six (6) |
| Data Security | Blowfish encryption algorithm (over 1 trillion codes) |
| Addressing | Each device has a 48-bit MAC address that is used to establish a connection with another device. |

The wireless network can accommodate a maximum of 127 nodes. The nodes are of four basic types:

- A connection point to support voice and data services.
- A voice terminal that only uses the TDMA service to communicate with a base station.
- A data node that uses the CSMA/CA service to communicate with a base station and other data nodes.
- An integrated node, which can use both TDMA and CSMA/CA services.

The SWAP system can operate either as an ad hoc network or as a managed network under the control of a connection point. In an ad hoc network, where only data communication is supported, all stations are equal, and control of the network is distributed among the stations. For time-critical communications such as interactive voice, a connection point is required to coordinate the system and provide the gateway to the PSTN. The SWAP system can also use the connection point to support power management for prolonged battery life by scheduling device wakeup and polling.

HomeRF is aimed at a PC-card implementation, while the Bluetooth specification adheres to an embedded component model. Also, HomeRF is designed for a 50-meter transmission range, and needs a 100-milliwatt powered transmitter. A Bluetooth device, in contrast, has a 10-meter range, and only needs one milliwatt of power. This means HomeRF products will always be more expensive than Bluetooth products. Moreover, Bluetooth devices won't drain precious battery life. The radio chip consumes only 0.3 milliamps (mA) in standby mode, which is less than 3 percent of the power used by a standard mobile phone. The chips also have excellent power-saving features, automatically shifting to a low-power mode as soon as traffic volume decreases or stops.

To achieve this low-power requirement and to avoid draining device batteries, the Bluetooth wireless technology is implemented on a single chip as part of a module (Figure 1.5). Designers combine both the radio frequency and logic components on the same chip, eliminating many "off-chip" components. These are among the largest contributors to a radio's cost. Such trimming is possible because of the short range of Bluetooth devices, which allows a receiver to be an order of magnitude less sensitive than a cordless phone, requiring only 1 milliwatt of power. The transmission rate—about 720,000 bits per second—is also modest, less than a tenth that of wireless LANs.

**Figure 1.5**
Size of a Bluetooth module from Ericsson compared to an ordinary matchstick.

It is inevitable that consumers will want to bring together HomeRF and Bluetooth devices. Normally the two types of devices are not compatible. Despite key differences between HomeRF and Bluetooth wireless technologies, however, there are enough synergies to justify the development of a dual-mode system that will allow a device to dynamically switch between HomeRF and Bluetooth, so that full communication can take place.

## Bluetooth Advantage

What we all need is an economical wireless solution that is convenient, reliable, easy to use, and operates over a longer distance than infrared without requiring a clear line of sight. Of the many emerging wireless solutions that attempt to address all of these needs, only Bluetooth wireless technology is promising enough to elicit the support of a broad base of vendors representing all segments of the computer and communications markets.

Simply put, the Bluetooth specification is a global technology specification for low-cost, small form-factor, wireless communication and networking between PCs, mobile phones, and other portable devices. As such, Bluetooth devices are intended to replace cable connections between computers, peripherals, and other electronic devices. To transfer data between your laptop and PC, for example, you typically need a proprietary cable, an infrared connection, or a special wireless adapter with a built-in antenna. With the Bluetooth technology embedded in microchips, you can connect devices by turning them on. Card slots in laptop and desktop computers remain available for other uses.

In allowing cluttered desktops to be cleared of cables, Bluetooth devices free up personal workspace. Since the wireless link has a range of 30 feet, you can place peripherals virtually anywhere in a room or office without consideration for cable lengths, which are typically less than 3 feet (1 meter). Although the range of each Bluetooth device is approximately 30 feet (10 meters), this distance can be extended to around 300 feet (100 meters) with optional amplifiers placed at strategic locations within a building. The elimination of cables also makes for a safer work environment, since there are no cables for people to trip over and unplug. In shedding cables, the work environment looks and feels more comfortable and inviting.

Besides eliminating the common tangle of wires, the Bluetooth wireless technology also enables devices to communicate with each other as soon as they come within range, rather than requiring the user to open an application or press a button to initiate a process. In fact, one of the main advantages of the Bluetooth specification is that it does not need to be set up—it is always on, running in the background. The devices do not even require a line of sight to communicate with each other. In contrast to infrared, Bluetooth devices can penetrate walls and briefcases. This means an electronic organizer in your pocket could transmit a phone number to the cell phone in your briefcase and initiate a phone call.

Since different devices can automatically link up with each other when they come into range, what you really have is your own personal area network (PAN), which is also capable of ad hoc networking. In this regard, Motorola sees Bluetooth wireless technology as enhancing the industry trade show experience. In preparation for your show visit, you can load preferences for product information into your palmtop device. As you walk through the exhibits, your palmtop detects other devices using Bluetooth wireless technology and selectively exchanges information with them. An exhibitor may collect contact information from your device, for example, so it can send product literature to your office or follow up with a sales call. Likewise, your device may collect contact information from the exhibitor's device, so you can follow up later if the exhibitor's information doesn't arrive in a timely manner.

## Origin of Bluetooth

In 1994, Ericsson Mobile Communications, the global telecommunications company based in Sweden, initiated a study to investigate the feasibility of a low-power, low-cost radio interface between mobile phones and their accessories. The aim of the study was to find a way to eliminate cables between mobile phones and PC cards, headsets, desktops, and other devices. The study was part of a larger project investigating how different communications devices could be connected to the cellular network via mobile phones. The company determined that the last link in such a connection should be a short-range radio link. As the project progressed, it became clear that the applications for a short-range radio link were virtually unlimited.

Ericsson's work in this area caught the attention of IBM, Intel, Nokia, and Toshiba. The companies formed the Bluetooth Special Interest Group (SIG) in May 1998, which grew to over 1500 member companies by April 2000—faster than any other wireless consortium has grown. The companies jointly developed the Bluetooth 1.0 specification, which was released in July 1999. The specification consists of two documents: the foundation core, which provides design specifications; and the foundation profile, which provides interoperability guidelines. The core document specifies components such as the radio, baseband, link manager, service discovery protocol, transport layer, and interoperability with different communication protocols. The profile document specifies the protocols and procedures required for different types of Bluetooth applications.

The five founding companies of the Bluetooth SIG were joined by 3Com, Lucent, Microsoft, and Motorola to form the Promoter group. The charter of the Promoter group is to lead the efforts of the Bluetooth SIG by creating a forum for enhancing the Bluetooth specification and providing a vehicle for interoperability testing.

## What's With the Name?

The engineers at Ericsson code named the new wireless technology Bluetooth to honor a 10th century Viking king in Denmark. Harald Bluetooth reigned from 940 to 985 and is credited not only with uniting that country, but with establishing Christianity there as well.[1]

At that time, the Danes lived in small communities under the authority of local chieftains, some of whom terrified European coastal towns with piratical Viking raids for slaves and treasure. For centuries the Danes had worshipped the gods Thor and Odin. As Christianity took hold throughout Europe, the struggle between Christians and pagans spilled over into the areas occupied by the Danes.

As the story goes, Harald was the son of King Gorm the Old of Denmark and of Thyra (or Tyra), said to be the daughter of an English nobleman. About 25 years into his reign, the German priest Poppos impressed Harald by holding a glowing hot piece of metal in his bare hands without producing any wounds. Poppos explained that his faith in God protected him, which convinced Harald of the powers of Christianity. King Harald's acceptance of Christianity and his ensuing baptism did much to alleviate the religious strife in Denmark.

---

[1] Harald's name was actually Blåtand, which roughly translates into English as "Bluetooth." This has nothing to do with the color of his teeth—some claim he neither brushed, nor flossed. Blåtand actually referred to Harald's very dark hair, which was unusual for Vikings. Other Viking states included Norway and Sweden, which is the connection with Ericsson (literally, Eric's son) and its selection of Bluetooth as the code name for this wireless technology.

The goals of Bluetooth wireless technology are unification and harmony as well—specifically, enabling different devices to communicate through a commonly accepted standard for wireless connectivity. It is a bit of a stretch, but this is how the marketing people at Ericsson rationalize their selection of the code name "Bluetooth."

## Bluetooth Technology

The Bluetooth specification comprises a system solution consisting of hardware, software and interoperability requirements. The set of Bluetooth specifications developed by Ericsson and other companies (see Appendix A) answers the need for short-range wireless connectivity for ad hoc networking. The Bluetooth baseband protocol is a combination of circuit and packet switching, making it suitable for both voice and data.

Bluetooth wireless technology is implemented in tiny, inexpensive, short-range transceivers in the mobile devices that are available today, either embedded directly into existing component boards or added into an adapter device such as a PC card inserted into a notebook computer. Potentially, this will make devices using the Bluetooth specification the least expensive wireless technology to implement. Once the chipsets reach mass production, they are expected to add only $5 (U.S.) to the cost of the products they are embedded into.

You will not necessarily have to buy new devices to take advantage of Bluetooth wireless technology. For example, those who have purchased Handspring's Visor—a device much like the Palm Pilot, but cheaper and more functional—will be able to snap on a module called Blue-Connect from Acer NeWeb. The Blue-Connect module beams applications from Visor to Visor or from Visor to a desktop and notebook using a synchronization scheme called Blue-Share. You can also beam address-book entries and transfer images from Visor to digital cameras.

Bluetooth wireless technology uses the globally available unlicensed ISM radio band of 2.4 GHz. The ISM (industrial, scientific, and medical) bands include the frequency ranges at 902—928 MHz and 2.4—2.484 GHz, which do not require an operator's license from the Federal Communications Commission (FCC) or any international regulatory authority. The use of a common frequency band means that you can bring devices using the Bluetooth specification virtually anywhere in the world and they will be able to link up with other such devices, regardless of what country you happen to be visiting.

## Types of Links

Two types of links have been defined in the Bluetooth specification for support of voice and data applications: an asynchronous connectionless (ACL) link and a synchronous connection-oriented (SCO) link. ACL links support data traffic on a best-effort basis. The information carried can be user data or control data. SCO links support real-time voice and multimedia traffic using reserved bandwidth. Both data and voice are carried in the form of packets and the Bluetooth specification can support ACL and SCO links at the same time.

Asynchronous connectionless links support symmetrical or asymmetrical, packet-switched, point-to-multipoint connections, which are typically used for data. For symmetrical connections, the maximum data rate is 433.9 Kbps in both directions, send and receive. For asymmetrical connections, the maximum data rate is 723.2 Kbps in one direction and 57.6 Kbps in the reverse direction. If errors are detected at the receiving device, a notification is sent in the header of the return packet, so that only lost or errored packets need to be retransmitted.

Synchronous connection-oriented links provide symmetrical, circuit-switched, point-to-point connections, which are typically used for voice. Three synchronous channels of 64 Kbps each are available for voice. The channels are derived through the use of either Pulse Code Modulation (PCM) or Continuously Variable Slope Delta (CVSD) modulation. PCM is the standard for encoding speech in analog form into the digital format of "1s" and "0s" for transmission through the PSTN. CVSD is another standard for analog-to-digital encoding but offers more immunity to interference and therefore is better suited than PCM for voice communication over a wireless link. The appropriate voice-coding scheme is selected after negotiations between the Link Managers of each Bluetooth device.

## Ad Hoc Networking

When it comes to ad hoc networking for data, a device equipped with a radio using the Bluetooth specification establishes instant connectivity with one or more other similarly equipped radios as soon they come into range. Each device has a unique 48-bit Medium Access Control (MAC) address, as specified in the IEEE 802 standards for LANs. For voice, when a mobile phone using Bluetooth wireless technology comes within range of another mobile phone with built-in Bluetooth wireless technology conversations occur over a localized point-to-point radio link. Since the connection does not involve a telecommunications service provider, there is no per-minute usage charge.

## Voice over Bluetooth

The Bluetooth specification allows compliant telephone handsets to be used in three different ways. First, telephones in the home or office may act as cordless phones connecting to the PSTN and incurring a per-minute usage charge. This scenario includes making calls via a voice base station, making direct calls between two terminals via the base station, and accessing supplementary services provided by an external network.

Second, telephones using the Bluetooth wireless technology can connect directly to other telephones for the purpose of acting as a "walkie-talkie" or handset extension. Referred to as the intercom scenario, the connection incurs no usage charges from a carrier. Third, the telephone may act as a mobile phone connected to the cellular infrastructure and incurring cellular charges.

The Bluetooth specification supports three voice channels. In addition to short-range voice connections of the walkie-talkie kind, a voice channel can be used for a radio link between a head set and a mobile phone, enabling you to keep your hands free for more important tasks, like driving, without having to connect a wire, which often gets in the way.

Ericsson, for example, offers a headset (Figure 1.6) with a built-in Bluetooth radio chip that weighs only 0.75 ounces (20 grams). The radio signal acts as a connector between the headset and the Bluetooth plug on an Ericsson phone. This means you can wear your phone on your belt or put it down and walk around as you converse. When the phone rings, you can answer by simply pressing a key on the headset. If you want to make a call, you just press the key on the headset and use voice recognition to initiate the call. The phone can be up to 30 feet away, in a briefcase, in your coat pocket, or even in another room while you're speaking with the other party. You have complete mobility without a dangling wire.

# The Case for Bluetooth

Infrared technology can also support voice. A component of the Infrared for Mobile Communications (IrMC) specification developed by the Infrared Data Association (IrDA) includes RTCON (for real-time connection), a method for transmitting digital voice over an infrared link. An RTCON module added to a portable infrared device provides real-time audio over an infrared link. However, this works well only if both sides of the link are in fixed positions relative to each other. The most common application involves a mobile phone placed in a hands-free car cradle, which allows hands-free cell phone operation in a car.

**Figure 1.6**
Bluetooth wireless headset from Ericsson.

## Video over Bluetooth

In addition to voice, the Bluetooth specification is capable of supporting video transmissions between devices. An integrated circuit developed by Toshiba, one of the five founding members of the Bluetooth SIG, supports video signal encoding and decoding in the MPEG-4 format. Toshiba's set-up involves transferring images grabbed by a digital camcorder, compressing them using the MPEG-4 format, and beaming them via Bluetooth wireless technology to another device, such as a workstation where they can be edited.

Inter-device communication is handled by TCP/IP[2] running over the Bluetooth specification's link-layer protocol. TCP/IP then provides the basis for supporting the Real-time Transfer Protocol (RTP), which ensures that the video's packets are correctly synchronized. At this writing, the transmission rate is only ten frames per second in the quarter common intermediate format (QCIF), so the image quality is much less than that offered by television at 30 frames per second. However, Toshiba is working on improving both picture quality and frame rate. Even now, the technology is good enough to allow a mobile terminal to allow the user in one room of a house to see what's currently being broadcast on TV so he or she can set the VCR remotely—the kind of role to which Bluetooth devices are ideally suited in a home environment.

Toshiba's integrated circuits provide the high-level processing performance required for MPEG-4 video encoding and decoding but with significant reductions in power consumption, making them suited for power-constrained wireless applications, including those run on the third generation of mobile computing and communications products and devices using bluetooth wireless technology.

MPEG (Moving Picture Experts Group) is the international organization, jointly organized by the International Organization for Standardization (ISO) and the International Electrotechnical Commission (IEC), which proposes formats for the compression of audio and video signals to assure their efficient storage and transmission. MPEG-1[3] covers recording to such media as CD-ROMs, while the MPEG-2 format sets standards for broadcasting and other audio and video equipment like DVD. MPEG-4 primarily targets wireless communication applications of the kind that can be supported by the Bluetooth specification, as well as Web-based multimedia applications.

---

[2] Transmission Control Protocol/Internet Protocol (TCP/IP) is the suite of networking protocols that makes the Internet work. TCP/IP is valued for its ability to interconnect diverse computing platforms—from palmtops and PCs, to Macintoshes and UNIX systems, to mainframes and supercomputers.

[3] MPEG-1 is the standard on which such products as Video CD and MP3 (MPEG-1, layer III) music files are based. MPEG-2 is the standard on which such products as Digital Television set top boxes and DVD are based. There is no MPEG-3. MPEG-4 is the standard for multimedia on the Web and for the wireless environment. The current thrust of the Moving Picture Experts Group is MPEG-7, "Multimedia Content Description Interface." In addition, work on MPEG-21, "Multimedia Framework," began in June 2000. To stay updated on these standards, visit the Web page of the Moving Picture Experts Group at www.cselt.it/mpeg/.

A key element of MPEG-4 is a video signal compression format suitable for applications characterized by unstable data transmission, including applications running over wireless links and the Internet. Toshiba has added to this an error-correction function to prevent image degradation that results from data communication errors.

## Radio Link

The radio link itself is very robust, using frequency-hopping spread-spectrum technology to mitigate the effects of interference and fading. As noted, spread spectrum is a digital coding technique in which the signal is taken apart or "spread" so that it sounds more like noise to the casual listener. The coding operation increases the number of bits transmitted and expands the bandwidth used.

Using the same spreading code as the transmitter, the receiver correlates and collapses the spread signal back down to its original form. With the signal's power spread over a larger band of frequencies, the result is a more robust signal that is less susceptible to impairment from electromechanical noise and other sources of interference. It also makes voice and data communications more secure. With the addition of frequency hopping—having the signals hop from one frequency to another—wireless transmissions are made even more secure against eavesdropping.

## Interference

Spread spectrum combats interference from other devices that also operate in the unlicensed 2.4 GHz part of the radio spectrum, including microwave ovens and other appliances used in the home, as well as some wireless LANs used in the office. Instead of staying on one frequency, each spread-spectrum device hops 1,600 times a second among 79 frequencies. The device initiating the connection will tell the other device what sequence of hops to use. If there is too much interference at one frequency, the transmission is lost for only a millisecond. To increase reliability, the system can send each data bit in triplicate. The result is that several dozen people in a room can use Bluetooth devices without significant interference.

Interference is of special concern in the corporate environment where wireless LANs may be in use. Bluetooth wireless technology uses the same spread-spectrum technology as wireless LANs based on the 802.11 standard and both operate in the same 2.4 GHz radio spectrum. While the wireless link used by Bluetooth devices operates over a shorter range than that used in 802.11 LANs, both occasionally can get into the same space. When a Bluetooth connection collides with a wireless LAN connection, either or both connections can jam, resulting in a transmission error. When this happens, error correction schemes on both the LAN and Bluetooth links will restore bit errors. The use of different frequency-hopping schemes minimizes the chance of interference, as does spreading over the entire frequency band.

### Safety

Radiation emissions from devices using Bluetooth wireless technology are no greater than emissions from industry-standard cordless phones. The Bluetooth module will not interfere or cause harm to public or private telecommunication network equipment, nor jeopardize the safety of the consumer using the equipment or those who come within the operating distance of Bluetooth devices.

## Personal Area Networks

One of the common goals shared by the IEEE and the Bluetooth SIG is the global use of wireless personal area networks (PANs). The IEEE's 802.15 working group is looking at creating standards that will provide the foundation for a broad range of interoperable consumer devices by establishing universally adopted standards for wireless digital communications.

The goal of the 802.15 working group is to create a consensus standard that has broad market applicability and deals effectively with the issues of coexistence with other wireless networking solutions. While the IEEE 802.11 wireless LAN technologies are specifically designed for devices in and around the office or home, devices using the IEEE 802.15 wireless PAN and Bluetooth wireless technology will provide country-to-country usage for travelers in cars, airplanes, and boats. Table 1.4 summarizes the performance characteristics of Bluetooth products that operate in the 2.4-GHz range.

# The Case for Bluetooth

**TABLE 1.4**

Performance Characteristics of Bluetooth Products

| Feature/Function | Performance |
|---|---|
| Connection Type | Spread spectrum (frequency hopping) |
| Spectrum | 2.4 GHz ISM band |
| Transmission Power | 1 milliwatt (mW) |
| Aggregate Data Rate | 1 Mbps using frequency hopping |
| Range | Up to 30 feet (10 meters) |
| Supported Stations | Up to eight (8) devices per piconet |
| Voice Channels | Up to three (3) |
| Data Security | For authentication, a 128-bit key; for encryption, the key size is configurable between 8 and 128 bits. |
| Addressing | Each device has a 48-bit MAC address that is used to establish a connection with another device. |

## Bluetooth Topology

The devices within a piconet play two roles: that of master or slave (Figure 1.7). The master is the device in a piconet whose clock and hopping sequence are used to synchronize all other devices (i.e., slaves) in the piconet. The unit that carries out the paging procedure and establishes a connection is by default the master of the connection. The slaves are the units within a piconet that are synchronized to the master via its clock and hopping sequence.

The Bluetooth topology is best described as a multiple piconet structure. Since the Bluetooth specification supports both point-to-point and point-to-multipoint connections, several piconets can be established and linked together in a topology called a "scatternet" whenever the need arises (Figure 1.8).

Piconets are uncoordinated, with frequency hopping occurring independently. Several piconets can be established and linked together ad hoc, where each piconet is identified by a different frequency hopping sequence. All users participating on the same piconet are synchronized to this hopping sequence. Although synchronization of different piconets is not permitted in the unlicensed ISM band, units using Bluetooth wireless technology may participate in different piconets through time division multiplexing (TDM). This enables a unit to participate sequentially in different piconets by being active in only one piconet at a time.

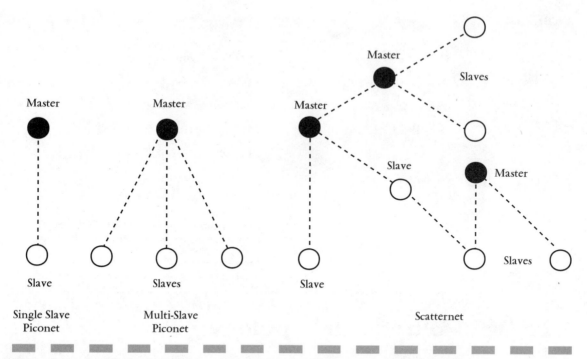

**Figure 1.7** Possible topologies of networked Bluetooth devices, where each device is either a master or slave.

**Figure 1.8**
Using Bluetooth wireless technology, devices can join in the collaborative environment by establishing a connection on a piconet.

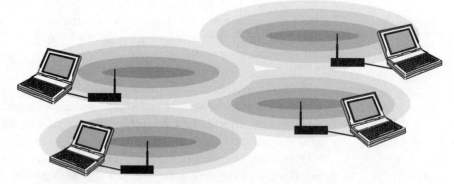

With its service discovery protocol, the Bluetooth specification enables a much broader vision of networking, including the creation of Personal Area Networks, where all the devices in a person's life can communicate and work together (Figure 1.9). Technical safeguards ensure that a cluster of Bluetooth devices in public places, such as an airport lounge or train terminal, would not suddenly start talking to one another.

The Bluetooth SIG intends to evolve the technology to provide greater bandwidth and distances, thus increasing the potential platforms and applications used in the emerging personal area network marketplace. Bluetooth wireless technology can scale upward to support wireless data transmission applications operating in the 5-GHz range, supporting connections to devices up to 300 feet away.

## Security

Worried about security? Don't be—the Bluetooth specification includes several security features. In addition to its limited range and its use of frequency hopping, which makes signal interception extremely difficult in the first place, the Bluetooth specification employs such link-level functions as authentication and encryption. Authentication prevents unwanted access to critical data and functions as well as protection against "spoofing" by hackers who try to impersonate authorized users. Encryption scrambles the data during transmission to prevent eavesdropping and maintain link privacy. In addition, Bluetooth wireless technology includes session-key generation that can be changed at any time during a connection. Even in the unlikely event that a hacker is able to grab a connection, he or she will not be able to stay on a piconet for any length of time.

Security is important not only to ensure the privacy of your messages and files as they fly through the air, but to ensure the integrity of electronic commerce transactions as well. Accordingly, the Bluetooth specification also offers a flexible security architecture that makes it possible to grant access to "trusted" devices and services without providing access to other "untrusted" devices and services.

Nokia is among the companies advancing the idea of using phones and palmtops equipped with Bluetooth wireless technology as "personal trusted devices" with which consumers can load "money" into an electronic wallet at an automated teller machine (ATM) and pay for merchandise at the point of sale, whether at a retail store or a vending machine.

Under the Bluetooth security architecture, untrusted or unknown devices may require authorization based on some type of user interaction before access is granted. Trusted devices are those that have been previously authenticated and allowed to have access based on their link-level key. For these devices, the link key may be stored in the device database, which identifies that device as trusted in terms of future access attempts.

The security architecture of the Bluetooth specification only authenticates devices, not users. This means that a trusted device that is stolen or borrowed can be used as if it were still in the possession of the rightful owner. If there is a need for user authentication, supplementary application-level security methods must be employed, such as the entry of a username and password, as would be the case if the device were to be used for mobile e-commerce transactions.

## What Can You Do With Bluetooth?

The Bluetooth specification enables you to connect a wide range of computing and telecommunications devices easily and simply, without the need to buy, carry, or connect cables. It will provide opportunities for rapid ad hoc connections, and make possible automatic connections between devices. It will virtually eliminate the need to purchase additional or proprietary cabling to connect individual devices. Because Bluetooth wireless technology can be used for a variety of purposes, it also replaces multiple types of cable connections with a single radio link. It will allow you to think about what you're working on, rather than how to connect everything to make the technology work.

### Presentations

Setting up a PowerPoint presentation, for example, would no longer require a spaghetti-like tangle of cables between projector, laptop, and printer. You would simply place the laptop near the projector, turn them on, and wait a few moments for them to communicate the necessary operating parameters. Via the same radio link, the laptop could send print requests to a nearby printer, allowing distribution of the latest reference materials to attendees. For small groups, the presentation can be delivered to each attendee's laptop that is using Bluetooth wireless technology. This would permit meetings to be held in any room, without requiring a projection screen, special lighting controls, or seating arrangements.

## Card Scanning

With a business-card scanner using the Bluetooth specification, you can scan cards into your own computer or any other computer within the 30-foot range without having to go through the hassle of connecting, disconnecting, and reconnecting cables between machines. Since the business-card scanner may not be used very frequently, significant cost savings can be achieved by sharing the device wirelessly among a group of users.

## Collaboration

With special software that turns notebooks, palmtops, or Windows CE devices using Bluetooth wireless technology into a sketchpad, you can use a stylus as the input device to annotate or draw on Word files, e-mail messages, JPEG photographs, or any Windows-based application. Instead of gathering people in one place and everyone fumbling with pen and paper, your marks made with the stylus on one device are transmitted over the persistent, wireless connections between the other devices, permitting collaboration among a group of users. The changes remain on screen, allowing the marked-up document to be saved on each device or e-mailed as an attachment.

## Synchronizing Data

A neat trick that Bluetooth devices can accomplish is sending messages to devices that are powered off or in sleep mode. For example, when a cell phone receives a message, it can send it to a laptop computer, even if the latter is packed in a briefcase and powered off. Of course, the technology can also be used for synchronizing data between the devices, ensuring that the most current version is available, regardless of what device you happen to select (Figure 1.9).

Automatic synchronization can be a real time saver. When you're finished adding information to your palmtop device at home, all you have to do is walk by your cubicle at work to get those files uploaded to your desktop PC. When you leave the office, any new files added to your desktop PC are automatically copied to the palmtop. When you arrive at home in the evening, the palmtop automatically loads the new information to your laptop as soon as the two devices come within range. You do not do anything—link-up just happens. With automatic synchronization, there's no longer any confusion about which file is on which machine—the Bluetooth specification ensures that you have the most updated information, regardless of what device you choose at any given time.

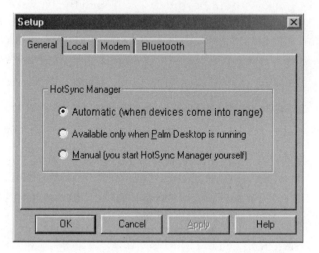

**Figure 1.9**
When Bluetooth devices are set up properly, they can communicate within a 30-foot range and synchronize files without explicit commands.

## Remote Synchronization

Synchronization also works between the devices of multiple users linked over the PSTN. Let's say that an important meeting has been rescheduled while you are on the road. Back at the office, your boss sends the changes from her workstation to your cellular phone, which automatically connects to your organizer and updates the schedule. The next time you power up your organizer, you are immediately alerted to the change.

By eliminating hard connections, the Bluetooth specification erases the difference between mobile and stationary computing—the devices are "plugged in" wherever they happen to be. On a train, for example, Bluetooth wireless technology would enable you to connect a laptop or palmtop to the Internet via a cell phone in your briefcase, while at the office, the link would be through an access point providing a wireless connection to the corporate LAN.

## Printing

Consider slightly more far-out scenarios, too: with Bluetooth wireless technology, your digital camera could send a photo straight to your printer. Or seconds after you snapped a photo of your kids at the zoo, the digital camera could send the image to the cell phone in your pocket, which could then send the photo as an e-mail attachment to relatives and friends back home.

## In-Car Systems

In the near future, the Bluetooth specification will allow a range of digital devices to share information wirelessly within an automobile—everything from cellular phones and pagers to hand-held computers and more.

Among the companies offering such systems is Johnson Controls. The company's TravelNote Connect is a modified TravelNote digital recorder that integrates Bluetooth wireless technology. TravelNote is a digital recorder/playback device that can be integrated into a vehicle's overhead console or sun visor. It enables the driver or front-seat passenger to record, store and play back "reminder" messages. By adding a Bluetooth wireless technology component, the device can do things like get a phone number from a handheld phone and automatically dial it so the driver doesn't have to take his or her hands off the wheel. When the connection is established, the Bluetooth compnent sets up a wireless voice link with the cellular phone, providing hands-free, speakerphone capabilities.

This and similar products from other manufacturers have the capacity to make every cellular phone a hands-free phone in the near future, without complex retrofits or costly installations in a vehicle interior. Since any device using Bluetooth wireless technology can talk to any other similarly equipped device, products can be mixed and matched for use in the car, irrespective of model, brand, manufacturer, or operating system.

## Communicator Platforms

Communicator platforms of the future will combine a number of technologies and features in one device, including mobile Internet browsing, messaging, imaging, location-based applications and services, mobile telephony, personal information management, and enterprise applications. With these integrated mobile information devices, you will be able fully to exploit mobile Internet and multimedia communication including voice, data, and images.

Ericsson is among several companies that will offer integrated communicator platforms capable of supporting high-speed data transfers, triple-band voice, and Internet access. The user interface is based upon the VGA format (Figure 1.10), which Ericsson believes is an ideal size for mobile communications devices and applications. The device has a color touch screen, which allows for easy navigation, pen input, and handwriting recognition. With a built-in GPS receiver, the platform can provide positioning information. With built-in Bluetooth and infrared components, the platform can connect wirelessly with other devices, networks, and third-party applications.

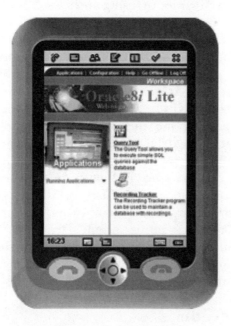

**Figure 1.10**
Ericsson has licensed its platform technology from Symbian, which supports Java, WAP, and the Bluetooth specification. This PDA-like device, available from Symbian, formerly the software division of Psion, runs Oracle8i Lite, an Internet database application from Oracle Corp.

## Electronic Books

With electronic books becoming popular, you can purchase titles on the Web from online booksellers and download them to your desktop or notebook computer. In essence, your computer becomes an electronic library from which you can select specific electronic volumes to take with you on business trips or vacations. Through a librarian program that manages the titles on your computer, you

# The Case for Bluetooth

can transfer any electronic volume (as well as your own documents) to a special reading device called an electronic book, or "ebook," which is attached to a cradle cabled to your computer. The ebook is a portable battery-powered device that weighs only 22 ounces (627 grams). It features a white back light for easy text viewing. Once these ebooks are enhanced with Bluetooth wireless technology, you will be able to transfer preselected titles between the devices merely by placing the ebook within range of the computer holding your library (Figure 1.11).

**Figure 1.11** In the future, you will be able to place an ebook using Bluetooth wireless technology components next to your computer to download preselected titles that can be taken with you. Shown here is the Rocketbook (right) from NuvoMedia.

## Travel

If you happen to be a frequent flyer, the Bluetooth specification offers some advantages for obtaining tickets. You can arrive at the airport and buy a ticket simply by walking past a wireless terminal, which confirms your identity, issues an electronic ticket, and bills your credit card. A flight attendant would no longer have to visit passengers and ask them to turn off their electronic devices. After the request is broadcast, the airplane's master Bluetooth device would shut down all electronic devices automatically for takeoff and landing.

After landing in your destination city, you get on the rent-a-car bus at the airport. Your reservation is automatically transferred to the rent-a-car database where it is verified, so the bus driver can let you off at your pre-assigned car. When you get in the rental car, which is equipped with Bluetooth wireless technology devices, your hotel reservation is automatically queried from your palmtop or cell phone and the Global Positioning System (GPS) provides you with on-screen directions to your hotel. As you enter the lobby, you are automatically checked in, and your room number and electronic key are automatically transferred to your Bluetooth device. As you approach the room with bags in hand, the door automatically opens. Later, if you can't fall asleep right away, you can catch up on some reading by breaking out your ebook.

## Home Entertainment

If you think all this is far-fetched, Microsoft envisions the day when you can buy a Bluetooth-equipped digital TV that will call you when your favorite football team is about to kick off and ask you if you want to record the game.

In the home, there are myriad applications for Bluetooth wireless technology. Imagine a simple data pad equipped with a Bluetooth transceiver and a touch-screen display. The data pad is slim, lightweight, and, with an advanced icon-driven menu, easy to use. The data pad will not only control all the entertainment devices in the home, but also control new ones that may be purchased in the future, finally putting an end to the proliferation of infrared remote controls.

According to Microsoft you might even be able to enhance your bathroom with special Bluetooth chips that monitor toothpaste and soap, and remind you to buy more when levels run low.

## Payment Systems

Using Bluetooth wireless technology, it will also be possible to link mobile phones and other kinds of handheld devices wirelessly to gasoline pumps so that when a driver fills up the tank, the cost of the fuel is automatically deducted from a credit-card account through the customer's handset. Bluetooth wireless technology could be used in a similar way for rail fares, cinema tickets, parking fees, and other every-

# The Case for Bluetooth

day purchases at kiosks—making for less hassle for consumers and lower transaction costs for companies (Figure 1.12).

**Figure 1.12**
Today, infrared-equipped cell phones like this one from BT Cellnet are being used in Europe to give users the ability to check their bank balances and transactions, and transfer funds or pay bills while on the move. With the addition of Bluetooth wireless components, such phones will provide users with even more flexibility.

## Scanners

Bluetooth wireless technology also encourages the development of entirely new products. For example, Swedish manufacturer C Tech offers a C Pen, a so-called "mobile information collector," which is a cross between a palmtop device and a text scanner. The device captures text or graphics via a tiny scanner and stores information in its 8 MB of onboard memory. The C Pen offers a range of palmtop-like features. For example, one feature allows the user to scan in a business card, line by line, and then upload it to a contacts database in Microsoft Outlook. Two-way language dictionaries let you translate, for instance, from English to German and vice versa. The company expects to release a version of the C Pen that uses both infrared and Bluetooth wireless technology. One version of the C Pen would have a built-in Bluetooth wireless port, while another version would have an infrared port for communicating with other computers. The gadget boasts a 100-MHz Intel processor plus a 3-hour battery, yet it is still small enough to fit in the hand.

## Behavior Enforcement

Bluetooth wireless devices can be combined with other technologies to offer wholly new capabilities, such as automatically lowering the ring volume or shutting off cell phones as users enter quiet zones. The convenience and potential benefits of cell phones are indisputable, but the technology's near ubiquity is provoking heated discussion about when and where they should be used. With more than 80 million cell phone subscribers in the U.S. alone, it is inevitable that some will ring during meetings, religious services, and entertainment events, causing disruptions. Bluetooth wireless technology can be used to enforce courteous cell phone usage in designated quiet areas, allowing users to keep their phones handy without compromising their good manners while theaters and meeting planners can provide the quiet atmosphere their patrons have come to expect.

BlueLinx Inc., a wireless innovator based in Charlotte, NC, offers a tool that prevents phones from ringing loudly in churches, theaters, and restaurants. The company's patented Q-zone uses Bluetooth wireless technology to create specific areas where electronic devices ring, beep, and ding at lower volume levels. Q-zone automatically changes settings on cell phones and other small electronic devices to use lower ring volumes or silent vibration upon entering specified areas, and reapplies the previous settings after users leave. It operates through a series of small nodes installed throughout the quiet zone that form a short-range wireless network, which allows seamless communication between the different devices.

## Mobile E-commerce

Bluetooth wireless technology will play a key role in electronic commerce. Soon you'll be able to avoid the line at retail stores to pay for items, or access the Internet anywhere via your mobile phone to order and pay for goods and services. While some mobile phones already have smart cards (i.e., subscriber identity modules, or SIMs) that can hold electronic cash (Figure 1.13), no one really wants to go through the trouble of removing them from the phone in order to have it read by a point-of-sale terminal. Bluetooth components will allow the smart card to be read while it remains in the phone or palmtop device (Figure 1.14).

# The Case for Bluetooth

**Figure 1.13**
SIM issued to subscribers of BT Cellnet, a cellular service provider in the UK. Within the larger card is a detachable postage stamp-sized SIM (left side). Both sizes are offered together to fit any kind of mobile phone the user happens to have.

**Figure 1.14**
Socket Communications offers a range of plug-in cards for select Windows CE-based palmtop computers. In addition to Bluetooth components, there are plug-in cards for wireless Web access, bar code scanning, serial communications, and Ethernet connectivity. The card is compatible with the Bluetooth specification and includes the Bluetooth protocol stack for Windows CE.

Ericsson and Visa International are among the companies developing payment solutions for the purchase of goods and services over the Internet through mobile devices, including cell phones and palmtops. Specifically, the two companies want to provide a variety of payment solutions that use several complementary open standards such as the Bluetooth specification and the Wireless Application Protocol (WAP), as well as the Secure Electronic Transactions (SET) and Europay-MasterCard-Visa (EMV) protocols, to address different market requirements.

Ericsson's wireless wallet, incorporating Bluetooth wireless technology, is a payment solution for mobile e-commerce. The wireless wallet, which can serve as a replacement to a conventional wallet for bills and coins, contains multiple smart card readers. A smart card inserted into the wallet can communicate with a mobile device using Bluetooth wireless components. The mobile device can be used for Internet shopping so that the appropriate smart card in the wireless wallet is used for payment.

SET was developed by Visa and other payment industry and technology companies as an open, global standard for secure electronic commerce. Based on encryption technology developed by RSA Data Security, it allows cardholders and merchants to use special encoding and identification software, called digital certificates, to authenticate themselves to each other and to allow cardholders to safely send credit card numbers over the Internet as payment.

EMV is a joint industry initiative to facilitate the introduction of chip technology into the international payment systems environment by developing joint specifications for Integrated Circuit Cards (ICC) and terminals for payment systems. EMV serves as the global framework for chip card and terminal manufacturers worldwide.

Industry studies predict that within the next 10 years, consumers will increasingly rely on non-PC devices such as the mobile phone to access the Internet. This partnership will ensure that Visa member banks will be able to offer cardholders secure payment solutions for the purchase of on-line goods and services with a mobile phone.

The advent of Web-enabled mobile phones allows users to access the Internet and receive messages over the air. This provides Visa and its member banks with a new opportunity to offer convenient and exciting services to cardholders on the go.

An example of a new service made possible by this technology is the targeted promotion of concert or theater tickets. This service will make it possible for mobile-phone subscribers to be notified of an event before the tickets are sold out. The user can purchase the tickets immediately over the Internet using a Web-enabled mobile phone that supports WAP. The tickets are then sent electronically to the phone, where they are stored on a smart card. When the user arrives at the theater, the electronic tickets on the smart card are presented via the Bluetooth wireless connection established with the theater's point-of-sale terminal.

WAP is an open global specification that enables mobile devices to access and interact with Web-based information and services. It provides the middleware necessary that runs on top of the Internet Protocol (IP), enabling the delivery of text content to wireless terminals. The standard specifies that the content must use the extensible markup language (XML)-compliant wireless markup language (WML). All WML content is accessed over the Internet using standard HyperText Transfer Protocol (HTTP) requests. Among the common information services that are delivered in this way to WAP-compliant portable devices are news items, sports scores, airline schedules, weather and traffic reports, local restaurant and movie listings, and stock quotes.

With Bluetooth wireless technology, merchants as well as the banking and finance industry will be able to generate increased revenues. In exploiting the full market potential of mobile commerce and expanding secure, mobile e-commerce solutions, participating companies of all types and sizes can open up new business opportunities. In this regard, the Bluetooth specification represents a completely new dimension to e-commerce that will radically change how portable and wireless devices are viewed and used. It provides endless opportunities for applications such as vending and ticketing machines, point-of-sale terminals, banking machines, and parking meters.

## Java and Bluetooth

Most current mobile devices are based upon proprietary platforms such as Microsoft's Windows CE or PalmOS, and cannot share application drivers or peripherals. This situation does not bode well for the acceptance of the Bluetooth specification, since one of the key advantages of the technology is ad hoc networking with devices that may be encountered spontaneously. Sun's Java is the more portable platform because the code runs on all systems and network appliances that support the Java Virtual Machine (JVM).[4] Java compilers generate code for the JVM, not any specific computer or operating system. This automatically makes all Java programs cross-platform applications. This method of code deployment also makes Java better suited for the memory-constrained embedded devices equipped with Bluetooth wireless technology.

There are two other compelling reasons for using Java in embedded systems: to achieve connectivity with a target system at the application level and to improve the software development process. In the first case, Java provides a secure environment for service delivery. The ability to download code securely from the Internet and run it locally is the most compelling reason for using Java. In this age of e-commerce, services can be anywhere on the Web; having the ability to meet the demands of customers for a very low cost is an essential ingredient to a company's success in this competitive arena.

In the second case, the company's software development team benefits from an elegant programming language for all code, from drivers to applications. While Java will not turn anyone into an embedded software expert, it can make both senior and junior firmware designers more productive. Rapid development is facilitated through code reuse, so embedded systems are easier to test and faster to deploy.

Wireless devices natively executing Java code will be able to communicate spontaneously and interact with other Java-enabled devices through various wireless protocols. In addition to the Bluetooth specification, HomeRF and 802.11 are also expected to extend wireless capability to these types of devices.

## Jini and Bluetooth

In addition to Java, Sun offers a service discovery technology called Jini. Built on Java technology, Jini is designed to enable users to make simple connections to other devices. When two Jini-enabled devices come within range, they automatically detect each other, exchange capabilities information, and establish a network connection.

As described by Sun, devices using Jini technology employ a discovery process to seek out other devices that also use Jini technology. The devices then post objects—representing the services they provide as

---

[4] Sun offers a small footprint version of the Java Virtual Machine for severely memory-constrained environments of 128 KB to 512 KB. The K Virtual Machine (KVM) is designed specifically for consumer and embedded devices and is part of a larger effort on the part of Sun to provide a modular, scalable architecture for the development and deployment of portable, dynamically downloadable, and secure applications in consumer and embedded devices. This larger effort, called the Java 2 Platform Micro Edition (Java 2 ME), is optimized for small-memory, limited-resource connected devices such as cellular phones, pagers, PDAs, television set-top boxes, and point-of-sale terminals.

# The Case for Bluetooth

well as their defining characteristics and attributes—to a Jini technology lookup service. When a device wants to use a service offered to the community, it can download the required objects, including any code such as applications, device drivers, or user interfaces, from the Jini technology lookup service.

To understand how Jini technology complements Bluetooth wireless technology, consider how a picture taken with a digital camera might be printed (Figure 1.15). To print a picture, the camera needs a service performed; specifically, it needs a printing service. The printing service typically will be performed locally; but sometimes, as shown on the right, the service may either need to communicate with an external device, such as an older printer that does not support the Java programming language, or else bridge to a foreign device (i.e., non-Jini service), such as a database of images available on the Web.

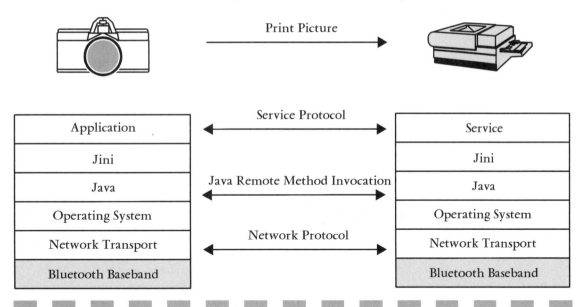

**Figure 1.15** The Bluetooth specification provides network transport, enabling a wireless connection to be established so communications can occur between different devices for service delivery. In this case, the service consists of printing a picture from a camera.

The arrows indicate the different levels of communication that are required to print the picture. First, the camera locates the printer service by using Jini technology, which in turn uses whatever network protocol is supported by the operating system. Jini technology does not require any particular operating system or network transport: the camera may use either Bluetooth or infrared wireless technology, for example, or be physically plugged into the local network.

Whatever the means used to establish itself on the network, once connected, the camera would be able instantly to tell a printer to print pictures, without the involvement of a PC. The printer would be loaded with the Java Virtual Machine and a package of Jini files, as well as its own drivers and user interface. The printer would send a Java object describing itself to the network's lookup service, which would in turn send the object to the camera when the camera requests the printer's services. To print the picture, the camera locates the printer service, then downloads and runs the Java code supplied by the printing service. This code uses the underlying network transport—that of the Bluetooth architecture or any other network transport protocol—to implement the printing service protocol needed to transmit the picture to the printer.

Jini-enabled devices were supposed to plug seamlessly into a network, find each other, and use each other's services. But Jini has not lived up to early expectations for a variety of reasons. It is poorly marketed and poorly understood. Technical, marketing, and licensing problems all have contributed to the lack of a committed developer community. And at 3 MB, the code is still too bloated for memory-constrained devices. At this writing, Jini's future is in doubt.

## Other Connectivity Solutions

In addition to Bluetooth and infrared wireless technologies, there are other emerging network technologies that have their own protocols to locate and communicate with devices over wireline or wireless links. Examples include JetSend and HAVi.

### JetSend

JetSend technology, developed by Hewlett-Packard and available since 1997, is an example of a service protocol that allows devices like printers, digital cameras, and palmtop computers to negotiate information exchange intelligently, over wireless links, with no user intervention. With digital cameras, for instance, real photos can be as close as your computer—just point and shoot—and your photos are transferred to your desktop computer, or sent to a printer for output on photo-quality paper. Or you can use JetSend to send electronic photos wirelessly to your palmtop computer or other Windows CE device.

The JetSend protocol allows the devices to identify a common data format and exchange data, eliminating the need for special drivers to make all devices work together. Currently, JetSend devices can communicate wirelessly over infrared links at 4 Mbps, but there is no reason why the protocol cannot support Bluetooth wireless links, as the technology becomes more popular in the future. JetSend also works over cabled networks using TCP/IP.

## HAVi

HAVi (Home Audio-Video interoperability) is a specification for home networks comprised of consumer electronics devices such as CD players, televisions, VCRs, digital cameras, and set-top boxes. In essence, HAVi will let these devices recognize one another and link up in a home network—without the use of a PC. The network configuration is automatically updated as devices are plugged in or removed. Applications are expected to coordinate the control of several devices and to simplify the operation of devices by the user.

Matsushita hopes to offer a set-top box that allows you to sift through hundreds of channels and record the programs you select. In a few years, the boxes may be able to guide the storing of 200 hours of content and allow you to use a search engine to navigate through it. IEEE 1394 cable, also known as FireWire, is used to connect devices at 100, 200, or 400 Mbps on the HAVi network. With HAVi, a bridge protocol would be required to provide a way to share services between HAVi devices and devices using Jini technology. Applications using Jini software can gain access to HAVi devices such as VCRs. In similar fashion, home devices on the HAVi network, specifically televisions, could connect to remote Jini-enabled services such as video-on-demand.

Microsoft has another scheme to take control of your living room. The company is pushing Windows CE and a HAVi-incompatible alternative known as HAPi (Home Application Programming interface). HAVi advocates claim Microsoft operating systems are not suited to consumer electronics because they are too complex, bug prone, and unable to cope with large continuous data streams. Yet Microsoft's plans are a bit more ambitious, aiming to work for all appliances, from stereos to refrigerators, and to allow PCs to edit videos or print out TV images.

These emerging network technologies play key roles in making impromptu digital networking become a truly universal, instantly accessible, and reliable method of device interconnectivity. Jini's approach to networking Java-enabled devices in combination with the Bluetooth specification and other standards like JetSend promises to enhance our computing experience by allowing it to become far more pervasive and personal, while making it much simpler. The list of innovations doesn't stop here. At the macro level, there is the emerging global 3G network to consider.

## Global 3G Wireless Framework

The Bluetooth wireless technology specification is one of the technologies being developed to optimize the use of third-generation (3G) mobile multimedia communications systems being developed by the International Telecommunication Union (ITU) and regional standards bodies. Under this "family of systems" concept, various wireless technologies are unified at a higher level to provide users with true global roaming and voice-data convergence, leading to enhanced services and innovative multimedia applications. This initiative, called International Mobile Telecommunications 2000 (IMT-2000), encompasses both satellite and terrestrial systems, serving fixed and mobile users in public and private networks.

The initiative aims to facilitate the evolution from today's national and regional second-generation (2G) systems, which are incompatible with one another, toward 3G systems that will provide users with interoperability, expanded coverage, and new service capabilities. The role of the ITU is to provide direction to and coordinate the many related technological developments in this area to assist the convergence of competing national and regional wireless access technologies. With a set of universally accepted specifications, all national and regional equipment manufacturers can build systems and products that interoperate. As a global standard, the Bluetooth specification feeds into this 3G framework very nicely.

Among the possible 3G support applications for which Bluetooth wireless technology might be suited is local intercommunication as well as wide area connectivity, to provide a greater level of service than either could achieve separately.

Take vending machines at a shopping mall, for example. Through a Bluetooth access system, a cluster of automatic vending machines can be connected to a central vending machine administration unit, that in turn uses a 3G access system to call for maintenance or supplies. Minor problems can be relayed to the mall technician directly through his or her Bluetooth wireless communicator. Pricing changes can be sent from the central administration unit and locally broadcast to all Bluetooth vending machines.

Another application where Bluetooth wireless technology can complement 3G systems is electronic mail. A Bluetooth/3G mobile phone can receive e-mail as a data transmission and forward it, via Bluetooth, to the notebook computer, assuming it is within close proximity. When the reception is complete, the notebook can notify the user via a short message over the Bluetooth link to his mobile phone that he has e-mail, and if an item is urgent, this fact can be forwarded as well.

Under this concept, the 3G terminal acts as a local "headend" for a variety of applications that are locally interconnected via Bluetooth wireless technology. If, for example, an important e-mail message were received while waiting for a train or plane, the user could approach a Bluetooth services booth and, for a fee chargeable to his credit card or e-wallet, instruct the notebook computer to print the e-mail of interest right at the kiosk using the 3G/Bluetooth mobile phone to control it, without having to unpack the notebook computer from the briefcase.

Systems equipped with Bluetooth wireless components in the home will permit remote control. Such systems may include central heating/air conditioning, lighting, and outdoor lawn sprinklers, among others. When you're out of range, maybe working late at the office, the 3G mobile phone or palmtop device would detect this and automatically dial the 3G/Bluetooth gateway in your home to gain access to all the household appliances within range. The things you can do while in the office or stuck in traffic include:

- Interrogate the refrigerator for its contents, so needed items can be picked up on the way home.
- Set the house temperature to a personal preference before arriving home.
- Despite getting home late, that favorite TV program is not missed; it is recorded by remotely activating the VCR.

# Problems with Bluetooth

The idea behind the Bluetooth specification is to replace the cables that tie devices to one another with a single short-range radio link. It is expected that, within a few years, about 80 percent of mobile phones will carry a Bluetooth chip that can provide a wireless connection to similarly equipped notebook computers, printers and, potentially, any other digital device within about 30 feet.

As well as defining how these devices find and talk to each other, the Bluetooth specification also ties into existing data networks, including the Internet. In the future, you might tap out an e-mail on your palmtop, tell it to make the Internet connection through your mobile phone, print a copy on the printer upstairs, and store the original on your desktop PC.

As promising as Bluetooth wireless technology is, it is severely limited in terms of its data rate of about 1 Mbps. As noted, the Bluetooth specification can scale upward to support wireless data transmission applications operating in the 5-GHz range, supporting connections to devices up to 300 feet away.

This type of technology evolution is not without precedent. When the Infrared Data Association (IRDA) introduced its specification for Serial Infrared (SIR) data link technology in 1994, it supported connections between devices at only 115.2 Kbps. A year later, the IRDA extended the SIR to transfer information at 4 Mbps and within the next four years, it released its specification for infrared connectivity at 16 Mbps. There is no reason why the Bluetooth SIG cannot greatly extend the speed and range of its technology within a similar timeframe to encompass more applications.

If there is one problem with Bluetooth wireless technology, it is that it may have been over hyped. For example, Bluetooth devices were promoted well before the standards were ready for release. It was announced early to create momentum because the Bluetooth SIG wanted many companies involved in developing products and applications. Originally, the Bluetooth SIG had hoped that the technology would be built into hundreds of millions of devices before 2002. A report by U.S. research firm Cahners In-Stat Group now forecasts that it will be 2005 before the technology has that kind of market penetration.

One reason for the delay is that the chips will have to cost $5 or less before they penetrate the mass market. That potential is still a way off. It will be at least another three years before Bluetooth chips will be cheap enough to be included in mainstream devices such as mobile phones and handheld computers. Despite the delay, the sheer weight of industry support for the Bluetooth specification—now over 2,000 companies—should eventually ensure that the standard will become commonplace. Bluetooth wireless technology is not the first technology to appear late, nor will it be the last.

## Bluetooth Qualification Program

The Bluetooth SIG has set up a qualification process to ensure that products comply with the Bluetooth specification. Upon passing of this qualification process, products can display the Bluetooth brand mark, signifying to consumers that they will interoperate as expected. Any product that displays the Bluetooth brand mark must be licensed to use the mark and only products that pass the qualification test can be issued a license.

This qualification process entails product testing by the manufacturer and by a Bluetooth Qualification Test Facility (BQTF), which issues test reports that are reviewed by a Bluetooth Qualification Body (BQB). All hardware or software modifications to a qualified product are documented and reviewed by the BQB that issued the qualification certificate for the product.

The qualification requirements are not the same for all products. Those that are specifically designed and marketed as development tools or demonstration kits are exempt from testing requirements, and qualification is possible for these products by filing a simple declaration of conformance. Products that integrate a Bluetooth component that has been prequalified may be exempt from repeating tests for which the component is prequalified.

## Market for Bluetooth

Since its introduction, Bluetooth wireless technology has been greeted by industry watchers as the most significant development in wireless communications in 20 years. Soon people all over the world will enjoy the convenience, speed, and security of instant wireless connections. To meet these expectations, Bluetooth components are likely to be embedded in hundreds of millions of mobile phones, PCs, laptops, and a whole range of other electronic devices in the next few years.

According to the research firm IDC, Bluetooth wireless technology will be embedded in more than 100 million devices in the U.S. and almost 450 million devices worldwide by 2004. In addition, the technology will move beyond the obvious devices such as laptops and handhelds to a variety of other devices such as printers, digital cameras, and other consumer devices. IDC expects smart phones to be among the earliest adopters of the Bluetooth specification. Printers will begin integrating Bluetooth wireless technology in the 2001 to 2002 timeframe. Cahners In-Stat Group predicts Bluetooth wireless technology will be a built-in feature in more than 670 million products worldwide by 2005.

## Summary

The Bluetooth specification can ease connection not only to the phone system or the Internet but also between devices. Indeed, the focus of Bluetooth wireless technology on low-cost, high levels of integration and ease of configuration has the potential to change current mobile computing and network connectivity paradigms. Moreover, because Bluetooth wireless technology supports both voice and data and a wide range of applications—from file synchronization and business card exchange to Internet access—it will strengthen the mobile computing offerings of all vendors. These include not only devices like notebook computers but also wireless connectivity solutions such as Wireless Domino Access and IBM Mobile Connect.

The choice of wireless LANs using Bluetooth, infrared, HomeRF, or 802.11 specifications will depend on the applications you have. Often, you will find that more than one of these technologies is required to meet your needs. Table 1.5 summarizes the typical applications of these technologies.

There will be points of convergence between Bluetooth and other wireless technologies. Infrared and Bluetooth wireless technologies, for example, provide complementary implementations for data exchange and voice applications. The capabilities of Bluetooth wireless devices complement infrared's point-and-shoot ease of use with omni-directional signaling, longer-distance communications, and capacity to penetrate solid surfaces. For some devices, having both Bluetooth and

## The Case for Bluetooth

infrared components will provide the optimal short-range wireless solution. For other devices, the choice of adding Bluetooth or infrared components will be based on the applications and intended usage models. The story on short-range wireless communications is still being written; both infrared and Bluetooth wireless technology will be major driving forces for development in this area.

**TABLE 1.5**
Typical Applications of Wireless Technologies

| Technology | Typical Applications |
|---|---|
| Bluetooth | Cable elimination, inter-device communication for voice as well as data, PANs, remote device control, mobile e-commerce |
| Infrared | Cable elimination, high-speed file transfer between devices, local device control |
| HomeRF | Cable elimination, data communication between computers, and computers and peripherals, in the home or small office |
| 802.11 Wireless LAN | Cable elimination, data communication between computers, and computers and peripherals, in corporate offices |

CHAPTER 2

# Basic Concepts

As noted in the previous chapter, the objective of the Bluetooth standard is to enable seamless communications of data and voice over short-range wireless links between both mobile and stationary devices. The standard specifies how mobile phones, wireless information devices (WIDs), handheld computers, and personal digital assistants (PDAs) using Bluetooth wireless components can interconnect with each other, with desktop computers, and with office or home phones. With its use of spread-spectrum technology, the first generation of the Bluetooth specification permits the secure exchange of data up to a rate of about 1 Mbps—even in areas with significant electromagnetic activity. With its use of continuously variable slope delta modulation (CVSD) for voice encoding, the Bluetooth specification allows speech to be carried over short distances with minimal disruption.

This chapter explores basic transmission concepts with the objective of establishing how the Bluetooth wireless specification operates to handle data and voice. To arrive at a thorough understanding of Bluetooth wireless technology's place in the communications continuum, comparisons will be made with alternative technologies used in both the wireline and wireless environments. In the process, the capabilities of Bluetooth wireless technology will become apparent and the foundation will be laid for the more detailed discussions of Bluetooth wireless technology in subsequent chapters.

## Serial versus Parallel

Like other communication technologies, Bluetooth wireless technology uses serial communication to transmit data in binary form (0s and 1s). A Bluetooth wireless device does this over a radio frequency (RF) link, but serial data communication is used over infrared as well as copper and fiber links. Serial communication entails the transmission of data in sequential fashion, with the 1s and 0s following each other down the link to the remote device. This mode of information transfer applies to both asynchronous and synchronous data communication. Parallel communication, on the other hand, conveys multiple bits at the same time, with each bit traveling over its assigned wire within a cable.

## Serial Transmission

As the term implies, serial transmission is a method of sending/receiving data one bit at a time in sequential fashion. Most readers are familiar with the RS-232 serial ports on their computers that are used for making the connection to an external modem at speeds up to 20 Kbps, or the newer high-speed USB (Universal Serial Bus) ports for making the connection to other peripherals, such as page scanners, at speeds of up to 12 Mbps. But serial transmission is not used just for connecting local devices to each other. It is the basis of all data communication that occurs over local area networks (LANs) and wide area networks (WANs) and everything in between—whether the medium is copper wire, optical fiber, or an infrared- or radio-based wireless technology.

## Parallel Transmission

While the serial method of transmission involves sending/receiving one bit at a time, the parallel method of transmission involves sending/receiving an entire byte at a time, with each bit carried over a separate channel or wire. If 8 bits are sent at a time, this will require 8 channels or wires, one for each data bit. To transfer data in this way, a separate channel or wire is used for the clock signal, which serves to inform the receiver when data are available. Another channel or wire would be used by the receiver to acknowledge receipt of the data, to indicate that it is ready to accept more data.

Parallel transmission mainly takes place within a computer or between a computer and a locally attached peripheral device such as a printer. Data going between a disk drive and a disk controller within your PC, for example, are transferred in parallel fashion through a ribbon cable.

The most familiar application of parallel transmission is printing, in which a cable connects a computer's parallel port to a printer's parallel port. The cable's 25-pin D-shaped connector, known as a DB-25 connector, plugs into the back of the computer, while the other end of the cable, which uses a 36-conductor Amphenol connector, also known as the Centronics connector, plugs into the back of the printer.[1]

Whatever the application, parallel connections are usually problem free because there is no need for packaging the data into packets on the send side and unwrapping the packets at receive side. The standard parallel port on a PC typically sends data at 115,200 bits per second, while newer enhanced parallel ports are up to 10 times faster, in excess of 1 Mbps. However, parallel transmission is good only for short distances, such as linking components within the computer, making a local connection from the computer to a printer or to some other device such as an external ZIP drive. The serial and parallel data transmission methods are compared in Figure 2.1.

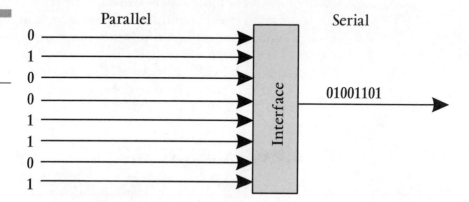

**Figure 2.1** Comparison of parallel and serial transmission.

Bluetooth devices transmit data to other devices in serial fashion over wireless connections—synchronously or asynchronously, depending on the application. But internal processes in the Bluetooth devices themselves use a combination of parallel and serial transmission. For example, to encrypt data for serial transmission over a wireless link, the Bluetooth device employs a stream ciphering process that loads four inputs in parallel to a payload key generator, which uses this information to formulate a payload key (Figure 2.2). The key stream generator sends the payload key in serial fashion to the encoder, where it is encrypted before being sent out over the wireless link, also in serial fashion. The sending and receiving devices use this key to establish mutual authentication so that subsequent message flow between them will be secure. Only with the proper key can the two stations unlock received data.

---

[1] The use of different connectors at each end of the cable resulted from the non-standard way equipment was built in the early days of desktop computing. When IBM built its PC, it opted to not use the 36-conductor connector from the dominant printer manufacturer of the time, which was Centronics. Instead, IBM equipped its computers with the 25-pin D shell connector—the DB-25 connector. Over the years, printer manufacturers have stuck with the Centronics connectors, while PC manufacturers have standardized on DB-25 connectors. This is the reason why special adapter cable is needed to make the connection between the two devices.

# Basic Concepts

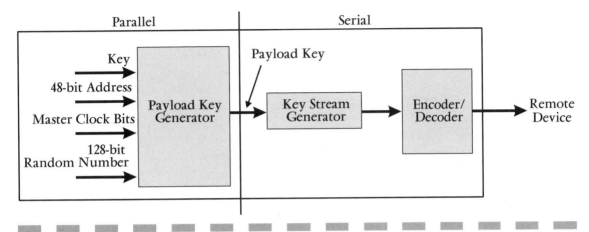

**Figure 2.2** At the component level, Bluetooth wireless technology uses a combination of parallel and serial processes to generate and send a payload key, which is used to establish mutual authentication between two devices and assure secure data transmission over the wireless link.

## Asynchronous versus Synchronous

The problem with serial data transmission is how to synchronize the receiver with the sender so that the receiver can detect the beginning of each new character in the bit stream being presented to it. Without synchronization, the receiver will not be able to interpret the incoming bit stream correctly. There are two approaches to serial data transmission that solve the problem of synchronization: asynchronous and synchronous transmission. The Bluetooth specification supports both.

In asynchronous transmission, synchronization is established by bracketing each set of 8 bits by a start and stop bit. A start bit is sent by the sending device to inform the receiving device that a character is being sent. The character is then sent, followed by a stop bit, which indicates that the transfer of that character is complete. This process continues for the duration of the session.

In synchronous transmission, data goes out over the link as a continuous stream of bits. Start and stop bits are not used to bracket each character. Instead, the devices at each end of the link rely on timing as the means of determining where characters begin and end in the bit stream. For this method of transmission to work properly, however, the devices at each end of the link must be in perfect synchronization with each other using a common clock source. They accomplish this by sending special characters, called synchronization, or syn, characters before sending any user data. When the clocks of each device are in synchronization, the user data are sent.

## Asynchronous

Asynchronous means that the bits in the serial data stream are not locked to a specific clock at the receiving end. This makes the asynchronous method of serial data transmission ideal for PC or simple terminal connections, where characters are generated at irregular intervals from a keyboard. The advantage of asynchronous transmission in this case is that each individual character is self-contained, so that if it becomes corrupted along the way, the characters before it and after it will be unaffected. Only the lost or corrupt character needs to be retransmitted.

In the PC environment, 7- or 8-bit characters are often used. Seven bits are enough to encode all upper- and lowercase characters, symbols and function keys, which number 128, in conformance with the American Standard Code for Information Interchange (ASCII). An optional eighth bit, called the *parity bit*, is used to check data integrity. When used, it is inserted between the last bit of a character and the first *stop bit*. As will be discussed later, a stop bit is used in asynchronous communication to indicate the end of a sequence of bits forming a character (Figure 2.3).

**Figure 2.3**
This 10-bit character frame consists of 2 bits for start-stop, a parity bit, and 7 bits of user data.

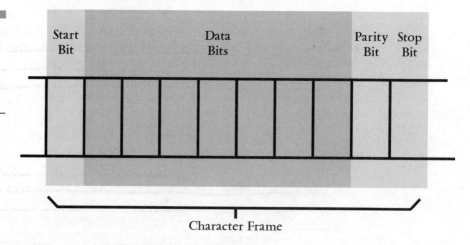

# Basic Concepts

The parity bit is included as a simple means of error checking. There is *even* and *odd parity*. The devices at each end of the connection must have the same parity setting. The idea is that you specify ahead of time whether the parity of the transmission is to be even or odd. This is done from within Windows when configuring the connection preferences the modem will use for dialup networking (Figure 2.4).

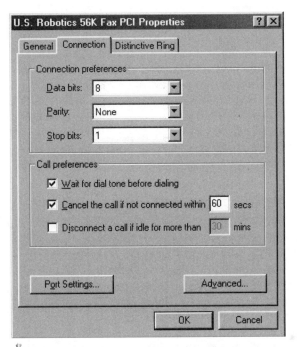

**Figure 2.4**
The connection preferences for a U.S. Robotics 56Kbps Fax PCI modem, as configured from within Windows 98. The settings of 8-None-1 cover just about every system you are likely to call, including Bulletin Board Systems (BBSs) and the Internet. Sometimes the settings of 7-Even-1 may be required when dialing up UNIX systems and mainframes.

Suppose the parity chosen is odd. The transmitter will then set the parity bit in such a way as to make an odd number of 1s among the data bits and the parity bit. For example, if there are five 1s among the data bits, already an odd number, the parity bit will be set to 0.

While this discussion of the parity bit applies to the modem connections most people are familiar with, it is also useful for understanding how this type of bit is used in wireless data communications between Bluetooth devices to implement rudimentary error correction. The data packets can be transferred with parity enabled or disabled. If it is enabled, users of Bluetooth devices will be able to configure their connection preferences for even or odd parity, depending on the host system they are connecting with. When it comes to voice, however, even though Bluetooth devices package digitally encoded speech into packets, parity is not enabled.

With asynchronous communication, the transmitter and receiver only have to approximate the same clock rate. For a 10-bit sequence, the last bit will be interpreted correctly even if the sender and receiver clocks differ by as much as 5 percent. This makes asynchronous a relatively simple and, therefore, inexpensive method of serial data transmission.

However, asynchronous transmissions include high overhead since each byte carries at least two extra bits for the start-stop functions, which results in a 20 percent loss of bandwidth (2/10 = .20) For large amounts of data, this adds up quickly. For example, to transmit 1,000 characters, or 8,000 bits, 2,000 extra bits must be transmitted for the start and stops, bringing the total number of bits sent to 10,000. The 2,000 extra bits is equivalent to sending 250 more characters over the link.

Bluetooth wireless technology supports one asynchronous channel that provides a data rate of almost 1 Mbps. The asynchronous connectionless link (ACL) supports data traffic on a best-effort basis. The ACL link supports packet-switched, point-to-multipoint connections, which typically are used for data. If errors are detected at the receiving device, a notification is sent in the header of the return packet, so that only lost or corrupt packets need to be retransmitted.

The ACL link can operate in symmetrical or asymmetrical fashion. Symmetrical means that the link offers the same data rate in both the send and receive directions. Asymmetrical means that the link offers a different data rate in the send and receive directions. For symmetrical connections, the maximum data rate is 433.9 Kbps in both directions, send and receive. For asymmetrical connections, the maximum data rate is 723.2 Kbps in one direction and 57.6 Kbps in the reverse direction. An asymmetrical connection would be used for Internet access, for example, where it is more important to have the higher data rate in the receive direction to retrieve Web content. Specifying the location of the Web page, on the other hand, requires far less bandwidth in the send direction.[2]

---

[2] The use of asymmetrical and symmetrical connections is not unique to the Bluetooth specification. There are Digital Subscriber Line (DSL) services, which operate over ordinary telephone lines, that are either asymmetrical or symmetrical. Asymmetrical Digital Subscriber Line (ADSL) runs at up to 8 Mbps in the downstream direction (toward the user) and up to 640 Kbps in the upstream direction (toward the Internet), with the actual speeds in each direction dependent upon the distance from the customer to the carrier's Central Office (CO). Symmetrical Digital Subscriber Line (SDSL) offers up to 1.1 Mbps in each direction, with the actual speeds in each direction dependent upon the distance from the customer to the carrier's Central Office (CO). Rate Adaptive Digital Subscriber Line (RADSL) supports both symmetrical and asymmetrical data transmission.

## Synchronous

The Bluetooth wireless technology specification supports a synchronous connection-oriented (SCO) link. This type of link provides circuit-switched, point-to-point connections, which typically are used for voice, data, and multimedia traffic using reserved bandwidth. SCO links are symmetrical, providing the same amount of bandwidth in both the send and receive directions.

### DATA

In dispensing with the need for start-stop bits, synchronous transmission relies on accurate timing between the sending and receiving devices to make sense of the 1s and 0s in the bit stream during decoding. If both devices use the same clock source, then transmission can take place with assurance the bit stream will be accurately interpreted by the receiver. To guard against the loss of synchronization, the receiver is periodically brought into synchronization with the transmitter through the use of control bits embedded somewhere in the bit stream.

In the synchronous method of transmission, data are not sent as individual bytes bracketed by start-stop bits, but as packets in reserved time slots that are set up between specific sending and receiving devices. Synchronous transmission is usually much more efficient in the use of bandwidth than asynchronous transmission, if only because the data field is usually much larger than the overhead fields. Another advantage of synchronous transmission is that the packet structure allows for easy handling of control information. There is a natural position, usually at the start of the packet, or through the use of a setup message as in the case of the Bluetooth architecture, for any special codes needed for conveying timing parameters and specifying the reserved slots.

### VOICE

The Bluetooth specification specifies three synchronous channels of 64 Kbps each, the same amount of bandwidth used to convey voice conversations over digital T-carrier lines or ISDN services over the Public Switched Telephone Network (PSTN). Unlike the data setup, packets carrying voice can never be retransmitted because the resulting delay would be disruptive to both speaker and listener. As noted in Chapter 1, the Bluetooth specification specifies the use of two voice-encoding schemes: Pulse Code Modulation (PCM) and Continuously Variable Slope Delta (CVSD) modulation. Of the two, CVSD is more immune to interference and therefore better for voice communication over a wireless link. The choice of PCM or CVSD is made by the Link Managers of each Bluetooth device, which negotiate the most appropriate voice-coding scheme for the application.

To appreciate CVSD, it is useful to compare it with PCM. Briefly, PCM samples the changing amplitudes of the analog waveform 8,000 times a second and assigns each point along the sampled waveform a value that is expressed as an 8-bit word, or byte (Figure 2.5).[3] By sampling the analog waveform in this manner, a person's voice can be captured and represented in digital form. The resulting bit stream is sent down the link in the form of electrical pulses that represent 0s and 1s. At the receiving device, this information is decoded to represent a very close approximation of the original analog waveform, which results in speech intelligible to the listener.

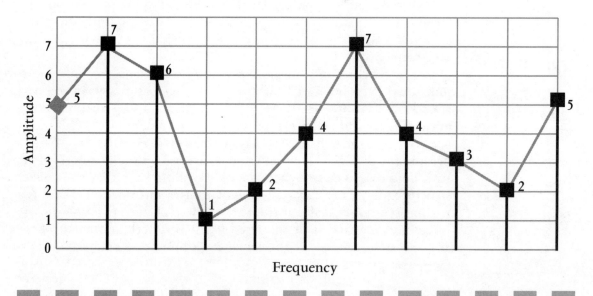

**Figure 2.5** Pulse Code Modulation (PCM) samples the constantly changing analog waveform 8,000 times a second. The resulting values are then encoded into 8-bit words.

---

[3] Under PCM, sampling of the analog waveform is done 8,000 times a second, with each value represented by 8 bits. This results in a transmission rate of 64,000 bits per second, or 64 Kbps, which constitutes the size of a basic voice channel on a digital link, whether the link is a wired or wireless connection.

# Basic Concepts

CVSD works differently in that the length of the samples is just 1 bit, rather than the 8 bits used in PCM. With a digital word this small, many more samples can be sent in the same amount of bandwidth, so that the transmission is more immune to interference. But 1-bit words cannot measure volume or loudness. Rather than representing the change in height of the analog signal, the sampling bit used in CVSD refers only to a change in the slope, or steepness, of the curve.

CVSD modulation is a method of digitizing a speech signal that takes advantage of the fact that voice signals do not change abruptly. Instead of sampling the amplitude of a speech signal 8,000 times a second and encoding the resultant values as 8-bit words, as in PCM, CVSD uses only 1-bit words, each of which refers to a change in the slope or steepness of the analog signal's "curve."

In essence, the CVSD modulator is a 1-bit analog-to-digital converter. The output of this 1-bit encoder is a serial bit stream, with each bit representing an incremental increase or decrease in signal amplitude, which is determined as a function of recent sample history. The number of past samples (i.e., bits) used to make a prediction is three. With only a 1-bit word, many more samples can be sent in the same amount of bandwidth, making CVSD very robust.

The CVSD modulator, or converter, is found in various vendors' Bluetooth chip sets, which are embedded into consumer devices like cell phones and PDAs. The CVSD converter consists of an encoder-decoder pair, with the decoder connected as part of a feedback loop. The encoder receives a band-limited audio signal and compares it to the analog output of the decoder. The result of the comparison is a serial string of ones and zeros, each indicating whether the band-limited audio sample's amplitude is above or below the decoded signal. When a run of three identical bits is encountered, the slope of the generated analog approximation is increased in its respective direction until the string of identical bits is broken. The CVSD decoder performs the inverse operation of the encoder and regenerates the audio signal.

CVSD has several attributes that make it well suited for digital coding of speech. One-bit words eliminate the need for complex framing schemes. With its robust performance in the presence of bit errors, error detection and correction functions are unnecessary. Other speech encoding schemes may require a digital signal processing (DSP) engine and extra components for analog-to-digital/digital-to-analog conversion. Another advantage of CVSD is that the entire coder/decoder algorithm, including input and output filters, can be integrated on a single silicon substrate. Even with this simplicity, CVSD has enough flexibility to allow digital encryption for secure conversations.

Finally, CVSD can operate over a wide range of data rates. Conceived in 1970, CVSD has been successfully used from 9.6 Kbps to 64 Kbps. While audio quality at 9.6 Kbps is noticeably degraded, it is still intelligible. At data rates of 24 Kbps to 48 Kbps audio quality is acceptable. And above 48 Kbps it is comparable to toll quality voice. All of these attributes make CVSD attractive to wireless communication systems, notably Bluetooth.

Although errors can occur even with CVSD, voice can never be retransmitted because of the delay that would ensue. So when errors occur, they show up at the receiving device as merely background noise. This noise intensifies as bit errors increase. But for limited distance, point-to-point voice communication in the Bluetooth environment, CVSD performs very well.

## Spread Spectrum

The Bluetooth wireless technology uses a digital coding technique called spread spectrum, a method of wireless communications that takes a narrowband signal and spreads it over a broader portion of the available radio frequency band. Among other advantages, the resulting signal is highly resistant to interference and more secure against interception. The same technology is used in cordless phones and wireless local area networks (LANs). In addition, many cellular services

# Basic Concepts

use Code Division Multiple Access (CDMA), a modulation and access technique that is based upon the spread-spectrum concept.

In recent years, CDMA systems have gained widespread global acceptance by wireless operators. This technique differs from that used to transmit voice and data over Time Division Multiple Access (TDMA) networks, which assigns each user a time slot in a narrow band of spectrum. While Bluetooth wireless technology makes use of spread-spectrum technology, it also uses a derivative of TDMA, called Time Division Duplexing (TDD), to provision the time slots used for voice and data communication.

The U.S. patent for spread-spectrum technology was held jointly by actress Hedy Lamarr (Figure 2.6) and music composer George Antheil. Their patent for a "Secret Communication System," issued in 1942, was based on the frequency-hopping concept, with the keys on a piano representing the different frequencies and frequency shifts used in music.[4]

**Figure 2.6**
Actress Hedy Lamarr (1914–2000), co-developer of spread-spectrum technology.

---

[4] In 1942, the technology did not exist for a practical implementation of spread spectrum. When the transistor finally did become available, the Navy used the idea in secure military communications. When transistors became really cheap, the idea was used in cellular phone technology to keep conversations private. By the time the Navy used the idea, the original patent had expired and Lamarr and Antheil never received any royalty payments for their idea.

During World War II, Lamarr had become intrigued with radio-controlled missiles and the problem of how easy it was to jam the guidance signal. She realized that if the signal could be made to jump from one frequency to another very quickly—like changing stations on a radio—and both the sender and receiver changed in the same order at the same time, then the signal could never be blocked without knowing exactly how and when the frequency changed. Although the frequency-hopping idea could not be implemented due to technology limitations at that time, it eventually became the basis for cellular communication.

## Spreading

Spread spectrum is a digital coding technique in which a narrowband signal is taken apart and "spread" over a spectrum of frequencies (Figure 2.7). The coding operation increases the number of bits transmitted and expands the amount of bandwidth used. Using the same spreading code as the transmitter, the receiver correlates and collapses the spread signal back down to its original form. The result is a highly robust wireless data transmission technology that offers substantial performance advantages over conventional narrowband radio systems.

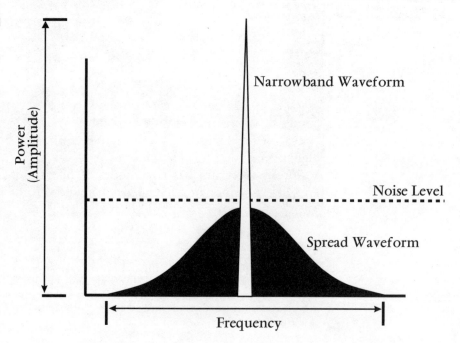

**Figure 2.7**
Spread spectrum transmits the entire signal over a bandwidth much greater than that required for standard narrowband transmission. Increasing the frequency range allows more signal components to be transmitted, which results in a more accurate reconstruction of the original signal at the receiving device.

# Basic Concepts

One of the advantages of spread spectrum is that the spread signal has a much lower power density. This low power density, spread over the expanded transmitter bandwidth, provides resistance to a variety of conditions that can plague narrowband radio systems, including:

- **Interference**—A condition in which a transmission is being disrupted by external sources, such as the noise emitted by various electromechanical devices, or internal sources such as cross talk
- **Jamming**—A condition in which a stronger signal overwhelms a weaker signal, causing a disruption to data communications
- **Multipath**—A condition in which the original signal is distorted after being reflected off a solid object
- **Interception**—A condition in which unauthorized users capture signals in an attempt to determine their content.

Conventional narrowband radio systems transmit and receive on a specific frequency that is just wide enough to pass the information, whether voice or data. In assigning users different channel frequencies, confining the signals to specified bandwidth limits, and restricting the power that can be used to modulate the signals, undesirable cross talk—interference between different users—can be avoided. These rules, enforced by regulatory agencies in each country, are necessary because any increase in the modulation rate widens the radio signal bandwidth, which increases the chance for cross talk.

The main advantage of spread spectrum radio waves is that the signals can be manipulated to propagate fairly well through the air, despite electromagnetic interference, virtually eliminating cross talk. In spread-spectrum modulation, a signal's power is spread over a larger band of frequencies. This results in a more robust signal that is less susceptible to interference from similar radio-based systems, since they too are spreading their signals, but with different spreading algorithms.

Spread spectrum has two modes of operation: frequency hopping and direct sequencing. _Frequency hopping_ spreads its signals by "hopping" the narrowband signal over the entire radio band as a function of time. _Direct sequencing_ spreads its signal by expanding the signal all at once over the entire radio band. Although Bluetooth wireless technology uses the frequency-hopping mode of spread spectrum, it is useful to contrast it with direct sequence to appreciate its advantages.

## Direct Sequence

In direct sequence spreading, the radio energy is spread across a larger portion of the band than is actually necessary for the data. This is done by breaking each data bit into multiple sub-bits called *chips* to create a higher modulation rate. The higher modulation rate is achieved by multiplying the digital signal with a chip sequence. If the chip sequence is 10, for example, and it is applied to a signal carrying data at 300 Kbps, then the resulting bandwidth will be 10 times wider. The amount of spreading is dependent upon the ratio of chips to each bit of information.

Because data modulation widens the radio carrier to increasingly larger bandwidths as the data rate increases, this chip rate of 10 times the data rate spreads the radio carrier to 10 times wider than it would otherwise be for data alone. The rationale behind this technique is that a spread spectrum signal with a unique spread code cannot create the exact spectral characteristics as another spread-coded signal. Using the same code as the transmitter, the receiver can correlate and collapse the spread signal back down to its original form, while other receivers using different codes cannot.

This feature of spread spectrum makes it possible to build and operate multiple networks in the same location. By assigning each one a unique spreading code, all transmissions can share the same frequency band yet remain independent of each other. The transmissions of one network appear to the other as random noise and are filtered out because the spreading codes do not match.

This spreading technique would appear to result in a weaker signal-to-noise ratio, since the spreading process lowers the signal power at any one frequency. Normally, a low signal-to-noise ratio would result in damaged data packets requiring retransmission. However, the processing gain of the receiver's despreading correlator recovers the loss in power when the signal is collapsed back down to the original data bandwidth, but is not strengthened beyond what would have been received had the signal not been spread.

In the United States, the FCC has set rules for direct sequence transmitters. Each signal must have 10 or more chips. This rule limits the practical raw data throughput of transmitters to 2 Mbps in the 902 MHz band and 8 Mbps in the 2.4 GHz band. The number of chips is directly related to a signal's immunity to interference. In an area with a lot of radio interference, users will have to give up throughput to limit interference successfully.

# Frequency Hopping

Bluetooth wireless technology uses the frequency-hopping version of spread spectrum, which entails the transmitter's jumping from one frequency to the next at a specific hop rate in accordance with a pseudorandom code sequence. The order of frequencies selected by the transmitter is taken from a predetermined set as dictated by the code sequence. For example, the transmitter may have a hopping pattern of going from the third frequency, to the twelfth frequency, to the fifth frequency, and so on across an entire range of frequencies. The receiver tracks these changes. Since only the intended receiver is aware of the transmitter's hopping pattern, only that receiver can make sense of the data being transmitted.

The FCC mandates that frequency-hopped spread-spectrum systems not spend more than 0.4 seconds on any one channel each 20 seconds, or 30 seconds in the 2.4-GHz band. Furthermore, they must hop through at least 50 channels in the 900-MHz band, and 75 channels in the 2.4-GHz band. These rules reduce the chance of repeated packet collisions in areas with multiple frequency-hopping transmitters. The Bluetooth specification specifies a rate of 1,600 hops per second among 79 frequencies.

All Bluetooth units participate in a piconet, with each unit sharing a common channel. Up to eight interconnected devices can be supported by a piconet, with one master and up to seven slaves. This relationship remains in place for the duration of the piconet connection. The units that participate in a piconet are time and hop synchronized to the same channel. Every Bluetooth unit has an internal system clock, which determines the timing and hopping used by its transceiver. Timing and the frequency hopping on the channel of a piconet are determined by the clock of the master. When the piconet is established, the master clock is communicated to the slaves. Each slave adds an offset to its native clock to get in step with the master clock. Since the clocks are free running, the offsets have to be updated regularly.

Other frequency-hopping transmitters in the vicinity will be using different hopping patterns and much slower hop rates than Bluetooth devices. Should transmitters that do not use Bluetooth wireless technology coincidentally attempt to use the same frequency at the same moment, the data packet transmitted by one or both devices will become garbled in the collision, and a retransmission of the affected data packets will be required. A new data packet will be sent again on the next hopping cycle of each transmitter.

Although the chance of devices that use Bluetooth wireless technology interfering with those that do not, but that share the same 2.4 GHz band is minimal, some manufacturers are concerned enough that they have petitioned the FCC for new guidelines that would provide more separation among competing protocols that operate in the 2.4-GHz band.[5] Both the Bluetooth SIG and IEEE recognize the potential for such signal interference and are working together to figure out how to enable their technologies to coexist. The impact of any change must be determined for portable phones and other devices such as microwave ovens, wireless speakers, and security systems, which also share this band.

## Circuit and Packet Switching

The Bluetooth protocol uses a combination of circuit and packet switching.* Circuit switching is familiar to anyone who has placed a telephone call on the PSTN. When a number is dialed, a dedicated path is set up through the network to handle the call (Figure 2.8). The path remains in place for the duration of the call. At the conclusion of the call, the path is torn down and the network resources used to build the connection become free to handle another call from someone else. Circuit switching can apply to wireless networks as well as wireline networks, and to data as well as voice.

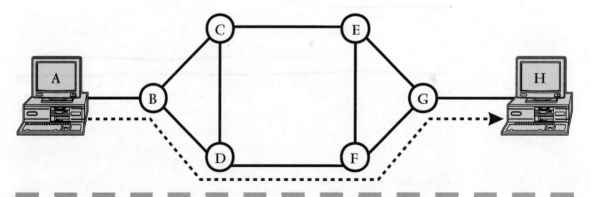

**Figure 2.8**  In circuit switching, a dedicated path is set up between endpoints to handle a voice or data call, which stays in place until the call is terminated.

---

[5] The FCC has opened Docket 99-231 for consideration of this matter. At this writing, no decision has been announced. Readers interested in staying updated on this issue can go to the FCC's Web page at www.fcc.gov.

# Basic Concepts

In packet switching, data travel in the form of individually addressed packets over multiple shared paths on the way to its destination. There is no connection setup and teardown—the paths are always available.

Packet switching is familiar to anyone who has accessed information on the Internet. When you download a Web page, for example, text and graphical content are assembled into packets for delivery to your computer. On the way through the Internet, however, the packets may take many different paths over connections that are always available and shared by others, who also may be downloading Web pages. Ultimately, all these individually addressed packets arrive at their proper destination. Packets destined for your computer, which arrive out of order, are first put back in order for delivery to your computer. There, a Web browser like AOL's Netscape Navigator or Microsoft's Internet Explorer renders the requested page according to instructions that are encoded with the HyperText Markup Language (HTML).

Circuit switching is effective for voice conversations, while packet switching is efficient for data transmission. For a voice call, the connection to another device must be specifically requested and set up. For data transmission, no connection needs to be set up through the network because the individually addressed packets travel over shared paths that are always available. However, because packets share the paths with other packets going to other, unrelated, destinations, there is always some amount of delay to contend with. Therefore, circuit-switching networks are ideal for communications that require voice and data to be transmitted in real time, while packet-switching networks are more efficient if some amount of delay is acceptable.

For both circuit and packet switching, the Bluetooth wireless technology uses time slots, which constitute channels. For circuit-switched voice, up to three time slots are reserved for synchronous packets so that up to three simultaneous synchronous voice channels are supported, each over its own time slot. For packet-switched data, Bluetooth wireless technology supports an asynchronous data channel. The devices in a piconet "listen" for packets addressed to them. The Bluetooth specification also supports a hybrid channel that can simultaneously support asynchronous data and synchronous voice.

Bluetooth systems can support a point-to-point connection between two units, or a point-to-multipoint connection with more than two units where the channel is shared among all participating Bluetooth units. Two or more units sharing the same channel form a piconet. The unit initiating the call becomes the master and the other units the slaves. The timing for synchronous communications comes from the master unit's internal clock.

## Time Division Duplexing

There are different ways of dividing up the available spectrum to provide users with organized access to it. In most communication systems it is desirable to be able to communicate in both directions at the same time. This is an obvious requirement for voice communication, but it applies to data as well. On the Internet, for example, you might enter the address of a Web page in the command line of your browser. This request goes out over the link in packet form to the appropriate Web server, which returns the appropriate content to your computer. Since devices using Bluetooth wireless technology will provide access to the Internet, as well as voice communication, two-way communication is a fundamental capability. Apart from Internet access, file transfers between portable devices and desktop computers require the two-way capability, especially during the file synchronization process.

When a device is capable of sending and receiving at the same time, the two-way capability is known as *full-duplex* operation. With a voice conversation this is desirable because it lets one party interrupt the other with a question or one device immediately request a retransmission of a block of information received in error during a data communications session. There are two basic ways of providing for full-duplex operation in a radio system: Frequency Division Duplexing (FDD) and Time Division Duplexing (TDD). Bluetooth devices use the latter.

Of the two, FDD is the older, having been in use since the 1920s. It is still the most common duplex method employed in wireless communication systems. Transmit and receive functions are separated in the frequency domain, and sufficient isolation of the two frequencies is required to prevent them from interfering with each other (Figure 2.9). Systems based on FDD must have two antennas, each tuned to a different frequency: one for transmission, the other for reception. Since both operations occur simultaneously, the radio system provides true full-duplex capability.

While FDD was designed to carry analog voice traffic, which is typically symmetrical and predictable, TDD was designed to carry digital data traffic, which is typically asymmetrical and unpredictable. TDD uses a single channel for both sending and receiving information, with the transmission direction alternating between sending and receiving. It uses one antenna that divides its time between transmitting and receiving signals. Separation between transmit and receive is achieved in the time domain. In addition, the amount of bandwidth allocated to each direction is flexible.

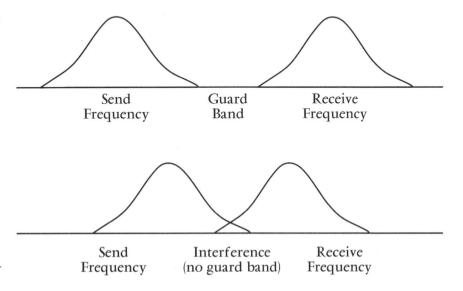

**Figure 2.9**
In Frequency Division Duplexing (FDD), two separate frequencies are used for sending and receiving. A guard band prevents the two frequencies from interfering with each other (top). If no guard band were present, there would be the possibility of interference (bottom).

Although not a true full-duplex capability, TDD comes very close to it, making it suitable for voice as well as data. The operation is more accurately described as *half-duplex*. However, the process of alternating between send and receive occurs so rapidly that the end user does not perceive any gaps or delays in what is heard. To the end user it appears as a true full-duplex connection. The Bluetooth specification even refers to TDD as "full-duplex."

The channel is divided into time slots, each 625 microseconds in length. The time slots are numbered according to the Bluetooth clock of the piconet master. In the time slots, master and slave can transmit packets. In the TDD scheme, master and slave transmit alternatively (Figure 2.10). The master starts its transmission in even-numbered time slots only, and the slave start its transmission in odd-numbered time slots only. The start of the packet is aligned with the slot start. Packets transmitted by the master or the slave may extend over as many as five time slots.

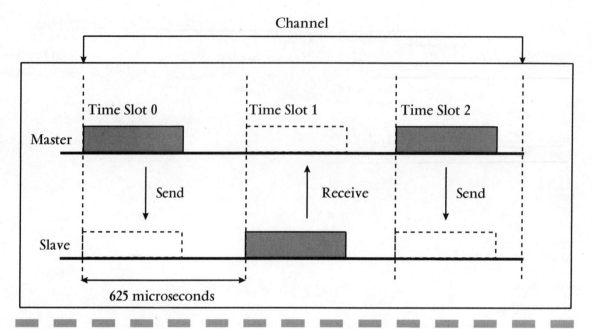

**Figure 2.10** With the TDD scheme used in Bluetooth wireless technology, packets are sent over time slots 625 microseconds in length between the master and slave units within a piconet.

The RF hop frequency is fixed for the duration of the packet (Figure 2.11). For a single packet, the RF hop frequency to be used is derived from the current Bluetooth clock value. For a multi-slot packet, the RF hop frequency to be used for the entire packet is derived from the Bluetooth clock value in the first slot of the packet. The RF hop frequency in the first slot after a multi-slot packet uses the frequency as determined by the master device's current clock value. If a packet occupies more than one time slot, the hop frequency applied is the same as the time slot where the packet transmission was started.

TDD is valued for its ability to handle the asynchronous demand of uploading and downloading better than FDD. With TDD, bandwidth can be allocated on an as-needed basis, changing the makeup of the traffic flow as demand warrants. For example, if the user wants to download a large data file, as much bandwidth as is needed will be allocated to the transfer. Then, at the next moment, if a file is being uploaded, that same amount of bandwidth can be allocated to that transfer. FDD technology, because of its division of frequencies for sending and receiving information, has a fixed allocation of bandwidth in each direction and therefore is not as spectrally efficient as TDD.

# Basic Concepts

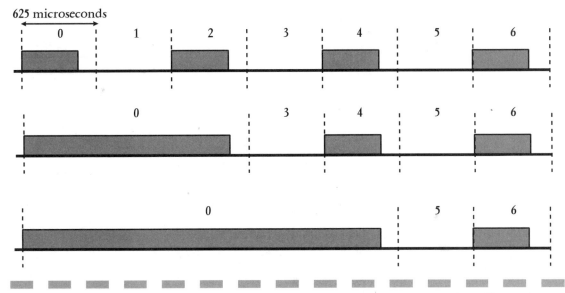

**Figure 2.11**  The hop definition on single-slot packets (top) and multi-slot packets (middle and bottom).

While FDD and TDD define different methods for establishing bidirectional radio links, they do not define a particular access technique that would allow multiple users to share the available bandwidth, enabling the system to operate more efficiently. Examples of technologies that employ multiple access schemes are Frequency Division Multiple Access (FDMA), Time Division Multiple Access (TDMA), and Code Division Multiple Access (CDMA). These technologies are widely used in cellular networks.

FDMA is still used on some first-generation cellular analog networks, such as AMPS (Advanced Mobile Phone Service) and TACS (Total Access Communications System).[6] TDMA is used on second-generation digital cellular networks, such as North American Digital Cellular and Global System for Mobile (GSM) communications. CDMA is also used on second-generation digital cellular networks, such as PCS 1900. Both TDMA and CDMA have been enhanced to support emerging third-generation networks that fall under the International

---

[6] There is a digital version of AMPS called D-AMPS (Digital-Advanced Mobile Phone Service). D-AMPS adds TDMA to AMPS to get three sub-channels for each AMPS channel, tripling the number of calls that can be handled on a channel.

Mobile Telecommunications-2000 (IMT-2000) initiative sponsored by the International Telecommunication Union (ITU).

FDMA divides the available bandwidth into a range of radio frequencies, each of which constitutes a channel. Only one subscriber is assigned to a channel at a time. Other conversations can access this channel only after the subscriber's call has terminated. TDMA divides conventional radio channels into time slots to obtain higher capacity. As with FDMA, no other conversations can access an occupied TDMA channel until it has been vacated, either by call termination or when the subscriber moves out of the cell's coverage area, in which case the call is handed off to another base station. Either way, the original time slot becomes free and may be reused.

CDMA uses a radically different approach. It is both a modulation and an access technique that is based upon the spread-spectrum concept. As noted, a spread-spectrum system is one in which the bandwidth occupied by the signal is much wider than the bandwidth of the information signal being transmitted. For example, a voice conversation with a bandwidth of 3 KHz or so would be spread over 1 MHz or more of spectrum.

In spread-spectrum systems, multiple conversations simultaneously share the available spectrum in both the time and frequency dimensions. The available spectrum is not "channelized" in frequency or time as in FDMA and TDMA systems, respectively. Instead, the individual conversations are distinguished through coding; that is, at the transmitter, each conversation is processed with a unique spreading code that is used to distribute the signal over the available bandwidth. The receiver uses the unique code to accept the signal associated with a particular conversation. The other signals present are each identified by a different code and simply resemble background noise. In this way, many conversations can be carried simultaneously within the same block of spectrum.

The following analogy is commonly used to explain how CDMA technology works. Four speakers are simultaneously giving a presentation, and they each speak a different native language: Spanish, Korean, English, and Chinese (Figure 2.12). If English is your native language, you only understand the words of the English speaker and tune out the Spanish, Korean, and Chinese speakers. You hear only what you know and recognize. The rest sounds like background noise. The same is true for CDMA. Each conversation is specially encoded and decoded for a particular user. Multiple users share the same frequency band at the same time, yet each user hears only the conversation he/she can interpret.

# Basic Concepts

**Figure 2.12**
In this analogy of CDMA functionality, each conversation is specially encoded and decoded for each particular user. Thus, the English-speaking person will only hear another English-speaking person and tune out the other languages, which are heard as background noise.

CDMA assigns each subscriber a unique code to put multiple users on the same wideband channel at the same time. These codes are used to distinguish between the various conversations. The result of this access method is increased call-handling capacity. Under ideal conditions, CDMA can provide 10 to 20 times the capacity of FDMA, and 4 to 8 times the capacity of TDMA. Figure 2.13 summarizes the distinctions between the three technologies.

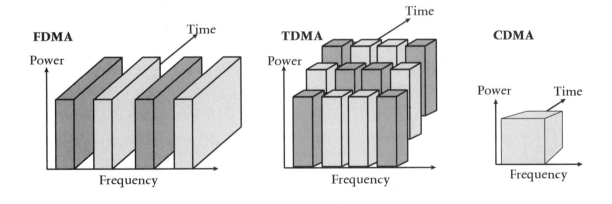

**Figure 2.13** Simple comparison of Frequency Division Multiple Access (FDMA), Time Division Multiple Access (TDMA), and Code Division Multiple Access (CDMA).

## Physical Links

As noted in Chapter 1, two kinds of links can be established between a master and one or more slaves: connection-oriented and connectionless. *Connection-oriented links* require that a session be established before any data can be sent, just like a voice call through the circuit-switched PSTN. With such networks, data are guaranteed to arrive in the order sent. *Connectionless networks* do not require that a session be established between a sender and receiver. The sender simply starts sending packets to the receiver. With such networks, data may not arrive in the same order it was sent. ATM and frame relay traffic, for example, use connection-oriented virtual circuits between a sender and receiver. With IP, packets may travel multiple paths through the network before arriving at their final destination, making IP a connectionless service.

Bluetooth wireless devices make use of both connection-oriented and connectionless links; specifically, Synchronous Connection-Oriented (SCO) and Asynchronous Connectionless (ACL) links.

### SCO Links

The SCO link is a point-to-point link between a master and a single slave in the piconet. The master maintains the SCO link by using reserved slots at regular intervals. The ACL link is a point-to-multipoint link between the master and all the slaves participating on the piconet. In the slots not reserved for one or more SCO links, the master can establish an ACL link on a per-slot basis to any slave, including the slave(s) already engaged in an SCO link.

The SCO link is a symmetric, point-to-point link between the master and one or more specific slaves. The SCO link typically supports time-bounded information like voice. Since the SCO link reserves slots, it is considered a circuit-switched connection between the master and the slave. The SCO link is established by the master, which sends a setup message via the Link Management (LM) protocol. This message contains the timing parameters and specifies the reserved slots.

The master can support up to three SCO links to the same slave or to different slaves within the piconet. A slave can support up to three SCO links from the same master, or two SCO links if the links originate from different masters. Since the packets carried over SCO links contain time-sensitive information, they are never retransmitted when errors occur.

The master sends packets over the SCO links at regular intervals, counted in slots, to the slave in the reserved master-to-slave slots. The slave is always allowed to respond with a packet in the following slave-to-master slot unless a different slave was addressed in the previous master-to-slave slot. If the slave fails to decode the slave address in the packet header, it is still allowed to return a packet in the reserved slot.

### ACL Links

In the slots not reserved for SCO links, the master can exchange packets with any slave on a per-slot basis. The ACL link provides a packet-switched connection between the master and all active slaves participating in the piconet. Both asynchronous and isochronous services are supported on the ACL link, but between a master and a slave only a single ACL link can be active. For most ACL packets, packet retransmission is applied to assure data integrity.

A slave is permitted to return an ACL packet in the slave-to-master slot only if it has been addressed in the preceding master-to-slave slot. If the slave fails to decode the slave address in the packet header, it is not allowed to transmit. ACL packets that are not addressed to a specific slave are considered broadcast packets intended for every slave. If there are no data to be sent on the ACL link and no polling is required, no transmission takes place.

## Peeking into Packets

Packets are used in many data-oriented communications services, and can also be used to convey digitized voice. Simply put, packets are packages of data that are exchanged between devices over a communications link. Large transmissions are divided into packets instead of being transmitted as one long string so that if a packet gets corrupted during transmission, only that packet needs to be retransmitted, instead of the entire transmission. If voice is transmitted in the form of packets, these are usually not retransmitted because of the delay that would be involved, which could disrupt a conversation.

Packets have a formal structure that is specified in the standards for each type of network. Standards organizations facilitate industry consensus on this and other matters so that manufacturers can build products that work together on the same network. Inside the packet are various fields that contain specific information that is interpreted and acted upon by devices in the network on the way to its destination. Depending on the protocol, a packet may include fields for the source and destination address, priority indication, flow control, error correction, and sequencing. These types of fields are often referred to as "overhead" to distinguish them from user data, which are also included in the packet.

There are two types of addresses: a device address (Layer 2) and a network address (Layer 3). Stations on conventional LANs like Ethernet and Token Ring use device addresses called Media Access Control (MAC) addresses. These are the 6-byte hardware-level addresses of the network interface cards (NICs) that provide workstations and other devices with the means to interconnect with each other through a hub or switch. An example of a MAC address is: 00 00 0C 00 00 01. The first three bytes contain a manufacturer code (the one above is for Cisco Systems), the last three bytes contain a unique station ID that are burned into the NIC's firmware. Manufacturer IDs are assigned by the Institute of Electronic and Electrical Engineers (IEEE).

Bridges on the LAN operate in "promiscuous" mode; that is, they listen to all traffic on all connected segments, which together comprise a single broadcast domain. A bridge reads the MAC addresses and filters them, so that information from Segment A addressed to a workstation on the same segment is not allowed to cross over the bridge. Information from Segment A addressed to a workstation on Segment B is allowed to cross over the bridge to that segment, where only the workstation with the right MAC address will recognize it.

However, if packets on the LAN are destined for another network, say the Internet, a Layer 3 address must be used; this is also known as an IP address. Most workstations and other LAN equipment are equipped with the TCP/IP protocol suite, which allows IP-formatted packets to ride over Ethernet, Token Ring, and other types of LANs. Instead of a bridge, which operates at Layer 2, a router is required, which operates at Layer 3. Routers read IP addresses to route packets to other routers on the Internet to get the packets to their proper destination over the most efficient path.[7]

Bluetooth devices can support traffic at both Layer 2 and Layer 3, depending on the application. When two Bluetooth devices establish a direct link with each other, the connection would be considered

---

[7] Often bridge and router functionality is found in the same device, which allows traffic to be bridged or routed as appropriate.

# Basic Concepts

Layer 2. When a Bluetooth device establishes a link with a gateway to retrieve information from the Internet, the connection would be considered Layer 3. In the first case, only device addresses are used; in the second case, IP addresses are used. An IP address might look like this: 192.100.168.1. Depending on what technologies are being used, packets may be fixed in length or variable in length. They vary in length up to about 12,000 bits for Ethernet and IP and up to about 32,000 bits for frame relay. Asynchronous Transfer Mode (ATM) services use the smallest size packets—fixed length cells of 424 bits each, or 53 bytes. Bluetooth devices convey data on a piconet channel in the form of packets that vary in length up to about 2,800 bits.

## Bluetooth Packets

The Bluetooth specification defines the use of two kinds of packets: SCO and ACL. SCO packets are used on the synchronous links for voice and are routed to the synchronous I/O (input/output) voice port. They do not include an error-checking mechanism and are never retransmitted because the resulting delay would diminish voice quality.

ACL packets are used on the asynchronous link. The information carried can be user data or control data. Since the data carried over an asynchronous link is not sensitive to delay, the packets may include an error-control mechanism and retransmission may be used to correct packets that have been corrupted during transmission.

The general packet format (Figure 2.14) used in Bluetooth wireless technology consists of three parts: access code, header, and payload.

**Figure 2.14**
General packet format specified for Bluetooth wireless technology.

| 72 Bits | 54 Bits | 0—2745 Bits |
|---|---|---|
| Access Code | Header | Payload |

### Access Code

Each packet starts with an access code, which is used for signaling purposes. The fields of the access code consist of a preamble, sync word, and trailer (Figure 2.15). The preamble indicates the arrival of a packet to the receiver. The sync word is used for timing synchronization with the receiver. The receiver correlates against the entire sync word in the access code, which results in a very robust signaling mechanism. The trailer is appended to the sync word as soon as the packet header follows the access code. The number of bits in the access code may vary, depending on whether a packet header follows. If a packet header follows, the access code is 72 bits long, otherwise it is only 68 bits.

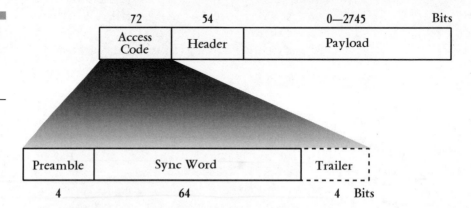

**Figure 2.15**
The access code is used for signaling between Bluetooth devices.

The functions provided by the access code may differ, depending on the operating mode of the Bluetooth device. Accordingly, there are three types of access codes:

- **Channel Access Code (CAC)**—The channel access code identifies a piconet. This code is included in all packets exchanged on the piconet channel. All packets sent in the same piconet start with the same channel access code.

- **Device Access Code (DAC)**—The device access code is used for special signaling procedures, such as paging and response to paging. Paging involves transmitting a series of messages with the objective of setting up a communication link to a unit active within the coverage area. When that unit responds to the page, the communication link can be set up.

- **Inquiry Access Code (IAC)**—There are two types of inquiry access code: general and dedicated. A general inquiry access code is common to all devices. It is used to discover other Bluetooth units that are in range. The dedicated inquiry access code is common for a dedicated group of Bluetooth units that share a common characteristic. It is used to discover only these dedicated Bluetooth units that are in range.

## Header

If used, the header contains link control (LC) information and consists of six fields totaling 18 bits (Figure 2.16):

- Active Member Address (3 bits)
- Type (4 bits)
- Flow (1 bit)
- Automatic Repeat Request (1 bit)
- Sequence Number (1 bit)
- Header Error Check (8 bits)

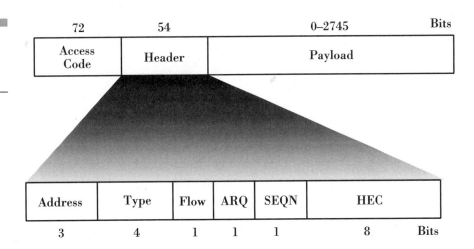

**Figure 2.16**
A packet's header format is comprised of six fields.

### ACTIVE MEMBER ADDRESS

This 3-bit field is used to distinguish between the active members participating on the piconet. In a piconet, one or more slaves are connected to a single master. To identify each slave separately, each slave is assigned a temporary 3-bit address to be used when it is active. Packets exchanged between the master and the slave carry the active member address of this slave. In other words, the address of the slave is used in both master-to-slave packets and in slave-to-master packets. An all-zero address is reserved for broadcasting packets from the master to the slaves. Slaves that are disconnected or *parked* give up their addresses and a new one must be assigned when they re-enter the piconet.

### TYPE

This 4-bit field is used for a code that specifies the packet type. The interpretation of this code depends on the type of link associated with the packet: either an SCO link or an ACL link. There are four different types of SCO packets and seven different types of ACL packets. The type code also indicates the number of slots the current packet will occupy. This allows non-addressed receivers to refrain from listening to the channel for the duration of the remaining slots.

### FLOW

This 1-bit field is used for flow control of packets over the ACL link. When the receiver's buffer for the ACL link is full, a "stop" indication is returned to temporarily halt the transmission of data. The stop signal only applies to ACL packets. Packets including only link-control information or SCO packets can still be received. When the receiver's buffer is empty, a "go" indication is returned. When no packet is received, or the received header is in error, a "go" indication is assumed.

### AUTOMATIC REPEAT REQUEST

This 1-bit field is used to inform the transmitting device of a successful transfer of payload data. The success of the reception is checked by means of a cyclic redundancy check (CRC). The return notification can be in the form of a positive acknowledgment (ACK) or a negative acknowledgment (NAK). If the payload data were received in good order, an ACK indication is returned, otherwise a NAK indication is returned. When no return message of any kind is received, a NAK is assumed. The ACK/NAK is piggy-backed in the header of the return packet.

The CRC is a computational means to ensure the accuracy of the data. The mathematical function is computed before the packet is transmitted at the originating device. Its numerical value is computed based on the content of the packet. This value is compared with a recomputed value of the function at the destination device. If the two values match, the ACK indication is returned. If a mismatch occurs, the NACK indication is returned.

### SEQUENCE NUMBER

This 1-bit field provides a sequential numbering scheme to put the data packet stream in the proper order when it reaches a receiving device. For each new transmitted packet that contains data with a CRC value, the sequence number bit is inverted to filter out retrans-

# Basic Concepts

missions at the destination. If a retransmission occurs due to a failing ACK, the destination receives the same packet twice. Comparison of sequence number of consecutive packets means correctly received retransmissions can be discarded.

**HEADER ERROR CHECK**
This 8-bit field is used to check the integrity of the header. After the HEC generator is initialized, a Header Error Check (HEC) value is calculated for the header bits. The receiver initializes its HEC circuitry so it can interpret the value. If the HEC does not check, the entire packet is ignored.

## Payload

The final part of the general packet format is the payload. In the payload, there are two types of fields: the (synchronous) voice field and the (asynchronous) data field. The ACL packets only have the data field and the SCO packets only have the voice field. The exception is a Data Voice (DV) packet, which has both. The data field consists of three segments: a payload header, a payload body, and possibly a CRC code (Figure 2.17).

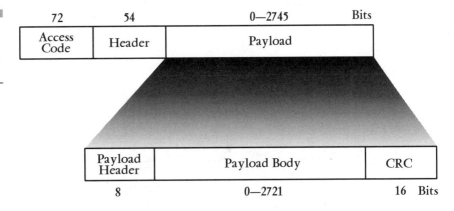

**Figure 2.17**
The ACL packet's payload is comprised of three fields.

**PAYLOAD HEADER**
Only data fields have a payload header. The payload header is one or two bytes long and specifies the logical channel, controls the flow on the logical channels, and has a payload length indicator. The length indicator indicates the number of bytes (i.e., 8-bit words) in the payload body, excluding the payload header and the CRC code.

## PAYLOAD BODY

The payload body includes the user information. The length of the payload body is indicated in the length field of the payload header.

## CRC CODE GENERATION

After the CRC generator is initialized, the 16-bit cyclic redundancy check code is calculated over the information to be transmitted and then appended to the information.

# Logical Channels

Logical channels refer to the different types of channels that run over a physical link; specifically, the ACL or the SCO link. In the Bluetooth system, five logical channels are defined for control and user information:

- Link Control (LC)
- Link Manager (LM)
- User Asynchronous (UA)
- User Isochronous (UI)
- User Synchronous (US)

The LC and LM are control channels that are used at the link control level and link manager level, respectively. The user channels UA, UI, and US are used to carry asynchronous, isochronous, and synchronous user information, respectively. The LC channel is carried in the packet header; all other channels are carried in the packet payload. The LM, UA, and UI channels are indicated in the Logical Channel field in the payload header. The US channel is carried by the SCO link only; the UA and UI channels are normally carried by the ACL link, although they can also be carried by the data in the DV packet on the SCO link. The LM channel can be carried by either the SCO or the ACL link.

## LINK CONTROL

The LC channel is mapped onto the packet header. This channel carries low-level link control information like the Automatic Repeat Request, flow control, and payload characterization. The LC channel is carried in every packet that has a packet header. All other channels are mapped onto the payload.

# Basic Concepts

**LINK MANAGER**
The LM control channel carries control information exchanged between the link managers of the master and one or more slaves.

**USER ASYNCHRONOUS**
The UA channel carries asynchronous user data. This data may be transmitted in one or more baseband packets.

**USER ISOCHRONOUS.**
The UI data channel is supported by properly timing the start packets at higher levels. This type of logical channel is used for time-bounded information like compressed audio over an ACL link.

**USER SYNCHRONOUS**
The US channel carries transparent synchronous user data. This channel is carried over the SCO link. The US channel can only be mapped onto the SCO packets. All other channels are mapped on the ACL packets, or possibly the SCO DV packet. The LM, UA, and UI channels may interrupt the US channel if they concern information of higher priority.

# Client-Server Architecture

For much of the 1990s, the client-server architecture has dominated corporate efforts to downsize, restructure, and otherwise reengineer for survival in an increasingly global economy. Frustrated with the restrictive access policies of traditional information systems managers and the slow pace of centralized, mainframe-centric applications development, client-server architecture grew out of the need to bring computing power and decision making down to the user, so businesses could respond faster to customer needs, competitive pressures, and market dynamics. This model of computing is also applicable to the Bluetooth wireless technology environment, where devices take on the identity of client and server, depending on the application or service being implemented.

## Architectural Model

The client-server architecture is not new. A more familiar manifestation of the architecture is the decades-old corporate telephone system, with the PBX acting as a server and the telephones acting as the clients. All the telephones derive their features and user access privileges from the PBX, which also processes incoming and outgoing calls. An example of a large-scale client-server data environment is the Internet, which is a global network of networks that is reducible to client-server information exchanges. What is relatively new is the application of this model to the LAN environment and more recently to the Bluetooth piconet.

Whatever the environment, the client-server architecture entails the use of an application program that is broken out into two parts: the client and server, which exchange information with each other over the network (Figure 2.18).

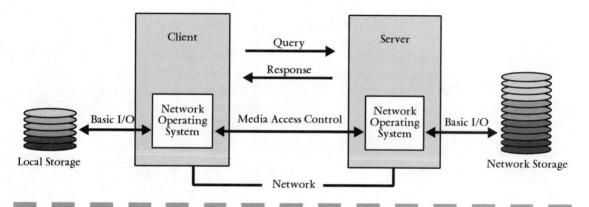

**Figure 2.18**  Simplified model of the client-server architecture.

**CLIENT**

In the conventional client-server environment, the client portion of the program, or front end, is run by individual users at their desktops and performs such tasks as querying a database, producing a printed report, or entering a new record. These functions are carried out through a database specification and access language, better known as Structured Query Language (SQL), which operates in conjunction with existing applications.

**SERVER**

The server portion of the program, or back end, is resident on a computer configured to support multiple clients, offering them shared access to numerous application programs as well as to printers, file storage, database management, communications, and other resources. The server not only handles simultaneous requests from multiple clients, it performs such administrative tasks as transaction management, security, logging, database creation and updating, concurrency management, and maintaining the data dictionary. The data dictionary standardizes terminology so that database records can be maintained across a broad base of users.

**NETWORK**
The network consists of the transmission facility—usually a LAN. When using Bluetooth wireless technology, the network is the piconet. Among the commonly used media for LANs is coaxial cable (thick and thin), twisted-pair wiring (shielded and unshielded), and optical fiber (single- and multimode). In some cases, wireless facilities such as infrared and spread spectrum are used to link clients and servers. Of course, Bluetooth wireless technology uses spread spectrum as well.

A media access control (MAC) protocol is used to regulate access to the transmission facility. Ethernet and Token Ring are the two most popular media access control protocols. Ethernet is based on a bus or star topology, which is contention-based, meaning that stations compete with each other for access to the network, a process that is controlled by a statistical arbitration scheme. Each station "listens" to the network to determine if it is idle. Upon sensing that no traffic is currently on the line, the station is free to transmit. If the network is already in use, the station backs off and tries again.

The media access control method used in token ring networks is the "token." The token, three octets in length, is passed from one station to another. A station seizes the token and replaces it with an information frame. Only the addressee can claim the message. At the completion of the information transfer the station reinserts the token on the ring. A token-holding timer controls the maximum amount of time a station can occupy the network before passing the token to the next station.

Bluetooth devices can assume the identity of either client or server. With regard to a procedure called service discovery, for example, the Service Discovery Profile (SDP) defines the roles of Bluetooth devices in terms of local device (LocDev) and remote device (RemDev). A LocDev initiates the service discovery procedure. To accomplish this, it must contain at least the client portion of the Bluetooth SDP architecture. A LocDev contains the service discovery application, which enables the user to initiate discoveries and display the results of these discoveries.

One or more RemDevs participate in the service discovery process by responding to the service inquiries generated by a local device. To accomplish this, they must contain at least the server portion of the Bluetooth SDP architecture. Each RemDev contains a service records database, which the server portion of SDP consults to create responses to service discovery requests.

The roles of LocDev or RemDev are not permanent or exclusive to one device or another. For example, a RemDev may have the service discovery application installed as well as an SDP client, and a LocDev may also have an SDP server. Thus, a device could be a LocDev for a particular SDP transaction, while acting as a RemDev for another SDP transaction.

## Service Discovery

Service discovery refers to the ability of network devices, applications, and services to search out and find other complementary network devices, applications, and services that are needed to properly complete specified tasks. In traditional networks, a central directory contains detailed records about the various network elements. In less formal networks, a lookup table may be constructed, with one node acting as the repository for the capabilities of the devices and services in a temporary network. Ad hoc, peer-to-peer networks do not require a directory at all, instead relying on direct interactions between the devices themselves to determine each other's capabilities. With the wireless, mobile paradigm, the same device may take part in all of these network types.

Bluetooth wireless technology makes extensive use of discovery processes that enable devices to identify each other when they come within range and establish appropriate connections so that the devices can run common applications and access various services. Through service discovery, personal networks are essentially self configuring; this in turn makes portable devices very easy to use. Connections are entirely transparent to the user, requiring no setup wizards or online settings. What is needed is a single service-discovery technology to support mobile devices as they move from traditional directory-centric and peer-to-peer networks and back again. To enable diverse appliances and equipment to communicate in this new permissive networking environment, there is widespread industry support for a service-discovery standard called the Salutation Architecture.

Developed by the Salutation Consortium, the architecture was created to help solve the problems of service discovery and utilization among the growing number of appliances and equipment becoming available from a broad range of vendors. The architecture is processor, operating system, and communication protocol independent. As such, the Salutation Architecture provides a standard method for applications, services, and devices to advertise and describe their capabilities to other applications, services, and devices, as well as to discover the capabilities of any other entity. For example, through service discovery, it is possible for a device to determine the capabilities of a printer with regard to its color capability, the various resolutions it provides, the availability of a document finisher and collator, and the contents of the input drawers. Once these capabilities are known, a session can be requested and established with the printer to utilize any of these capabilities.

The Salutation Architecture has implementations modeled in TCP/IP, InfraRed Data Association standards, the Bluetooth specificatio, and the Internet Engineering Task Force's (IETF) Service Location Protocol (SLP), a protocol for automatic resource discovery on IP networks. SLP would be used on a LAN, corporate intranet, or public Internet where users otherwise would not have a clear picture of the services and resources available to them. Consequently, many hours of a network administrator's time is spent configuring desktops so users can access necessary services.

For example, users have traditionally had to find services by knowing the name of a network host (a human-readable text string), which is an alias for a network address. SLP eliminates the need for a user to know the name of a network host supporting a service. Rather, the user supplies the desired type of service and a set of attributes that describe the service. Based on that description, the Service Location Protocol resolves the network address of the service for the user.

Through the use of tools that have been enabled with SLP, a clearer picture of the network-attached resources and services is available to all users. Users can browse through resources and select the most appropriate service to meet the task at hand based on any attribute. Finding the Human Resources corporate Web server, the nearest color printer, alternative file servers, or routing a print job to a printer in a remote sales office is easy and automatic using the Service Location Protocol.

The Salutation Architecture supports Windows, Windows CE, Java, VXWorks and Tornado, and offers ports to Palm. Starting with version 2.1 of the Salutation service-discovery architecture, a single API (application programming interface) is provided to allow a device to access two service-location protocols simultaneously. In addition, version 2.1 provides a single service manager to support the Salutation protocol locally as well as SLP on the Internet or corporate intranet simultaneously.

## Summary

Devices adhering to the Bluetooth specification are well positioned to handle virtually any short-range communication need, enabling seamless data and voice communication over wireless links between mobile and stationary devices, including mobile phones, wireless information devices, handheld computers, and personal digital assistants that can interconnect with each other, as well as with desktop computers, home and office phones, and new gateway devices that will permit dialup access to virtually any appliance in the home for remote control.

With its use of spread-spectrum technology the Bluetooth specification permits the effective exchange of data even in areas with significant noise levels. With the signal's power spread over a larger band of frequencies, the original signal can be reconstructed with more accuracy, resulting in a more robust signal that is less susceptible to impairment from electromechanical noise and other sources of interference. In being more difficult to distinguish from random noise, the spread signal also makes voice and data communications more secure against interception.

CHAPTER 3

# Bluetooth Protocol Architecture

As noted in Chapter 1, the Bluetooth Special Interest Group (SIG) developed the Bluetooth specification. The specification includes the protocols that allow for the development of interactive services and applications that operate over wireless links established with interoperable radio modules. Various usage models for Bluetooth wireless technology were also envisioned by the SIG, and the specification includes a description of the protocols for the implementation of these models.

## What Are Protocols?

Protocols are an agreed-upon way that devices exchange information. For every type of network technology, including that of the Bluetooth specification, there are sets of protocols or rules that define exactly how messages are passed over the link. As you might guess from reading Chapter 2, the protocol defines the formatting of these messages, including which parts are reserved for such things as the address, error control, and user data.

The protocols that define how traffic is passed over the link are a lot like the protocols we all observe in daily life. When driving a car toward a busy intersection, for example, we know what to do by observing the protocol displayed by the traffic light: green for go, yellow for slow down, and red for stop. In the rare instances when we encounter intersections with no traffic lights or signs, there is general confusion among drivers about which vehicle should go next. When this happens, traffic flow can be very slow, especially if collisions occur. If you travel on this road, you risk getting to your destination too late for a meeting.

Without generally accepted protocols, a network would not work properly because hardware manufacturers and software developers would all be doing things their own way. The result would be proprietary products unable to communicate with each other over the same network. Until work began in the late 1960s on what was to become the Transmission Control Protocol/Internet Protocol (TCP/IP), computers were based on proprietary platforms, with the result that they could not exchange information across different platforms over a network. Today, virtually all computers of any make and model include a TCP/IP stack so they can become interoperable when connected to the Internet. Through the common TCP/IP stack installed on all computers, they can exchange information with each other and access various services easily and quickly.

# Bluetooth Protocol Architecture

In the mid-1970s, international standards organizations took up the task of developing a generic reference model called Open Systems Interconnection (OSI). The model provides a useful framework for visualizing the communications process and comparing products in terms of standards conformance and interoperability potential. This layered structure not only aids users in visualizing the communications process, it provides vendors with the means for segmenting and allocating various communications requirements within a workable format. This can reduce much of the confusion normally associated with the complex task of supporting successful communications.

## Open Systems Interconnection

The OSI Reference Model was issued by the International Organization for Standardization (ISO) in 1974. The purpose of the seven-layer model was to separate the various network functions to promote interoperability between the products of different vendors (Table 3.1). Each layer contributes protocol functions that together ensure the smooth error-free exchange of information between various interconnected devices.

**TABLE 3.1**

Summary of the Seven-layer OSI Reference Model

| OSI Layer | Description | Included Functions |
|---|---|---|
| Application | Defines the manner in which applications interact with the network | Electronic mail, file transfer, terminal emulation |
| Presentation | Defines the way in which data are formatted, presented, converted, and encoded | Character code translation, data conversion, data compression, data encryption |
| Session | Defines the procedures for establishing, maintaining and disconnecting a communications link between devices on a network | Data synchronization, name lookup, authentication, logging |
| Transport | Defines the procedures for ensuring reliable data transmission | Packet assembly/disassembly, packet error checking, packet sequencing, retransmission requests |
| Network | Defines the procedures for routing data through intermediate systems to the correct destination node | Physical/logical address translation, quality of service, route selection |

*continued on next page*

**TABLE 3.1**

Summary of the Seven-layer OSI Reference Model (continued)

| OSI Layer | Description | Included Functions |
|---|---|---|
| Data Link | Validates the integrity of the flow of data from one node to another by synchronizing blocks of data and controlling the flow of data | Frame assembly/disassembly, frame error-checking, frame retransmission |
| Physical | Defines the physical and electrical characteristics of the medium over which data in the form of 1s and 0s are transmitted | Wiring (including pin definitions for cable connectors), network interfaces, transmit/receive signaling, detection of signaling errors on the medium |

Since it was issued, the OSI reference model has influenced the development of every open network technology, including Bluetooth wireless technology. Even though the Bluetooth architecture has its own four-layer protocol stack, it reuses existing protocols at the higher layers. The reuse of protocols helps to adapt existing (legacy) applications to work with the Bluetooth wireless technology.

## Application Layer

The highest layer in the OSI reference model is the application layer, which is used for applications that are specifically written to run over the network. When programmers write applications that require network services, this is the layer the application program will access. Specifically, this layer provides services that directly support user applications, such as electronic mail, database access, and file transfers (Figure 3.1).

According to the OSI reference model, each type of application must employ its own Layer 7 protocol. With the wide variety of available application types, Layer 7 offers definitions for each, including:

- Resource sharing
- Remote file access
- Remote printer access
- Inter-process communication

# Bluetooth Protocol Architecture

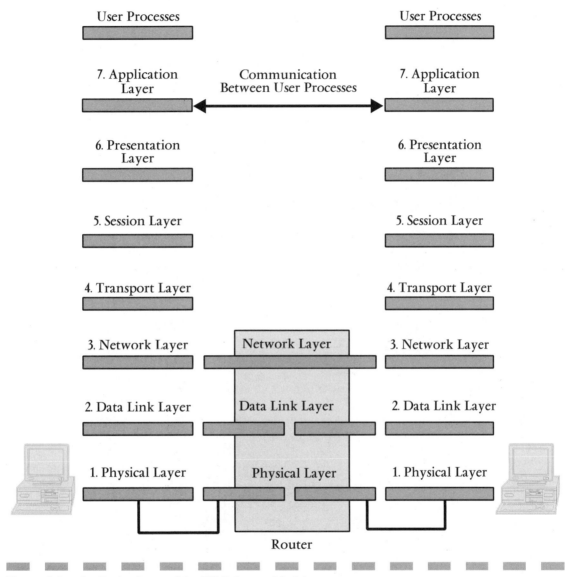

**Figure 3.1** Application Layer of the OSI Reference Model.

- Network management
- Directory services
- Electronic messaging (such as e-mail)
- Network virtual terminals

## Presentation Layer

Layer 6 deals with the format and representation of data that the applications use; specifically, it controls the formats of screens and files (Figure 3.2). Layer 6 defines such things as syntax, control codes, special graphics, and character sets. This level also determines how variable alphabetic strings will be transmitted, how binary numbers will be presented, and how data will be formatted.

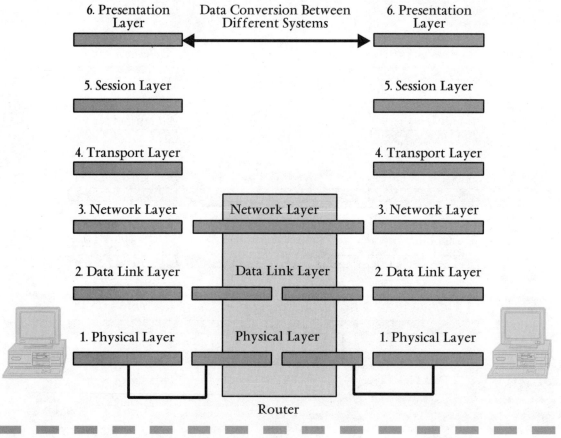

**Figure 3.2** Presentation Layer of the OSI Reference Model.

The presentation layer formats the data to be presented to the application layer. It can be viewed as the translator for the network. This layer may translate data from a format used by the application layer into a common format at the sending station, and then translate

# Bluetooth Protocol Architecture

the common format to a format known to the application layer at the receiving station. The Presentation Layer includes:

- **Character code translation**—For example, ASCII to EBCDIC.
- **Data conversion**—Bit order, carriage return (CR) or carriage return/line feed (CR/LF), integer-floating point, and other functions.
- **Data compression**—This reduces the number of bits that need to be transmitted on the network.
- **Data encryption**—This encrypts data for security purposes.

## Session Layer

The session layer manages communications; for example, it sets up, maintains, and terminates virtual circuits between sending and receiving devices (Figure 3.3). It sets boundaries for the start and end of messages, and establishes how messages will be sent: half-duplex, with each computer taking turns sending and receiving, or full-duplex, with each computer sending and receiving at the same time. These details are negotiated during session initiation.

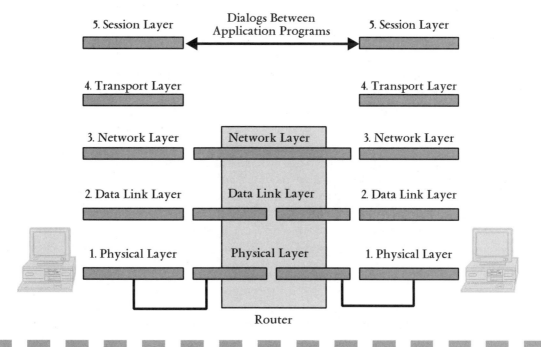

**Figure 3.3** Session Layer of the OSI Reference Model.

The session layer allows session establishment between processes running on different stations. Specifically, the session layer provides:

- **Session establishment, maintenance, and termination**—This allows two application processes on different machines to establish, use, and terminate a connection, called a *session*.
- **Session support**—This performs the functions that allow these processes to communicate over the network: authentication for security purposes, name recognition, logging, and other administrative functions. Data synchronization and checkpointing are done at this layer, so that when a link recovers from a failure, only the data sent after the link failed need to be retransmitted.

## Transport Layer

Layer 4 handles end-to-end transport (Figure 3.4). If there is a need for reliable, end-to-end sequenced delivery, then the transport layer performs this function. For example, each packet of a message might have followed a different route through the network toward its destination. The transport layer re-establishes packet order through a process called sequencing so that the entire message is received exactly the way it was sent. At this layer, lost data are recovered and flow control is implemented. With flow control, the rate of data transfer is adjusted to prevent excessive amounts of data from overloading network buffers.

Layer 4 may also support datagram transfers, transactions that need not be retransmitted if data are lost. This is required for voice and video, which may tolerate loss of information, but need to have low delay and low variance in transmittal time. This flexibility is the result of the protocols implemented in this layer, ranging from the five OSI protocols—TP0 to TP4—to TCP and UDP in the TCP/IP suite, and many others in proprietary suites. Some of these protocols do not perform retransmission, sequencing, checksums, and flow control. Specifically, the transport layer provides:

- **Message segmentation**—This accepts a message from the (session) layer above it, splits the message into smaller units (if not already small enough), and passes the smaller units down to the network layer. The transport layer at the destination station reassembles the message.

# Bluetooth Protocol Architecture

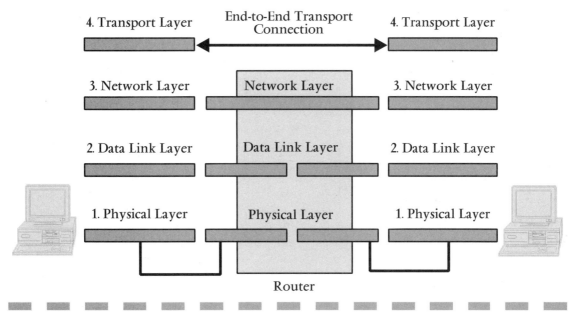

**Figure 3.4** Session Layer of the OSI Reference Model.

- **Message acknowledgment**—This provides reliable end-to-end message delivery with acknowledgments.
- **Message traffic control**—This tells the transmitting station to "back-off" when no message buffers are available.
- **Session multiplexing**—This interleaves several message streams, or sessions, onto one logical link and keeps track of which messages belong to which sessions (see session layer).

## Network Layer

The network layer formats the data into packets, adds a header containing the packet sequence and the address of the receiving device, and specifies the services required from the network (Figure 3.5). The network does the routing to match the service requirement. Sometimes a copy of each packet is saved at the sending node until it receives confirmation that it has arrived at the next node undamaged, as is done in X.25 packet-switched networks. When a node receives the packet, it searches a routing table to determine the best path for that packet's destination without regard for its order in the message.

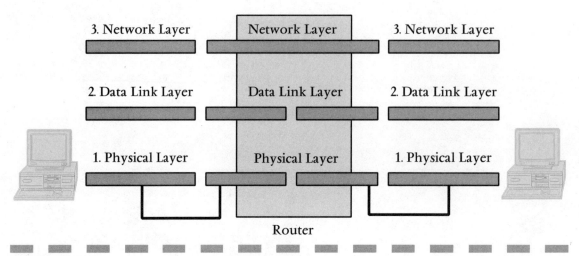

**Figure 3.5** Network Layer of the OSI Reference Model.

In a network where not all nodes can communicate directly, the network layer takes care of routing packets through the intervening nodes. Intervening nodes may reroute the message to avoid congestion or node failures. Specifically, the network layer provides:

- **Routing**—This routes frames among network nodes.
- **Subnet traffic control**—Routers (network layer intermediate systems) can instruct a sending station to scale back its frame transmission when the router's buffer fills up.
- **Frame fragmentation**—If it determines that a downstream router's maximum transmission unit (MTU) size is less than the frame size, a router can fragment a frame for transmission and reassembly at the destination station.
- **Logical-physical address mapping**—This translates logical addresses, or names, into physical addresses.
- **Subnet usage accounting**—This has accounting functions to keep track of frames forwarded by subnet intermediate systems, to produce billing information.

## Data-Link Layer

All modern communications protocols, whether used over wire-line or wireless communication links, use the services defined in Layer 2, including the Bluetooth protocol (Figure 3.6).

# Bluetooth Protocol Architecture

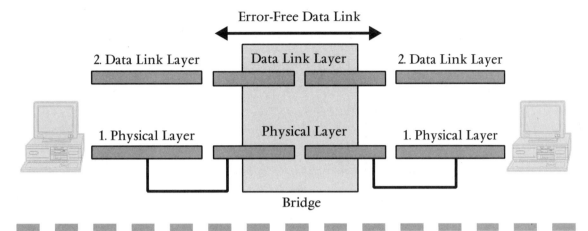

**Figure 3.6** Date-Link Layer of the OSI Reference Model.

Layer 2 does not know what the information or packets it encapsulates mean or where they are headed. Networks that can tolerate this lack of information are rewarded by low transmission delays. Specifically, the functions provided by the data link layer include:

- **Link establishment and termination**—This establishes and terminates the logical link between two nodes.
- **Frame traffic control**—This tells the transmitting node to back off when no frame buffers are available.
- **Frame sequencing**—This transmits/receives frames sequentially.
- **Frame acknowledgment**—This provides/expects frame acknowledgments. Detects and recovers from errors that occur in the physical layer by retransmitting non-acknowledged frames and handling duplicate frame receipt.
- **Frame delimiting**—This creates and recognizes frame boundaries.
- **Frame error checking**—This checks received frames for integrity.
- **Media access management**—This determines when the node has the right to use the physical medium.

## Physical Layer

The lowest OSI layer is the physical layer. This layer represents the actual electrical and mechanical interface which connects a device to a transmission medium (Figure 3.7).

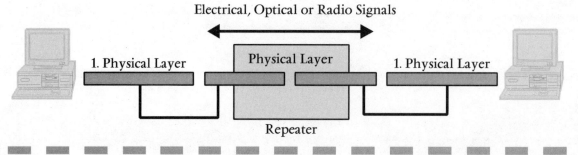

**Figure 3.7** Physical Layer of the OSI Reference Model.

Because the physical interface has become so standardized, it is usually taken for granted. Yet physical connections—cables and connectors—with their pin assignments and transmission characteristics can still be a problem in designing a reliable network if they do not conform to a common model. Specifically, the functions provided at the physical layer include:

- **Data encoding**—This modifies the simple digital signal pattern (1s and 0s) used by the PC, the better to accommodate the characteristics of the physical medium, and to aid in bit and frame synchronization.
- **Physical medium attachment**—This accommodates various possibilities in the medium, such as the number of pins a connector has and what each pin is used for.
- **Transmission technique**—This determines whether the encoded bits will be transmitted via digital or analog signaling.
- **Physical medium transmission**—This transmits bits as electrical or optical signals appropriate for the physical medium, and determines how many volts/decibels should be used to represent a given signal state, using a given physical medium.

Throughout the 1980s, the prediction was often made that OSI would replace TCP/IP as the preferred technique for interconnecting multi-vendor networks. Obviously, this did not happen. There are several reasons for this, including the slow pace of OSI standards progress in the 1980s, as well as the expense of implementing complex OSI software and having products certified for OSI interoperability. Furthermore, TCP/IP was already widely available and plug-in protocols continue to be developed to add functionality. The situation is different in Europe where OSI compliance was mandated early on by the regulatory authorities in many countries.

## Bluetooth Protocol Stack

Like OSI, the Bluetooth specification makes use of a layered approach in its protocol architecture. And like OSI, the ultimate aim of the Bluetooth specification is to allow applications written to the specification to interoperate with each other. Interoperability is achieved when the applications in remote devices run over identical protocol stacks. Different protocol stacks are used for different applications. Regardless of the specific application, the protocol stacks involved use a common Bluetooth data-link and physical layer (Figure 3.8).

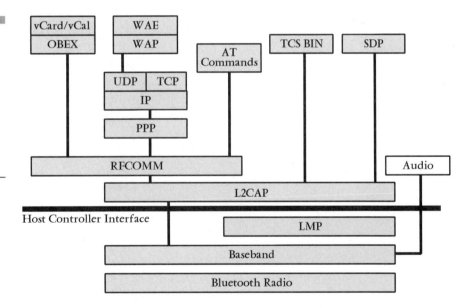

**Figure 3.8**
Interoperable applications supporting the Bluetooth usage models are built on top of this Bluetooth protocol stack.
Source: Bluetooth Specification 1.0.

Not every application makes use of all the protocols in the Bluetooth protocol stack; instead, they may run over one or more vertical slices from the stack, drawing upon a particular service needed to support the main application. The protocols may also have other relations with each other. For example, protocols like Logical Link Control and Adaptation Layer (L2CAP) and Telephony Control Specification Binary (TCS BIN) may use LMP (Link Manager Protocol) when there is need to control the link.

The complete protocol stack comprises both protocols that are specific to Bluetooth wireless technology like LMP and L2CAP, and those protocols like OBEX (Object Exchange Protocol) UDP (User Datagram Protocol) and WAP (Wireless Application Protocol) that can be used with many other platforms. In designing the protocols for the Bluetooth protocol stack, instead of reinventing the wheel, existing protocols were simply reused for different purposes at the higher layers. This way of doing things not only speeded development of the Bluetooth specification, it facilitated adaptation of legacy applications to work with Bluetooth wireless technology and helped to ensure the smooth operation and interoperability of these applications.

The openness of the Bluetooth specification allows many applications already developed by vendors to take immediate advantage of hardware and software systems that are compliant with the Bluetooth specification. This openness also makes it possible for vendors freely to implement their own (proprietary) or commonly used application protocols on top of the protocols specific to Bluetooth wireless technology. Thus, the open specification greatly expands the number of new and legacy applications that can take full advantage of the capabilities offered by Bluetooth wireless technology.

The Bluetooth protocol stack consists of four layers. The layers and the protocols that fit into them are summarized in Table 3.2.

**TABLE 3.2**

Summary of Protocols and Layers in the Bluetooth Protocol Stack

| Bluetooth Protocol Layer | Members of the Protocol Stack |
|---|---|
| Bluetooth Core Protocols | • Baseband |
| | • Link Management Protocol (LMP) |
| | • Logical Link Control and Adaptation Layer (L2CAP) |
| | • Service Discovery Protocol (SDP) |
| Cable Replacement Protocol | • Radio Frequency Communication (RFCOMM) |
| Telephony Control Protocols | • Telephony Control Specification Binary (TCS BIN) |
| | • AT-Commands |
| Adopted Protocols | • Point-to-Point Protocol (PPP) |
| | • User Datagram Protocol (UDP)/Transmission Control Protocol (TCP)/Internet Protocol (IP) |

*continued on next page*

# Bluetooth Protocol Architecture

**TABLE 3.2**

Summary of Protocols and Layers in the Bluetooth Protocol Stack (continued)

| Bluetooth Protocol Layer | Members of the Protocol Stack |
|---|---|
| | • Object Exchange Protocol (OBEX) |
| | • Wireless Application Protocol (WAP) |
| | • vCard |
| | • vCalendar |
| | • Infrared Mobile Communications (IrMC) |
| | • Wireless Application Environment (WAE) |

The Bluetooth specification also defines a Host Controller Interface (HCI), which provides a command interface to the baseband controller, link manager, and access to hardware status and control registers. In Figure 3.8, the HCI is positioned below L2CAP, but it can exist above L2CAP as well.

Together, the Cable Replacement layer, the Telephony Control layer, and the Adopted protocol layer form application-oriented protocols which enable applications to run over the Bluetooth Core protocols. Since the Bluetooth specification is open, additional protocols such as the HyperText Transfer Protocol (HTTP) and File Transfer Protocol (FTP) can be accommodated in an interoperable fashion on top of those transport protocols specific to the Bluetooth specification or on top of the application-oriented protocols shown previously in Figure 3.8.

## Bluetooth Core Protocols

The core protocols are specific to Bluetooth wireless technology, having been developed by the Bluetooth SIG. RFCOMM and the TCS BIN protocol were also developed by the Bluetooth SIG, but they are based on existing standards: ETSI TS 07.10 and the ITU-T Recommendation Q.931, respectively, which are discussed later. The core protocols, plus the Bluetooth radio (as shown in Figure 3.8), are required by most Bluetooth devices, while the rest of the protocols are used only when needed.

## Baseband

The baseband layer enables the physical RF link between Bluetooth units in a piconet. Since the Bluetooth RF system uses frequency-hopping spread-spectrum technology, in which packets are transmitted in defined time slots across defined frequencies, this layer uses inquiry and paging procedures to synchronize the transmission-hopping frequency and clock of different Bluetooth devices.

It provides the two different kinds of physical links with their corresponding baseband packets: Synchronous Connection-Oriented (SCO) and Asynchronous Connectionless (ACL), which can be transmitted in a multiplexed manner on the same RF link. ACL packets are used for data only, while the SCO packet can contain audio only or a combination of both audio and data. All audio and data packets can be provided with different levels of error correction and can be encrypted to ensure privacy. In addition, the link management and control messages are each allocated a special channel.

Packets containing audio data can be transferred between one or more Bluetooth devices, making possible various usage models. The audio data in SCO packets are routed directly to and from Baseband, and do not go through L2CAP. The audio model is relatively simple within the Bluetooth specification; any two Bluetooth devices can send and receive audio data between each other just by opening an audio link.

## Link Manager Protocol (LMP)

LMP is responsible for link setup and control between Bluetooth devices, including the control and negotiation of baseband packet sizes. It is also used for security: authentication and encryption; generating, exchanging, and checking link and encryption keys. The LMP also controls the power modes and duty cycles of the Bluetooth radio device, and the connection states of a Bluetooth unit in a piconet.

LMP messages are filtered out and interpreted by the link manager on the receive side, so they are never passed up to higher layers. LMP messages have higher priority than user data. If a link manager needs to send a message, it will not be delayed by L2CAP traffic. In addition, LMP messages are not explicitly acknowledged, since the logical channel provides a reliable enough link, making acknowledgments unnecessary.

## Logical Link Control and Adaptation Protocol

The Logical Link Control and Adaptation Protocol (L2CAP) supports higher level protocol multiplexing, packet segmentation and reassembly, and Quality of Service (QoS). L2CAP permits higher-level protocols and applications to transmit and receive data packets up to 64 kilobytes in length. Although the baseband protocol provides the SCO and ACL link types, L2CAP is defined only for ACL links and no support for SCO links is planned. Voice-quality channels for audio and telephony applications are usually run over baseband SCO links. However, audio data may be packetized and sent using communication protocols running over L2CAP.

## Service Discovery Protocol (SDP)

Discovery services are an important element in the Bluetooth framework since they provide the basis for all the usage models. Using SDP, device information, services, and the characteristics of the services can be queried. Having located available services within the vicinity, the user may select from any of them. After that, a connection between two or more Bluetooth devices can be established.

# Cable Replacement Protocols

The Bluetooth specification includes two protocols that deliver control signaling over wireless links, emulating the kind of signaling normally associated with wireline links.

## RFCOMM

RFCOMM is a serial line emulation protocol based on a subset of the European Telecommunications Standards Institute's Technical Standard (TS) 07.10, which is also used for Global System for Mobile (GSM) communication devices. The ETSI is a non-profit organization that produces the telecommunications standards used throughout Europe. The RFCOMM protocol provides emulation of RS-232 serial ports over the L2CAP protocol. This "cable replacement" protocol emulates

RS-232 control and data signals over the baseband, providing both transport capabilities for upper level services (e.g., OBEX) that use serial line as the transport mechanism.

RFCOMM supports applications that make use of a device's serial port. In a simple configuration, the communication segment is a Bluetooth link from one device to another (Figure 3.9). Where the communication segment is another network, Bluetooth wireless technology is used for the path between the device and a network connection device like a modem. RFCOMM is only concerned with the connection between Bluetooth devices in the direct connect case, or between the Bluetooth device and a modem in the network case. RFCOMM can support other configurations, such as modules that communicate via Bluetooth wireless technology on one side and provide a wired interface on the other side, as shown in Figure 3.10. These devices are not really modems but offer a similar service. An example would be a LAN access point that supports Bluetooth devices, enabling handheld devices to connect to the corporate network.

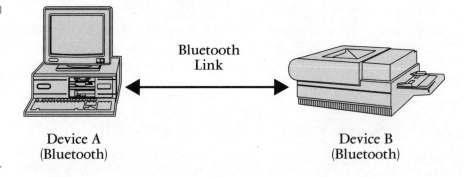

**Figure 3.9**
RFCOMM as used to support a direct connection between two Bluetooth devices, such as a computer and printer.
Source: Bluetooth Specification 1.0.

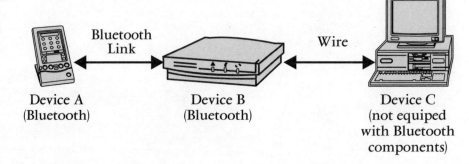

**Figure 3.10**
RFCOMM as used to support a Bluetooth device (PDA) and a device without Bluetooth wireless technology (corporate network hub) through an intermediate device (LAN access point).
Source: Bluetooth Specification 1.0.

## Telephony Control Protocols

Telephony Control Specification Binary (TCS Binary or TCS BIN) is a bit-oriented protocol that defines the call control signaling for setting up speech and data calls between Bluetooth devices. It also defines mobility management procedures for handling groups of Bluetooth TCS devices. TCS BIN is based on Recommendation Q.931 issued by the International Telecommunication Union-Telecommunications (ITU-T), an agency of the United Nations, which coordinates standards for global telecom networks and services. Q.931 is the ITU-T specification for basic call control under ISDN (Integrated Services Digital Network).

In addition to TCS BIN, the Bluetooth SIG has defined a set of AT commands that define how a mobile phone and modem can be controlled in various usage models. These are discussed later. The AT commands are based on ITU-T Recommendation V.250 and ETS 300 916 (GSM 07.07). When it comes to facsimile services, the commands are specified by one of the following implementations:

- Fax Class 1.0 TIA-578-A and ITU T.31 Service Class 1.0
- Fax Class 2.0 TIA-592 and ITU T.32 Service Class 2.0
- Fax Service Class 2—no industry standard

# Adopted Protocols

As previously noted, the Bluetooth specification makes use of several existing protocols that are reused for different purposes at the higher layers. This allows older applications to work with Bluetooth wireless technology and helps to ensure the smooth operation and interoperability of these applications with newer applications specifically designed for Bluetooth devices.

## PPP

The Bluetooth specification uses the Point-to-Point Protocol (PPP) developed by the Internet Engineering Task Force (IETF). This standard defines how IP datagrams are transmitted over serial point-to-point links. If you access the Internet with a modem on a dialup basis, or with a router on a dedicated line, you use such links. Datagrams are simply the units of data that are transported across the link on a best-effort basis. PPP has three main components:

**ENCAPSULATION**
PPP provides a method for encapsulating datagrams over serial links. It offers an encapsulation protocol over both bit-oriented synchronous links and asynchronous links with eight bits of data and no parity. These links are full duplex, but may be either dedicated or circuit-switched. PPP uses the High-level Data Link Control (HDLC) protocol as a basis for the encapsulation. The PPP encapsulation also provides for the multiplexing of different network-layer protocols simultaneously over the same link. It furnishes a common solution for easy connection between a wide variety of hosts, bridges, and routers.

**LINK CONTROL PROTOCOL (LCP)**
PPP provides a Link Control Protocol to ensure its portability across a wide variety of environments. The LCP is used for automatic agreement on the encapsulation format options, to handle varying limits on sizes of packets, to authenticate the identity of its peer on the link, to determine when a link is functioning properly and when it is defunct, to detect a looped-back link and other common configuration errors, and to terminate the link.

**NETWORK CONTROL PROTOCOLS (NCP)**
Point-to-Point links tend to exacerbate many problems with network protocols. For instance, assignment and management of IP addresses, a problem even in LAN environments, is especially difficult over circuit-switched point-to-point links, such as dial-up modem servers. These problems are handled by a family of Network Control Protocols, which manage the specific needs required by their respective network-layer protocols.

In Bluetooth wireless networks, PPP runs over RFCOMM to implement point-to-point serial links, say between a mobile device and a LAN access point. PPP-networking is the means of taking IP packets to/from the PPP layer and placing them onto the LAN, giving the user access to corporate e-mail, for example.

## TCP/UDP/IP

The Transmission Control Protocol (TCP), User Datagram Protocol (UDP), and Internet Protocol (IP) are all defined by the IETF and used for communication across the Internet. As such, they are now among the most widely used family of protocols in the world. These protocols are included in numerous devices, among them all makes and models of desktop and notebook computers, as well as minicomputers, mainframes, and supercomputers. Increasingly, printers, handheld computers, and mobile phones are also being equipped with these protocols.

## TRANSMISSION CONTROL PROTOCOL (TCP)

TCP is a connection-oriented, end-to-end reliable protocol that fits into a layered hierarchy of protocols which support multinetwork applications. TCP forwards data delivered in the form of IP datagrams or packets to the appropriate process at the receiving host. Among other things, TCP defines the procedures for breaking up the data stream into packets, reassembling them in the proper order to reconstruct the original data stream at the receiving end, and issuing retransmission requests to replace missing or damaged packets. Since the packets typically take different paths through the Internet to their destination, they arrive at different times and out of sequence. All packets are temporarily stored until the late packets arrive so they can be put in the correct order. If a packet arrives damaged, it is discarded and another one resent in response to a retransmission request.

## USER DATAGRAM PROTOCOL (UDP)

While TCP offers assured delivery, UDP merely passes individual messages to IP for transmission on a best-efforts basis. Since IP is not inherently reliable, there is no guarantee of delivery. Nevertheless, UDP is very useful for certain types of communications, such as quick database lookups. For example, the Domain Name System (DNS) consists of a set of distributed databases that provide a service that translates between plain-language domain names and their IP addresses. For simple messaging between applications and these network resources, UDP is adequate.

## INTERNET PROTOCOL (IP)

The Internet Protocol delivers datagrams between different networks through routers that process packets from one autonomous system (AS) to another. Each device in the AS has a unique IP address. The Internet Protocol adds its own header and checksum to make sure the data is properly routed. This process is aided by the presence of routing update messages that keep the address tables in each router current. Several different types of update messages are used, depending on the collection of subnets involved in a management domain. The routing tables list the various nodes on the subnets as well as the paths between the nodes. If the data packet is too large for the destination node to accept, it will be segmented into smaller packets by the higher-level TCP.

The implementation of these standards by the Bluetooth specification allows communication with any other device connected to the Internet. The Bluetooth device, whether a cellular handset or a LAN access point, for example, is then used as a bridge to the Internet. TCP, IP, and PPP are used for all the Internet Bridge usage scenarios (discussed later) and will be used for OBEX in future versions of the Bluetooth specification. UDP, IP, and PPP are also available as transport for the Wireless Application Protocol (WAP).

## OBEX Protocol

OBEX is a session-layer protocol originally developed by the Infrared Data Association (IrDA) as IrOBEX. Its purpose is to support the exchange of objects in a simple and spontaneous manner. For example, the OBEX protocol defines a folder-listing object, which is used to browse the contents of folders on a remote device. As such, OBEX provides the same basic functionality as the HyperText Transfer Protocol (HTTP), but in a much lighter fashion. Like HTTP, OBEX is based on the client-server model and is independent of the actual transport mechanism. In the first phase of implementation with the Bluetooth specification, only RFCOMM is used as the transport layer for OBEX, but future implementations will likely support TCP/IP as well.

In May 1999, OBEX became the first protocol common to both the Bluetooth and infrared wireless specifications. By adopting common usage models and then exploiting the unique advantages of each technology, the combination of Bluetooth and infrared create the only short-range wireless standards that can meet user needs varying from wireless voice transmission to high-speed (16 Mbps) robust data transfer.

An example of how the OBEX protocol can be put to practical use via a mobile phone that supports Bluetooth or infrared wireless technology comes from Nokia. The company's 9110 Communicator currently supports an infrared connection and the IrOBEX protocol for object exchange. (Future models will support Bluetooth links and OBEX.) Nokia's Object Exchange application makes it possible to capture documents from a handheld scanner like the Hewlett-Packard

# Bluetooth Protocol Architecture

Capshare 910, send them to the Communicator via an infrared IrOBEX connection, and forward them to other devices as a fax or e-mail attachment.

The user slides the handheld device (Figure 3.11) over any piece of paper, from a business card to a contract or even a flip chart. Then, at the push of a button, the electronic copy is sent to a computer or handheld PC, where it can be edited, or shared through e-mail or e-fax. The user can even obtain immediate paper copies by sending the documents wirelessly to an infrared-equipped printer. The HP Cap-Share has the capacity to store up to 50 letter-sized text pages, 15 graphics pages, or 150 flip chart pages.

**Figure 3.11**
Hewlett-Packard's Capshare 910 is a handheld scanner that can send documents to mobile phones via an IrOBEX connection, and forward them to other devices as a fax or an e-mail attachment.

## Wireless Application Protocol (WAP)

WAP is a specification for sending and reading Internet content and messages on small wireless devices, such as cellular phones equipped with text displays. Common WAP-enabled information services are news, stock quotes, weather reports, flight schedules, and corporate announcements. Special Web pages called WAP portals are specifically formatted to offer information and services. CNN and Reuters are among the content providers that offer news for delivery to cell phones, wireless PDAs, and handheld computers. Electronic commerce and e-mail are also among the WAP-enabled services that can be accessed from these devices.

Typically, these devices will have very small screens, so content must be delivered in a "no-frills" format. In addition, the bandwidth constraints of today's cellular services mean that the content must be optimized for delivery to handheld devices. To get the information in this form, Web sites are built with a light version of the HyperText Markup Language (HTML) called the Wireless Markup Language (WML).

The strength of WAP is that it spans multiple airlink standards and, in the true Internet tradition, allows content publishers and application developers to be unconcerned about the specific delivery mechanism. Like the Internet, the WAP architecture is defined primarily in terms of network protocols, content formats, and shared services. This approach leads to a flexible client-server architecture, which can be implemented in a variety of ways, but also provides interoperability and portability at the network interfaces. The WAP protocol stack is depicted in Figure 3.12.

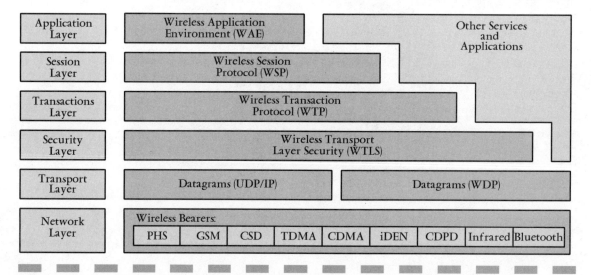

**Figure 3.12** The WAP protocol stack. Source: Wireless Application Protocol Forum, Ltd.

WAP solves the problem of using Internet standards such as HTML, HTTP, TLS, and TCP over mobile networks. These protocols are inefficient, requiring large amounts of mainly text-based data to be sent. Web content written with HTML generally cannot be displayed in an effective way on the small-size screens of pocket-sized mobile phones and pagers, and navigation around and between screens is not easy with one hand.

Furthermore, HTTP and TCP are not optimized for the intermittent coverage, long latencies, and limited bandwidth associated with wireless networks. HTTP sends its headers and commands in an inefficient text format instead of compressed binary format. Wireless services using these protocols are often slow, costly, and difficult to use. The TLS security standard, too, is problematic, since many messages need to be exchanged between client and server. With wireless transmission latencies, this back-and-forth traffic flow results in a very slow response for the user.

WAP has been optimized to solve all these problems. It makes use of binary transmission for greater compression of data and is optimized for long latency and low-to-medium bandwidth. WAP sessions cope with intermittent coverage and can operate over a wide variety of wireless transports using IP where possible and other optimized protocols where IP is not possible. The WML language used for WAP content makes optimum use of small screens and allows easy navigation with one hand without a full keyboard; it has built-in scalability from two-line text displays through to the full graphic screens on smart phones and communicators.

There are a couple of things WAP is good for in the Bluetooth environment: information delivery and *hidden computing*. With regard to information delivery, a WAP client using Bluetooth wireless technology discovers the presence of a WAP server using the Service Discovery Protocol (SDP). At the time of service discovery, the WAP server's address is determined. When the client gets the address, it establishes a connection to the server and can access the information or service provided by that server on a push/pull basis. Encryption and authentication for server-to-client security is provided by the Wireless Transport Layer Security (WTLS) protocol, which is important to safeguard the privacy of e-commerce and hidden computing applications.

Hidden computing is the ability to access and control the functionality of a computer from a peer mobile device. A hidden computing application might involve a kiosk at an airport, shopping center, or some other public place, which allows the mobile device to look up information, buy merchandise, or order tickets. The reason for the Bluetooth SIG's use of the WAP stack for hidden computing is to be able to reuse the software applications developed for the Wireless Application Environment.

## WAP Applications Environment (WAE)

WAP applications are built within the Wireless Application Environment (WAE), which closely follows the Web content delivery model, but with the addition of gateway functions. Figure 3.13 contrasts the conventional Web model with the WAE model. All content is specified in formats similar to the standard Internet formats, and is transported using standard protocols on the Web, while using an optimized HTTP-like protocol in the wireless domain (i.e., WAP). The architecture is designed for the memory and CPU processing constraints found in mobile terminals. Support for low bandwidth and high-latency networks are also included in the architecture. Where existing standards were not appropriate due to the unique requirements of small wireless devices, WAE has modified the standards, without losing the benefits of Internet technology.

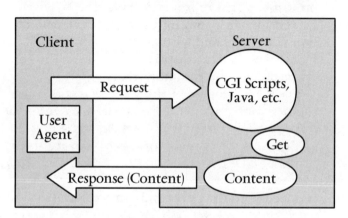

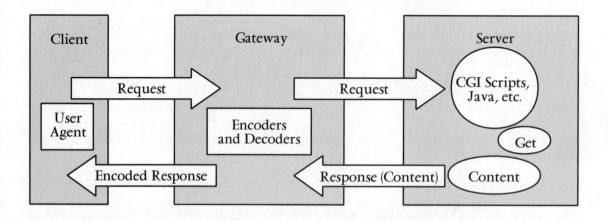

**Figure 3.13** The standard Web content delivery model (top) and the Wireless Application Environment model (bottom). Source: Wireless Application Protocol Forum, Ltd.

The major elements of the WAE model include:

- **WAE User Agents**—These client-side software components provide specific functionality to the end-user. An example of a user agent is a browser that displays content downloaded from the Web. In this case, the user agent interprets network content referenced by a Uniform Resource Locator (URL). WAE includes user agents for the two primary standard content types: encoded Wireless Markup Language (WML) and compiled Wireless Markup Language Script (WMLScript.)

- **Content Generators**—Applications or services on servers may take the form of Common Gateway Interface (CGI) scripts that produce standard content formats in response to requests from user agents in the mobile terminal. WAE does not specify any particular content generator, since many more are expected to become available in the future.

- **Standard Content Encoding**—This set of well-defined content encoding allows a WAE user agent (e.g., a browser) to conveniently navigate Web content. Standard content encoding includes compressed encoding for WML, bytecode encoding for WMLScript, standard image formats, a multi-part container format, and adopted business and calendar data formats (i.e., vCard and vCalendar).

- **Wireless Telephony Applications (WTA)**—This collection of telephony-specific extensions provides call and feature control mechanisms, allowing users to access and interact with mobile telephones for phonebooks and calendar applications.

WMLScript is a lightweight procedural scripting language based on JavaScript. It enhances the standard browsing and presentation facilities of WML with behavioral capabilities. For example, an application programmer can use WMLScript to check the validity of user input before it is sent to the network server, provide users with access to device facilities and peripherals, and interact with the user without a round-trip to the network server (e.g., display an error message).

In addition to WML and WMLScript, other supported content formats for WAP over Bluetooth wireless technology are vCard and vCalendar. These and other components are all part of the WAP applications environment.

## Content Formats

vCard and vCalendar are open specifications originally developed by the Versit Consortium[1] and now controlled by the Internet Mail Consortium (IMC) and being further developed by the IETF. These specifications define the format of an electronic business card and personal calendar entries and scheduling information, respectively. vCard and vCalendar do not define any transport mechanism but only the format under which data are transported between devices.

### VCALENDAR

vCalendar defines a transport- and platform-independent format for exchanging calendaring and scheduling information in an easy, automated, and consistent manner. It captures information about event and "to-do" items normally used by applications such as personal information managers (PIMs) and group schedulers. Programs that use vCalendar can exchange important data about events so that it is possible to schedule meetings with anyone who has a vCalendar-aware program (Figure 3.14).

vCalendar, which provides a consistent, simple format for the capture and exchange of events and action items, helps streamline the daily task of scheduling meetings or appointments. The core set of vCalendar properties includes advanced features such as attachments, audio and e-mail reminders, and classification of events. Among the items that can be attached to an event is the sender's electronic business card, called a vCard. In addition, the vCalendar specification provides interoperability among different calendaring and scheduling applications to facilitate the planning of meetings or appointments across the public Internet, private intranets, or shared extranets. With its adoption by the Bluetooth SIG, the functions of vCalendar can be carried out between devices in close proximity over a piconet.

### VCARD

vCard is the specification for the electronic business card originally developed by the Versit Consortium, along with vCalendar. The respon-

---

[1] The Versit Consortium is a global initiative of IBM, Siemens, and Lucent Technologies which seeks to eliminate barriers to communication and collaboration by developing and promoting open, cross-platform specifications for a wide range of computer and telephony products.

# Bluetooth Protocol Architecture

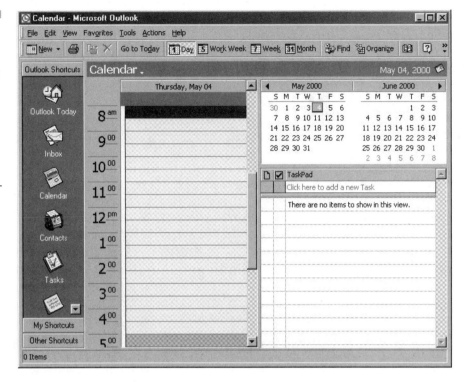

**Figure 3.14**
The vCalendar interface as viewed from within Microsoft Outlook, showing separate panes for daily tasks (left), monthly summary (top right), and TaskPad (bottom right).

sibility for developing and promoting vCard also now rests with the Internet Mail Consortium. vCard is used in applications such as Internet mail, voice mail, Web browsers, telephony, call centers, video conferencing, PIMs, PDAs, pagers, fax, office equipment, and smart cards. vCard information goes beyond simple text, and may include elements like pictures, company logos, and hyperlinks to Web pages.

vCards carry vital directory information such as name, addresses (business, home, mailing, parcel), telephone numbers (home, business, fax, pager, cellular, ISDN, voice, data, video), e-mail addresses, and links to Web content (Figure 3.15).

All vCards can also contain graphics and multimedia, including photographs, company logos, and audio clips. Geographic and time zone information in vCards lets others know when they can call. vCards also support multiple languages.

The vCard specification is transport- and operating system-independent so that the user can install vCard-ready software on any computer. Different programs have different ways of storing vCards. Some programs allow the vCard icon to be dragged and dropped into programs, others require that the vCard be saved to disk then import-

**Figure 3.15**
The vCard interface allows the user to enter a variety of useful information, which is categorized as personal, home, business, other, conferencing, and digital IDs.

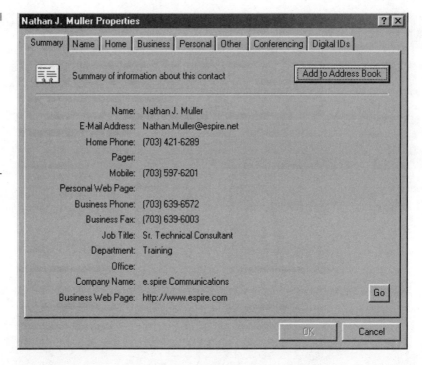

ed to the desired program; still others automatically detect when a vCard is being taken and will prompt the user on whether to open it immediately or save it to disk.

There is no more need for manual swapping of business cards; enter all the information in a portable computer, and then rekey it into one or more devices. The vCard makes it possible to exchange business card information by walking into a meeting and beaming it over Bluetooth or infrared links between hand-held organizers, PDAs, and notebook PCs from any manufacturer. Within seconds, all meeting participants can have this vital information automatically stored in their favorite directories. Later it can be used to place a phone call, send a fax or e-mail, or even to initiate a videoconference.

## Usage Models and Profiles

The Bluetooth SIG has identified various usage models, each of which is accompanied by a profile. These profiles define the protocols and features that support a particular usage model.

# Bluetooth Protocol Architecture

For example, in the Internet Bridge usage model identified by the Bluetooth SIG, a mobile phone or wireless modem acts as the modem for a desktop PC, providing dial-up networking and facsimile capabilities without the need for a physical connection to the PC. For dialup networking, a two-piece protocol stack is required, in addition to the SDP branch, as shown in Figure 3.16. The AT-commands are needed to control the mobile phone or modem, and another stack such as PPP over RFCOMM is needed to transfer payload data. For faxing, a similar protocol stack is used, but PPP and the networking protocols above PPP are not used. Instead, application software sends the facsimile directly over RFCOMM.

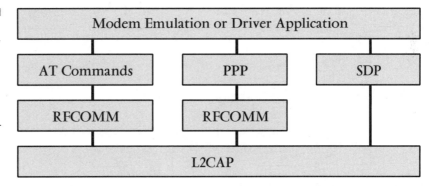

**Figure 3.16**
The protocol stack for the dialup networking usage model.
Source: Bluetooth Specification 1.0.

Other usage models and profiles determine the functionality of the telephone handset, in essence giving the handset multiple capabilities. Telephone handsets built to this profile may be used in three different ways. First, the handset may act as a cordless phone connecting to the public switched telephone network (PSTN) at home or in the office, with usage incurring the usual per-minute charges. This scenario includes making calls via a voice base station, making direct calls between two terminals via the base station, and accessing supplementary services provided by an external network. Second, the handset may act as a cellular phone connecting to the cellular infrastructure and incurring per-minute charges from the cellular carrier. Finally, the handset can be used for connecting directly to other handsets in the piconet. Referred to as the intercom scenario, this type of connection incurs no carrier charges. The cordless and intercom scenarios use the same protocol stack shown in Figure 3.17. The audio stream is directly connected to the baseband protocol, bypassing the L2CAP.

**Figure 3.17**
Protocol stack for cordless phone and intercom usage scenarios.
Source: Bluetooth Specification 1.0.

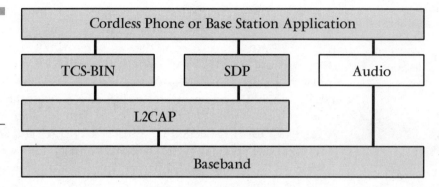

These usage models and profiles are discussed in more detail in the following chapter, along with additional usage models and profiles.

## Summary

The Bluetooth protocols are intended to facilitate the rapid development of applications that can take advantage of Bluetooth wireless technology. The lower layers of the Bluetooth protocol stack are designed to provide a flexible base for further protocol development. Other protocols, such as RFCOMM, are adopted from existing protocols and are only slightly modified for the purposes of the Bluetooth specification. Upper-layer protocols such as WAP are used without modifications. In this way, existing applications may be reused to work with the Bluetooth wireless technology without impeding interoperability.

The purpose of the Bluetooth specification is to promote the development of interoperable applications targeted at the highest priority usage models identified by the SIG's marketing team. However, the Bluetooth specification also serves as a framework for further development, thereby encouraging hardware and software vendors to create more usage models within the framework. Bluetooth wireless technology used with the capabilities of current computers and communications devices makes the possibilities for new future wireless applications virtually unlimited.

CHAPTER 4

# Link Management

When two Bluetooth devices come within range of each other, the Link Manager entity in each device discovers the other. Peer-to-peer communication between the Link Managers occurs through messages exchanged via the Link Manager Protocol (LMP). These messages perform link setup, including security mechanisms like authentication and encryption, which entails the generation, exchange, and checking of link and encryption keys, and the control and negotiation of baseband packet sizes. Through this message exchange, LMP also controls the power modes and duty cycles of the Bluetooth radio devices, and the connection states of Bluetooth units in a piconet. The relationship of the Link Manager Protocol to the other Bluetooth wireless entities is shown in Figure 4.1.

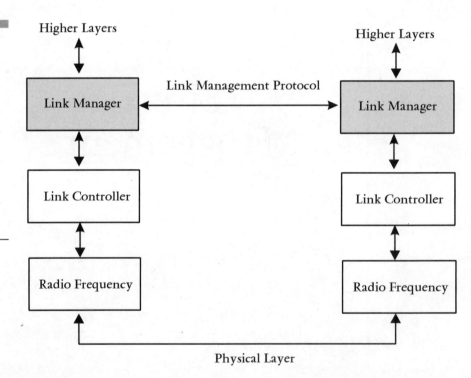

**Figure 4.1**
The Link Manager Protocol (LMP) operates between the Link Manager (LM) of each Bluetooth device to set up the link between them and make the arrangements for security.
Source: Bluetooth Specification 1.0.

The Link Manager is software that runs on a microprocessor in the Bluetooth unit to manage the communication between Bluetooth devices. Each Bluetooth device has its own Link Manager that discovers other remote link managers, and communicates with them to handle link setup, authentication, configuration, and other functions.

# Link Management

To perform its service provider role, the Link Manager draws upon the functions provided by the underlying Link Controller (LC), a supervisory function that handles all the Bluetooth baseband functions and supports the Link Manager. It sends and receives data, requests the identification of the sending device, authenticates the link, sets up the type of link (SCO or ACL), determines what type of frame to use on a packet-by-packet basis, directs how devices will listen for transmissions from other devices, or puts them on hold.

The messages exchanged between the Link Managers take the form of Packet Data Units (PDUs). These have a higher priority than user data, so that a message the Link Manager needs to send will not be delayed by L2CAP traffic. However, it is possible for PDUs to be delayed by many retransmissions of individual baseband packets. In any case, these messages are filtered out and interpreted by the Link Manager on the receiving side and are not propagated to higher layers.

There is no explicit message acknowledgment provision in LMP because the Link Controller provides such a reliable link that it makes such acknowledgments unnecessary. However, the time between receiving a baseband packet carrying an LMP PDU and sending a baseband packet carrying a valid response PDU must be less than the LMP response timeout threshold, which is 30 seconds.

## Types of PDUs

There are 55 different types of PDUs defined in the Bluetooth specification, each used to carry out a unique function. Each PDU is assigned a 7-bit op code for unique identification of its type. The source/destination of the PDUs is determined by the Active Member Address (AM_ADDR) in the packet header. A summary of PDUs is provided in Table 4.1, along with their length in bytes, op code, possible direction in terms of the master (M)-slave (S) relationship, and contents. The usage of these PDUs is discussed in more detail throughout the rest of this chapter.

**TABLE 4.1**

Summary of LMP Protocol Data Unit (PDU) Types, as Described in Bluetooth Specification 1.0

| LMP PDU | Bytes | Op code | Direction | Contents |
|---|---|---|---|---|
| LMP_accepted | 2 | 3 | M → S | Op code |
| LMP_au_rand | 17 | 11 | M ↔ S | Random number |
| LMP_auto_rate | 1 | 35 | M ↔ S | |
| LMP_clkoffset_req | 1 | 5 | M → S | |
| LMP_clkoffset_res | 3 | 6 | M ← S | Clock offset |
| LMP_comb_key | 17 | 9 | M ↔ S | Random number |
| LMP_decr_power_request | 2 | 32 | M ↔ S | Future use |
| LMP_detach | 2 | 7 | M ↔ S | Reason (for detach) |
| LMP_encryption_key_size_req | 2 | 16 | M ↔ S | Key size |
| LMP_encryption_mode_req | 2 | 15 | M ↔ S | Encryption mode |
| LMP_features_req | 9 | 39 | M ↔ S | Features |
| LMP_features_res | 9 | 40 | M ↔ S | Features |
| LMP_host_connection_req | 1 | 51 | M ↔ S | |
| LMP_hold | 3 | 20 | M ↔ S | Hold time |
| LMP_hold_req | 3 | 21 | M ↔ S | Hold time |
| LMP_incr_power_req | 2 | 31 | M ↔ S | Future use |
| LMP_in_rand | 17 | 8 | M ↔ S | Random number |
| LMP_max_power | 1 | 33 | M ↔ S | |
| LMP_max_slot | 2 | 45 | M → S | Maximum slots |
| LMP_max_slot_req | 2 | 46 | M ← S | Maximum slots |
| LMP_min_power | 1 | 34 | M ↔ S | |
| LMP_modify_beacon | 11 or 13 | 28 | M → S | Timing control |
| LMP_name_req | 2 | 1 | M ↔ S | Name offset |
| LMP_name_res | 17 | 2 | M ↔ S | Name offset<br>Name length<br>Name fragment |
| LMP_not_accepted | 3 | 4 | M ↔ S | Op code<br>Reason |

*continued on next page*

# Link Management

**TABLE 4.1**

Summary of LMP Protocol Data Unit (PDU) Types, as Described in Bluetooth Specification 1.0 (continued)

| LMP PDU | Bytes | Op code | Direction | Contents |
|---|---|---|---|---|
| `LMP_page_mode_req` | 3 | 53 | M ↔ S | Paging scheme<br>Paging scheme settings |
| `LMP_page_scan_mode_req` | 3 | 54 | M ↔ S | Paging scheme<br>Paging scheme settings |
| `LMP_park` | 17 | 26 | M → S | Timing control flags |
| `LMP_park_req` | 1 | 25 | M ↔ S | |
| `LMP_preferred_rate` | 2 | 36 | M ↔ S | Data rate |
| `LMP_quality_of_service` | 4 | 41 | M → S | Poll interval |
| `LMP_quality_of_service_req` | 4 | 42 | M ↔ S | Poll interval |
| `LMP_remove_SCO_link_req` | 3 | 44 | M ↔ S | SCO handle<br>Reason |
| `LMP_SCO_link_req` | 7 | 43 | M ↔ S | SCO handle<br>Timing control flags<br>SCO packet<br>Air mode |
| `LMP_set_broadcast_scan_window` | 4 or 6 | 27 | M → S | Timing control flags<br>Broadcast scan window |
| `LMP_setup_complete` | 1 | 49 | M → S | |
| `LMP_slot_offset` | 9 | 52 | M ↔ S | Slot offset<br>Bluetooth device address |
| `LMP_sniff` | 10 | 22 | M → S | Timing control flags<br>Sniff attempt<br>Sniff timeout |
| `LMP_sniff_req` | 10 | 23 | M ↔ S | Timing control flags<br>Sniff attempt<br>Sniff timeout |
| `LMP_sres` | 5 | 12 | M ↔ S | Authentication response |
| `LMP_start_encryption_req` | 17 | 17 | M → S | Random number |
| `LMP_stop_encryption_req` | 1 | 18 | M → S | |
| `LMP_supervision_timeout` | 3 | 55 | M ↔ S | Supervision timeout |
| `LMP_switch_req` | 1 | 19 | M ↔ S | |
| `LMP_temp_rand` | 17 | 13 | M → S | Random number |

*continued on next page*

**TABLE 4.1**

Summary of LMP Protocol Data Unit (PDU) Types, as Described in Bluetooth Specification 1.0 (continued)

| LMP PDU | Bytes | Op code | Direction | Contents |
|---|---|---|---|---|
| LMP_temp_key | 17 | 14 | M → S | Key |
| LMP_timing_accuracy_req | 1 | 47 | M ↔ S | |
| LMP_timing_accuracy_res | 1 | 48 | M ↔ S | Drift<br>Jitter |
| LMP_unit_key | 17 | 10 | M ↔ S | Key |
| LMP_unpark_BD_ADDR_req | varies | 29 | M → S | Timing control flags<br>AM_ADDR 1st unpark<br>AM_ADDR 2nd unpark<br>BD_ADDR 1st unpark<br>BD_ADDR 2nd unpark |
| LMP_unpark_PM_ADDR_req | varies | 30 | M → S | Timing control flags<br>AM_ADDR 1st unpark<br>AM_ADDR 2nd unpark<br>PM_ADDR 1st unpark<br>PM_ADDR 2nd unpark |
| LMP_unsniff_req | 1 | 24 | M ↔ S | |
| LMP_use_semipermanent_key | 1 | 50 | M → S | |
| LMP_version_req | 6 | 37 | M ↔ S | Version number<br>Company ID<br>Sub-version number |
| LMP_version_res | 6 | 38 | M ↔ S | Version number<br>Company ID<br>Sub-version number |

Each PDU is either mandatory (M) or optional (O). The LM does not need to be able to transmit a PDU that is optional, but it must recognize all optional PDUs it receives and, if a response is required, send a valid response, even in the case of an unsupported LMP feature. If the optional PDU received does not require a response, no response is sent. A list of which of the optional PDUs a device supports can be requested.

Some LMP PDUs are reserved for future use. The PDUs for requesting that power be decreased (LMP_decr_power_req) and that power be increased (LMP_incr_power_req), for example, are not supported in the first version of the Bluetooth specification, but may be supported in a subsequent version.

## General Response Messages

There are two general response messages used between Link Managers (Table 4.2). The PDUs LMP_accepted and LMP_not_accepted are used as response messages to other PDUs in a number of different procedures. The PDU LMP_accepted includes the op code of the message that is accepted. The PDU LMP_not_accepted includes the op code of the message that is not accepted and the reason why it was not accepted.

**TABLE 4.2**

General Response Messages
Source: Bluetooth Specification 1.0

| Mandatory/Optional | PDU | Contents |
|---|---|---|
| M | LMP_accepted | Op code |
| M | LMP_not_accepted | Op code<br>Reason |

## Authentication

For security, an authentication procedure is used; this is based on a challenge-response scheme. Both the master and the slave can be verifiers. The verifier sends an LMP_au_rand PDU which contains a random number (the challenge) to the claimant. The claimant calculates a response. The response is a function of the challenge, the claimant's BD_ADDR (Bluetooth Device Address), and a secret key. The response is sent back to the verifier, which checks if the response is correct or not. A proper calculation of the authentication response requires that the two devices (verifier and claimant) share a secret key. Table 4.3 lists the PDUs used in the authentication procedure.

**TABLE 4.3**

Authentication Messages
Source: Bluetooth Specification 1.0

| Mandatory/Optional | PDU | Contents |
|---|---|---|
| M | LMP_au_rand | Random number |
| M | LMP_sres | Authentication response |

If the claimant has a link key associated with the verifier, it calculates the response and sends it to the verifier with LMP_sres. The verifier checks the response. If the response is not correct, the verifier can end the connection by sending LMP_detach with the reason: "code authentication failure." If the claimant does not have a link key associated with the verifier, it sends `LMP_not_accepted` with the reason code "key missing" after receiving `LMP_au_rand`.

When an authentication fails, and to prevent an intruder from trying a large number of keys, a certain waiting interval must pass before a new authentication attempt can be made. For each successive authentication failure with the same Bluetooth address, the waiting interval is increased exponentially; that is, after each failure, the waiting interval before a new attempt can be sent is twice as long as the waiting interval prior to the previous attempt. The maximum waiting interval depends on the implementation. The waiting time exponentially decreases to a minimum when no new failed attempts are being made during a certain time period.

## Pairing

When two devices do not have a common link key, an initialization key must be created based on a PIN and a random number. The 128-bit initialization key is created when the verifier sends `LMP_in_rand` to the claimant. Authentication then needs to be done; the calculation of the authentication response is based on the initialization key. After a successful authentication, the link key is created. The PDUs used in the pairing procedure are summarized in Table 4.4.

**TABLE 4.4**

Pairing Procedure Messages
Source: Bluetooth Specification 1.0

| Mandatory/Optional | PDU | Contents |
|---|---|---|
| M | `LMP_in_rand` random number | Random number |
| M | `LMP_au_rand` random number | Random number |
| M | `LMP_sres` authentication response | Authentication response |
| M | `LMP_comb_key` random number | Random number |
| M | `LMP_unit_key` key | Key |

The verifier sends `LMP_in_rand` and the claimant replies with `LMP_accepted`. Both devices calculate the initialization key, and an

# Link Management

authentication based on this key is performed. The verifier checks the authentication response and if it is correct, the link key is created. If the authentication response is not correct the verifier can end the connection by sending `LMP_detach` with the reason code "authentication failure."

If the claimant has a fixed PIN, it may request a switch of the claimant-verifier role in the pairing procedure by generating a new random number and sending it back in `LMP_in_rand`. If the device that started the pairing procedure has a variable PIN it must accept this and respond with `LMP_accepted`. The roles are then successfully switched and the pairing procedure continues.

If the device that started the pairing procedure has a fixed PIN and the other device requests a role switch, the switch is rejected by sending `LMP_not_accepted` with the reason code "pairing not allowed." The pairing procedure is then stopped at this point. If the claimant rejects pairing, it sends `LMP_not_accepted` with the reason code "pairing not allowed" after receiving LMP_in_rand.

When the authentication is finished the link key must be created. This will be used in the authentication between the two units for all subsequent connections until it is changed. The link key created in the pairing procedure will be either a combination key or one of the unit's unit keys. If one unit sends `LMP_unit_key` and the other unit sends `LMP_comb_key`, the unit key will be the link key. If both units send `LMP_unit_key`, the master's unit key will be the link key. If both units send `LMP_comb_key`, the link key is calculated.

When the authentication during pairing fails because of a wrong authentication response, a waiting interval must pass before a new authentication attempt can be made. For each successive authentication failure with the same Bluetooth address, the waiting interval is increased exponentially. This prevents an intruder from trying a large number of different PINs in a relatively short time.

## Changing the Link Key

If two devices are paired and the link key is derived from combination keys, the link key can be changed. If the link key is a unit key, the units must go through the pairing procedure in order to change the link key. The contents of the PDU are protected with the current link key. Table 4.5 shows the PDUs used for the change of link key.

**TABLE 4.5**

PDUs Used for the Change of Link Key
Source: Bluetooth Specification 1.0

| Mandatory/Optional | PDU | Contents |
|---|---|---|
| M | LMP_comb_key | Random number |
| M | LMP_unit_key | Key |

If the change of link key is successful, the new link key is stored in non-volatile memory, and the old link key is discarded. The new link key will be used as the link key for all subsequent connections between the two devices until the link key is changed again. The new link key also becomes the current link key and remains so until the link key is changed again, or until a temporary link key is created.

If encryption is used on the link and the current link key is a temporary link key, the procedure for changing the link key must immediately be followed by a stop of the encryption. Encryption can then be started again. This procedure assures that encryption parameters known by other devices in the piconet are not used when the semi-permanent link key is the current link key.

## Changing the Current Link Key

The current link key can be a semi-permanent link key or a temporary link key. It can be changed temporarily, but the change is only valid for the session. Changing to a temporary link key is necessary if the piconet is to support encrypted broadcast.

To change to a temporary link key, the master starts by creating the master key. The master then issues a random number and sends it to the slave in LMP_temp_rand. Both sides can then calculate an overlay value. The master sends the master key to the slave in LMP_temp_key. The slave, which knows the overlay value, calculates the master key. After this, the master key becomes the current link key. It will remain the current link key until a new temporary key is created or until the link key is changed.

### Changing a Temporary Link Key

It is possible to make the semi-permanent link key the current link key. After the current link key has been changed to the master key,

# Link Management

this change can be undone so that the semi-permanent link key again becomes the current link key. If encryption is used on the link, the procedure of going back to the semi-permanent link key must be immediately followed by a stop of the encryption. Encryption can then be started again. This is to assure that the encryption parameters known by other devices in the piconet are not used when the semi-permanent link key is the current link key.

## Encryption

Encryption is an option. If at least one authentication has been performed, encryption may be used. If the master wants all slaves in the piconet to use the same encryption parameters, it must issue a temporary key and make this key the current link key for all slaves in the piconet before encryption is started. This is necessary if it is desired that broadcast packets be encrypted. Table 4.6 shows the PDUs used for handling encryption.

**TABLE 4.6**

PDUs Used for Handling Encryption
Source: Bluetooth Specification 1.0

| Mandatory/Optional | PDU | Contents |
|---|---|---|
| O | LMP_encryption_mode_req | Encryption mode |
| O | LMP_encryption_key_size_req | Key size |
| O | LMP_start_encryption_req | Random number |
| O | LMP_stop_encryption_req | |

The master and the slave must first agree whether to use encryption or not, and whether encryption should be applied only to point-to-point packets or to both point-to-point and broadcast packets. If the master and slave agree on the encryption mode, the master continues to give more detailed information about the encryption.

The next step is to determine the size of the encryption key. The master sends the message LMP_encryption_key_size_req, including the suggested key size. Messages are passed between the master and slave to negotiate the size of the encryption key until a key size agreement is reached or it becomes clear that no such agreement can be reached. If an agreement is reached, a unit sends LMP_accepted and the key size in the last LMP_encryption_key_size_req will be used.

After this, the encryption is started. If an agreement is not reached, a unit sends `LMP_not_accepted` with the reason code "unsupported parameter value" and the units are not allowed to communicate using Bluetooth link encryption.

On the other hand, if encryption is started, the master issues the random number and calculates the encryption key. The random number is the same for all slaves if the piconet supports encrypted broadcast. The master then sends `LMP_start_encryption_req`, which includes the random number. When this message is received, the slave calculates the current link key and sends back the acknowledgment `LMP_accepted`. On both sides, the current link key and random number are used as input to the encryption algorithm.

Before starting encryption, higher-layer data traffic must be temporarily stopped to prevent reception of corrupt data. The start of encryption is a three-step process:

1. The master is configured to transmit unencrypted packets, and to receive encrypted packets.
2. The slave is configured to transmit and receive encrypted packets.
3. The master is configured to transmit and receive encrypted packets.

Between steps 1 and 2, master-to-slave transmission is possible and this is when `LMP_start_encryption_req` is transmitted. Step 2 is triggered when the slave receives this message. Between step 2 and step 3, slave-to-master transmission is possible. This is when `LMP_accepted` is transmitted. Step 3 is triggered when the master receives this message.

Before stopping encryption, higher-layer data traffic must be temporarily stopped to prevent reception of corrupt data. Like the start of encryption, the stopping of encryption is a three-step process:

1. The master is configured to transmit encrypted packets, and to receive unencrypted packets.
2. The slave is configured to transmit and receive unencrypted packets.
3. The master is configured to transmit and receive unencrypted packets.

Between steps 1 and 2, master-to-slave transmission is possible and this is when `LMP_stop_encryption_req` is transmitted. Step 2 is triggered when the slave receives this message. Between steps 2 and 3 slave-to-master transmission is possible and this is when `LMP_accepted` is transmitted. Step 3 is triggered when the master receives this message.

# Link Management

If the encryption mode, encryption key, or encryption random number need to be changed, encryption must first be stopped and then restarted with the new parameters.

## Clock Offset Request

Clock offset is the difference between the slave's clock and the master's clock; its value is included in the payload of the Frequency Hop Synchronization (FHS) packet. When a slave receives this packet, the difference is computed between its own clock and the master's clock, and is included in the payload of the FHS packet. The clock offset is also updated each time a packet is received from the master. The master can request this clock offset any time during the connection. By saving this clock offset the master knows on what RF channel the slave wakes up to page scan after it has left the piconet. This can be used to speed up the paging time when the same device is paged again. Table 4.7 lists the PDUs used for the clock offset request.

**TABLE 4.7**
PDUs Used for the Clock Offset Request
Source: Bluetooth Specification 1.0

| Mandatory/Optional | PDU | Contents |
|---|---|---|
| M | LMP_clkoffset_req | |
| M | LMP_clkoffset_res | Clock offset |

## Slot Offset Information

Slot offset is the difference between the slot boundaries in different piconets. The PDU LMP_slot_offset carries two parameters: slot offset and the Bluetooth Device Address (BD_ADDR). Slot offset, expressed in microseconds, is the time between the start of the master's transmit (TX) slot in the piconet, where the PDU is transmitted, and the start of the master's TX slot in the piconet, where BD_ADDR is transmitted. Before doing a master-slave switch, this PDU is transmitted from the device that assumes the role of master in the switch procedure. If the master initiates the switch procedure, the slave sends LMP_slot_offset before sending LMP_accepted. If the slave initiates the switch procedure, the slave sends LMP_slot_offset before sending LMP_switch_req. Table 4.8 shows the PDU used for slot offset.

**TABLE 4.8**

PDU Used for the Slot Offset
Source: Bluetooth Specification 1.0

| Mandatory/Optional | PDU | Contents |
|---|---|---|
| O | LMP_slot_offset | Slot offset<br>BD_ADDR |

## Timing Accuracy Information Request

Timing information can be used to minimize the scan window for a given hold time during return from hold and to extend the maximum hold time. This information can also be used to minimize the scan window when scanning for the sniff mode slots or the park mode beacon packets. Accordingly, LMP supports requests for the timing. The timing accuracy parameters returned are the long-term drift as measured in parts per million (ppm) and the long-term jitter measured in milliseconds of the clock used during hold, sniff, and park mode.

These parameters are fixed for a certain device and must be identical when requested several times. If a device does not support the timing accuracy information it sends the PDU LMP_not_accepted with the reason code "unsupported LMP feature" when the request is received. The requesting device assumes the worst-case values: drift at 250 ppm and jitter at 1 ms. Table 4.9 lists the PDUs used for requesting timing accuracy information.

**TABLE 4.9**

PDUs Used for Requesting Timing Accuracy Information
Source: Bluetooth Specification 1.0

| Mandatory/Optional | PDU | Contents |
|---|---|---|
| O | LMP_timing_accuracy_req | |
| O | LMP_timing_accuracy_res | Drift<br>Jitter |

## LMP Version

LMP supports requests for the version of the LM protocol, in which case the answering device sends back a response with three parameters: version number (VersNr), company ID (CompId), and sub-version num-

# Link Management

ber (Sub-VersNr). VersNr specifies the version of the Bluetooth LMP specification that the device supports. CompId is used to track possible problems with the lower Bluetooth layers. All companies that create a unique implementation of the Link Manager will have their own CompId (Table 4.10).

**TABLE 4.10**

The LMP_CompId Parameter Codes
Source: Bluetooth Specification 1.0

| Code | Company |
|---|---|
| 0 | Ericsson Mobile Communications |
| 1 | Nokia Mobile Phones |
| 2 | Intel Corp. |
| 3 | IBM Corp. |
| 4 | Toshiba Corp. |
| 5—65534 | (reserved) |
| 65535 | Unassigned. For use in internal and interoperability tests before a Company ID has been assigned; may not be used in products. |

Each company is also responsible for the administration and maintenance of the SubVersNr. In addition, each company will also have a unique SubVersNr for each radio frequency, Bluetooth baseband, and Link Manager implementation. For a given VersNr and CompId, the values of the SubVersNr must increase each time a new implementation is released. There is no ability to negotiate the LMP version, only to exchange the parameters (Table 4.11).

**TABLE 4.11**

PDUs Used for Requesting LMP Version
Source: Bluetooth Specification 1.0

| Mandatory/Optional | PDU | Contents |
|---|---|---|
| M | LMP_version_req | VersNr<br>CompId<br>SubVersNr |
| M | LMP_version_res | VersNr<br>CompId<br>SubVersNr |

## Supported Features

Since the Bluetooth radio and link controller may support only a subset of the total number of packet types and features described in Baseband Specification and Radio Specification, there must be a way for devices to learn what other devices support. This information is exchanged through the PDUs LMP_features_req and LMP_features_res (Table 4.12). After the features request has been carried out, the supported packet types for both sides may also be exchanged.

**TABLE 4.12**
PDUs Used for Requesting LMP Features
Source: Bluetooth Specification 1.0

| Mandatory/Optional | PDU | Contents |
|---|---|---|
| M | LMP_features_req | Features |
| M | LMP_features_res | Features |

Whenever a request is issued, it must be compatible with the supported features of the other device. For instance, when establishing an SCO link the initiator may not propose to use a packet type not supported by the other device. Exceptions to this rule are the PDUs LMP_slot_offset and the optional LMP_switch_req, both of which can be sent as the first LMP messages when two Bluetooth devices have been connected and before the requesting side is aware of the other side's features.

## Switching of Master-Slave Role

Since the paging device always becomes the master of the piconet, a switch of the master-slave role is sometimes needed, but this is an optional capability. If device A is the slave and device B is the master, for example, the device that initiates the switch finalizes the transmission of the current L2CAP message and then sends the PDU LMP_switch_req, which requests the role switch. If the switch is accepted, the other device finalizes the transmission of the current L2CAP message and then responds with LMP_accepted. If the switch is rejected, however, the other device responds with LMP_not_accepted and no switch of master-slave roles is performed.

# Link Management

## Name Request

LMP supports name request to another Bluetooth device. The name is a user-friendly one associated with the Bluetooth device and may consist of up to 248 bytes. The name is fragmented over one or more packets. When the `LMP_name_req` is sent, a name offset indicates which fragment is expected. The corresponding `LMP_name_res` carries the same name offset, the name length indicating the total number of bytes in the name of the Bluetooth device, and the name fragment (Table 4.13).

**TABLE 4.13**

PDUs Used for Requesting the Name of a Bluetooth Device
Source: Bluetooth Specification 1.0

| Mandatory/Optional | PDU | Contents |
| --- | --- | --- |
| M | LMP_name_req | Name offset |
| M | LMP_name_res | Name offset<br>Name length<br>Name fragment |

## Detach

Either the master or the slave can close the connection between the Bluetooth devices at any time. The PDU for this capability is `LMP_detach`. A reason parameter is included in the message to inform the other party of why the connection is being closed.

## Hold Mode

The Asynchronous Connection-Less (ACL) link between two Bluetooth devices can be placed in hold mode for a specified time. The PDU for requesting hold mode is `LMP_hold_req` (Table 4.14). During this time no packets can be transmitted from the master.

**TABLE 4.14**

PDUs Used for Hold Mode
Source: Bluetooth Specification 1.0

| Mandatory/Optional | PDU | Contents |
| --- | --- | --- |
| O | LMP_hold | Hold time |
| O | LMP_hold_req | Hold time |

The hold mode is typically entered into when there is no need to send data for a relatively long period. During this period, the transceiver can then be turned off in order to save power. However, the hold mode can also be used if a device wants to discover or be discovered by other Bluetooth devices, or wants to join other piconets. What a device actually does during the hold time is up to each device user to decide, and is not controlled by the hold message.

The master can force hold mode if there has been a request for hold mode that has been previously accepted. The only limitation is that the hold time included in the PDU when the master forces hold mode cannot be longer than any hold time the slave has previously accepted. The same limitation applies to the slave: it can force hold mode if there has previously been a request for hold mode that has been accepted, but the hold time included in the PDU cannot be longer than any hold time the master has previously accepted.

Either the master or the slave can request to enter hold mode. Upon receipt of the request, the same request with modified parameters can be returned, or the negotiation can be terminated. If there is agreement, LMP_accepted terminates the negotiation and the ACL link is placed in hold mode. If no agreement is reached, LMP_not_accepted with the reason code "unsupported parameter value" terminates the negotiation and hold mode is not entered.

## Sniff Mode

Another power-saving mode of operation for Bluetooth devices is sniff mode. To enter sniff mode, master and slave negotiate a sniff interval and a sniff offset, which specifies the timing of the sniff slots. The offset determines the time of the first sniff slot. After that, the sniff slots follow periodically with the sniff interval. When the link is in sniff mode the master can only start a transmission in the sniff slot (Table 4.15).

Two parameters control the listening activity in the slave. The first is the "sniff attempt" parameter, which determines how many slots the slave must listen to, beginning at the sniff slot, even if it does not receive a packet with its own Active Member Address. The second is the "sniff timeout" parameter, which determines how many additional slots the slave must listen to if it continues to receive only packets with its own Active Member Address.

# Link Management

**TABLE 4.15**

PDUs Used for Sniff Mode
Source: Bluetooth Specification 1.0

| Mandatory/Optional | PDU | Contents |
|---|---|---|
| O | LMP_sniff | Timing control flags<br>Sniff offset<br>Sniff interval<br>Sniff attempt<br>Sniff timeout |
| O | LMP_sniff_req | Timing control flags<br>Sniff offset<br>Sniff interval<br>Sniff attempt<br>Sniff timeout |
| O | LMP_unsniff_req | |

The master can force the slave into sniff mode, but either the master or slave can request to enter sniff mode. Upon receipt of the request, the same request with modified parameters can be returned, or the negotiation can be terminated. If agreement is reached, LMP_accepted terminates the negotiation, and the ACL link is placed in sniff mode. If no agreement is reached, LMP_not_accepted with the reason code "unsupported parameter value" terminates the negotiation and sniff mode is not entered.

Sniff mode is ended by sending the PDU LMP_unsniff_req. The requested device must reply with LMP_accepted. If the slave requests, it will enter active mode after receiving LMP_accepted. If the master requests, the slave will enter active mode after receiving LMP_unsniff_req.

## Park Mode

If a slave does not need to participate in the channel, but still should be frequency-hop synchronized (FHS), it can be placed in park mode. In this mode the device gives up its Active Member Address (AM_ADDR). When a slave is placed in park mode it is assigned a unique Parked Member Address (PM_ADDR), which can be used by the master to unpark that slave.[1] Even without its AM_ADDR, the device can still resyn-

---

[1] If the slaves are identified with the PM_ADDR, a maximum of seven slaves can be unparked with the same message. If the slaves are identified with the Bluetooth Device Address (BD_ADDR), a maximum of two slaves can be unparked with the same message.

chronize to the channel by waking up in response to *beacons*, which occur at a predetermined interval. At a beacon, the parked slave can be reactivated by the master. The master can also change the park mode parameters, transmit broadcast information, or let the parked slave request access to the channel.

All PDUs sent from the master to the parked slaves are broadcast. These (`LMP_set_broadcast_scan_window`, `LMP_modify_beacon`, `LMP_unpark_BD_addr_req`, and `LMP_unpark_PM_addr_req`) are the only PDUs that can be sent to a slave in park mode and the only PDUs that can be broadcast. To increase reliability for broadcast, the packets are made as short as possible.

The master can force park mode. In this procedure, the master finalizes the transmission of the current L2CAP message and then sends `LMP_park`. When this PDU is received by the slave, it finalizes the transmission of the current L2CAP message and then sends `LMP_accepted`.

The master can also request that the slave enter park mode. In this procedure, the master finalizes the transmission of the current L2CAP message and then sends `LMP_park_req`. If the slave accepts the request to enter park mode, it finalizes the transmission of the current L2CAP message and then responds with `LMP_accepted`. Finally, the master sends `LMP_park`. If the slave rejects park mode, it sends back `LMP_not_accepted`.

The slave can request to be placed in park mode. In this case, the slave finalizes the transmission of the current L2CAP message and then sends `LMP_park_req` to the master. If the master accepts the park mode request, it finalizes the transmission of the current L2CAP message and then sends back `LMP_park`. If the master rejects park mode, it sends back `LMP_not_accepted`.

# Power Control

The power control is used for limiting the transmit power of the Bluetooth device to optimize power consumption and overall interference level. When a Bluetooth device measures the strength of the received signal, it reports back whether the power should be increased or decreased. If the Receiver Signal Strength Indicator (RSSI) value differs too much from the preferred value, the Bluetooth device can request an increase or a decrease of the other device's transmit power.

# Link Management

Equipment with power control capability optimizes the output power in a link with LMP commands (Table 4.16). At the master side, the transmit power is independent for different slaves, such that a request from one slave can only affect the master's transmit power for that particular slave.

**TABLE 4.16**

PDUs Used for Power Control Source: Bluetooth Specification 1.0

| Mandatory/Optional | PDU | Contents |
|---|---|---|
| O | LMP_incr_power_req | For future use (1 byte) |
| O | LMP_decr_power_req | For future use (1 byte) |
| O | LMP_max_power | |
| O | LMP_min_power | |

If the receiver of LMP_incr_power_req already transmits at maximum power, then LMP_max_power is returned. The device may then request an increase again only after having requested a decrease at least once. Similarly, if the receiver of LMP_decr_power_req already transmits at minimum power, then LMP_min_power is returned and the device may request a decrease again only after having requested an increase at least once.

One byte is reserved in LMP_incr/decr_power_req for future use. It could, for example, be the mismatch between preferred and measured RSSI. The receiver of LMP_incr/decr_power_req could then use this value to adjust to the correct power at once, instead of changing it only one step at a time in response to each request. The parameter is set at a default value for all versions of LMP where this parameter is not yet defined.

## Channel Quality-Driven Change of Data Rate

Bluetooth devices are configured always to use Data-Medium Rate (DM) packets or Data-High Rate (DH) packets, or automatically to adjust the packet type according to the quality of the channel. The difference between DM and DH is that the payload in a DM packet is protected with a Forward Error Correction (FEC) code, whereas the payload of a DH is not protected. If a device wants to adjust automatically between DM and DH, it sends `LMP_auto_rate` to the other device. Using as a basis quality measures in LC, the device determines if throughput will be increased by a change of packet type. If so, `LMP_preferred_rate` is sent to the other device (Table 4.17).

**TABLE 4.17**

PDUs Used for Quality-driven Change of the Data Rate
Source: Bluetooth Specification 1.0

| Mandatory/Optional | PDU | Contents |
|---|---|---|
| O | LMP_auto_rate | |
| O | LMP_preferred_rate | Data rate |

## Quality of Service (QoS)

The Link Manager provides QoS capabilities determined through a poll interval, which is the maximum amount of time between subsequent transmissions from the master to a particular slave. As such, the poll interval is used for bandwidth allocation and latency control. The poll interval is guaranteed, except when there are collisions with page, page-scan, inquiry, and inquiry scan. In addition, master and slave negotiate the number of repetitions for broadcast packets (NBC).

The master can notify a slave of new quality of service. In this case, the master notifies the slave of the new poll interval and NBC. The slave cannot reject the notification. Alternatively, the master and slave can attempt dynamically to negotiate the quality of service as needed. In this case, each device can accept or reject a new quality of service (Table 4.18).

**TABLE 4.18**

PDUs Used for Quality of Service (QoS)
Source: Bluetooth Specification 1.0

| Mandatory/Optional | PDU | Contents |
|---|---|---|
| M | LMP_quality_of_service | Poll interval<br>NBC |
| M | LMP_quality_of_service_req | Poll interval<br>NBC |

# SCO Links

The initial connection between two Bluetooth devices is an Asynchronous Connection-Less (ACL) link. Once the ACL link is in place, one or more Synchronous Connection-Oriented (SCO) links can then be established. The SCO link reserves slots separated by the SCO interval. The first slot reserved for the SCO link is defined by the SCO interval and the SCO delay. After that, the SCO slots follow periodically with the SCO interval. To avoid problems with a wrap-around of the clock during initialization of the SCO link, a flag indicating how the first SCO slot should be calculated is included in a message from the master. Each SCO link is distinguished from all other SCO links by an SCO handle (Table 4.19).

**TABLE 4.19**

PDUs Used for Managing SCO Links
Source: Bluetooth Specification 1.0

| Mandatory/Optional | PDU | Contents |
|---|---|---|
| O | LMP_SCO_link_req | SCO handle<br>Timing control flags<br>SCO interval<br>SCO delay<br>SCO packet<br>Air interface |
| O | LMP_remove_SCO_link_req | SCO handle<br>Reason |

To establish an SCO link, the master sends a request with parameters that specify the timing, packet type, and coding that will be used on the SCO link. For each of the SCO packets, the Bluetooth specification supports three different voice-encoding formats on the air interface: µ-law PCM, A-law PCM, and CVSD. The slots used for the SCO links are determined by three parameters controlled by the master: SCO interval, SCO delay, and a flag indicating how the first SCO slot should be calculated. After the first slot, the SCO slots follow periodically with the SCO interval.

If the slave does not accept the SCO link, but is willing to consider another possible set of SCO parameters, it can indicate what it does not accept in the error reason field of LMP_not_accepted. The master can then issue a new request with modified parameters. The SCO handle in the message must be different from any already existing SCO link(s).

The slave can also initiate SCO link establishment with `LMP_SCO_link_req`, but timing control flags and SCO delay are invalid as well as the SCO handle. If the master is not capable of establishing an SCO link, it replies with `LMP_not_accepted`. Otherwise it sends back `LMP_SCO_link_req`. This message includes the assigned SCO handle, SCO delay, and the timing control flags. For the other parameters, the master tries to use the same parameters as in the slave request; if the master cannot meet that request, it can use other values. The slave must then reply with `LMP_accepted` or `LMP_not_accepted`.

Both the master and the slave can request a change of SCO parameters. Regardless of which device originates the request, it is initiated with `LMP_SCO_link_req`, where the SCO handle is the handle of the SCO link for which the device wishes to change parameters. If the device accepts the new parameters, it replies with `LMP_accepted` and the SCO link will change to the new parameters. If the device does not accept the new parameters, it replies with `LMP_not_accepted` and the SCO link is left unchanged. When the device replies with `LMP_not_accepted`, it indicates in the error reason parameter what it does not accept. The device can then try to change the SCO link again with modified parameters.

The master or slave can remove the SCO link by sending a request, including the SCO handle of the SCO link to be removed and a reason indicating why the SCO link is removed. In this case, the receiving device must respond with `LMP_accepted`.

## Control of Multi-Slot Packets

The number of slots used by a slave in its return packet can be limited. The master allows the slave to use the maximum number of slots by sending the PDU `LMP_max_slot`, providing "max slots" as the parameter. Likewise, each slave can request to use a maximum number of slots by sending the PDU `LMP_max_slot_req`, providing "max slots" as the parameter. The default value is one slot. If the slave has not been informed about the number of slots, it may only use one-slot packets. Two PDUs are used for the control of multi-slot packets (Table 4.20).

# Link Management

**TABLE 4.20** PDUs for Controlling the Use of Multi-slot Packets
Source: Bluetooth Specification 1.0

| Mandatory/Optional | PDU | Contents |
|---|---|---|
| M | `LMP_max_slot` | Max slots |
| M | `LMP_max_slot_req` | Max slots |

## Paging Scheme

In addition to the mandatory paging scheme, which has to be supported by all Bluetooth devices, the Bluetooth specification allows for optional paging schemes (Table 4.21).[2] LMP provides the means to negotiate the paging scheme, which will be used the next time a Bluetooth device is paged. The key difference between an optional paging scheme and the mandatory scheme is the construction of the page train sent by the pager. In addition to transmission in the even-numbered master slots, the master is transmitting in the odd-numbered master slots. In the optional scheme, the same set of frequencies used for transmitting is also used for reception. This allows the slave unit to reduce the scan window.

**TABLE 4.21** PDUs Used to Request the Paging Scheme
Source: Bluetooth Specification 1.0

| Mandatory/Optional | PDU | Contents |
|---|---|---|
| O | `LMP_page_mode_req` | Paging scheme<br>Paging scheme settings |
| O | `LMP_page_scan_mode_req` | Paging scheme<br>Paging scheme settings |

The page mode procedure is initiated by device A, which negotiates the paging scheme used when it pages device B. Device A proposes a paging scheme, including the parameters for this scheme, and device B can accept or reject the request. On rejection of the request, the old setting is not changed. A request to switch back to the mandatory scheme may also be rejected.

The page scan mode procedure is initiated by device A, which negotiates the paging scheme used when device B pages device A. Device A proposes a paging scheme, including the parameters for this

---

[2] At this writing, the optional paging schemes are not completely defined in Bluetooth Specification 1.0.

scheme, and device B can accept or reject the request. On rejection of the request, the old setting is not changed. However, a request to switch to the mandatory scheme must be accepted.

## Link Supervision

Each Bluetooth link has a timer used for link supervision. This mandatory capability is used to detect link loss caused by devices moving out of range, a device's power-down, or some other type of failure. An LMP procedure is used to set the value of the supervision timeout. The PDU used to set the supervision timeout is LMP_supervision_timeout.

## Connection Establishment

After the paging procedure, the master must poll the slave by sending POLL or NULL packets, with a max poll interval. LMP procedures that do not require any interactions between the LM and the host at the paged unit's side can then be carried out.

When the paging device needs to create a connection involving layers above the Link Manager (LM), it sends LMP_host_connection_req (Table 4.22). When the other side receives this message, the host is informed about the incoming connection. The remote device can accept or reject the connection request by sending LMP_accepted or LMP_not_accepted. When a device does not require any further link setup procedures, it sends LMP_setup_complete; in this case, it can still respond to requests from the other device. When the other device is also ready with link setup, it sends LMP_setup_complete. After this, the first packet on a logical channel different from LMP can then be transmitted.

**TABLE 4.22**

PDUs Used for Connection Establishment
Source: Bluetooth Specification 1.0

| Mandatory/Optional | PDU | Contents |
|---|---|---|
| M | LMP_host_connection_req | |
| M | LMP_setup_complete | |

## Test Modes

In addition to the 55 PDUs previously described, LMP has PDUs to support different Bluetooth test modes, which are used for certification and compliance testing of Bluetooth radio and baseband devices. These functions are not seen by end-users of the Bluetooth devices. The test modes include transmitter tests (packets with constant bit patterns) and loop back tests. While in test mode, a device may not support normal operation. If the Bluetooth device is not in test mode, the control and configuration commands used for testing are rejected. In this case, an LMP_not_accepted is returned.

The test mode is activated by sending LMP_test_activate to the device under test, which is always the slave. The Link Manager must be able to receive this message at any time. If entering test mode is locally enabled in the device under test, it responds with LMP_accepted and the test mode is entered. Otherwise, the device under test responds with LMP_not_accepted and remains in normal operation. In this case, the reason code in LMP_not_accepted is "PDU not allowed."

The test mode can be deactivated in two ways. First, sending LMP_test_control with the test scenario set to "exit test mode" ends the test mode, in which case the slave returns to normal operation, but is still connected to the master. Second, sending LMP_detach to the device under test ends the test mode as well as the connection to the master.

When a device has entered test mode, the PDU LMP_test_control can be sent to it to start a specific test. This PDU is acknowledged with LMP_accepted. If a device that is not in test mode receives LMP_test_control it responds with LMP_not_accepted, where the reason code is "PDU not allowed."

## Error Handling

In the event the Link Manager receives a PDU with an unrecognized op code, it responds with LMP_not_accepted with the reason code "unknown LMP PDU." The op code parameter echoed back is the unrecognized op code. If the Link Manager receives a PDU with invalid parameters, it responds with LMP_not_accepted with the reason code "invalid LMP parameters." If the maximum response time is exceeded or if link loss is detected, the party that waits for the response will conclude that the procedure has terminated unsuccessfully.

Erroneous LMP messages can be caused by errors on the channel or systematic errors at the transmit side. To detect the latter case, the LM monitors the number of erroneous messages and disconnects if this exceeds a predefined threshold. Since LMP PDUs are not interpreted in real time, collision situations can occur where both LMs initiate the same procedure and both cannot be completed. In this situation, the master rejects the slave-initiated procedure by sending `LMP_not_accepted` with the reason code "LMP Error Transaction Collision." The master-initiated procedure is then completed.

## Summary

The Link Management Protocol provides the means to implement peer-to-peer communication between the Link Managers of two Bluetooth devices when they come within range of each other. It is the Link Manager entity in each device which discovers the other. The actual communication between Link Managers occurs through various messages (Packet Data Units, or PDUs) exchanged via the Link Manager Protocol. There are 55 types of messages (plus 2 that are used for certification and compliance testing of the Bluetooth radio and baseband), which are used to initiate various functions, such as link setup, security mechanisms like authentication and encryption, and the control and negotiation of baseband packet sizes. Through this message exchange, LMP also controls the power modes and duty cycles of the Bluetooth radio devices as well as the connection states of Bluetooth units in a piconet. Regardless of what function is being carried out in response to these messages, they are filtered out and interpreted by the Link Manager on the receiving side, and are not sent up to higher layers.

CHAPTER 5

# Logical Link Control

The Bluetooth specification includes the Logical Link Control and Adaptation Protocol (L2CAP)[1], which provides for higher-level protocol multiplexing, and packet segmentation and reassembly (SAR). L2CAP also conveys Quality of Service (QoS) information between Bluetooth endpoints. Like the Link Manager Protocol (LMP) discussed in the previous chapter, L2CAP is layered over the baseband protocol and resides within the data-link layer of the OSI reference model, as shown in Figure 5.1.

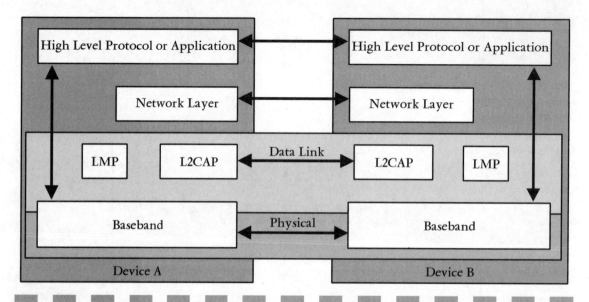

**Figure 5.1** Logical Link Control and Adaptation Protocol (L2CAP) resides at the data-link layer to support higher-level protocol multiplexing, packet segmentation and reassembly (SAR), and conveys Quality of Service (QoS) information. Source: Bluetooth Specification 1.0.

L2CAP permits higher-level protocols and applications to transmit and receive L2CAP data packets up to 64 kilobytes (KB) in length. While the Baseband specification of Bluetooth defines two link types—Synchronous Connection-Oriented (SCO) links and Asynchronous Connection-Less (ACL) links—the L2CAP specification is defined for only ACL links and no more than one ACL link can exist between any two devices. L2CAP relies on integrity checks at the baseband

---

1 The "L2" in L2CAP refers to OSI Layer 2, which is the data-link layer. The data-link layer consists of the logical link control and media access control sublayers.

# Logical Link Control

layer to protect the transmitted information over the ACL link. No support for SCO links is planned.

L2CAP interfaces with other communication protocols (Figure 5.2) such as the Service Discovery Protocol, RFCOMM, and Telephony Control Specification (TCS). While voice-quality channels for audio and telephony applications are usually run over baseband SCO links, packetized audio data, such as IP Telephony, may be sent using the communication protocols running over L2CAP, in which case, the audio is treated as just another data application.

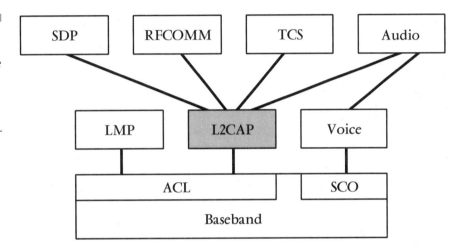

**Figure 5.2**
L2CAP in relation to other protocols in the Bluetooth protocol architecture.
Source: Bluetooth Specification 1.0.

Among the characteristics of L2CAP are simplicity and low overhead, which makes it suitable for implementation in devices with limited computing and memory resources, such as handheld computers, PDAs, digital cellular phones, wireless headsets, and other devices that support Bluetooth wireless technology. The protocol's low overhead enables it to achieve high bandwidth efficiency without consuming too much power, which is in keeping with the power efficiency objectives of the Bluetooth radio.

## L2CAP Functions

Among the functions performed at the L2CAP layer is protocol multiplexing. L2CAP must support protocol multiplexing because the baseband protocol does not support a "type" field to identify the higher-layer protocol being multiplexed above it. L2CAP must therefore be able to distinguish among upper-layer protocols such as the SDP, RFCOMM, and TCS.

Another function performed at the L2CAP layer is segmentation and reassembly, which is necessary to support protocols using packets larger than those supported by the baseband. Compared to the physical media in the wired environment, the data packets used in conjunction with the Bluetooth baseband protocol are limited in size. Exporting a maximum transmission unit (MTU) associated with the largest baseband payload (341 bytes for high-rate data packets) limits the efficient use of bandwidth for higher-layer protocols designed to use larger packets. Large L2CAP packets must be segmented into multiple smaller baseband packets prior to wireless transmission. On the receive side, baseband packets are reassembled into a single larger L2CAP packet after a simple integrity check.

The L2CAP connection establishment process allows the exchange of information regarding the QoS expected between two Bluetooth devices. The L2CAP implementation at each end monitors the resources used by the protocol and ensures that the QoS contracts are enforced.

Many protocols include the concept of address groups. The baseband protocol supports the concept of a piconet, a group of up to eight devices operating together according to the same clock. The L2CAP group abstraction permits implementations for efficient mapping of protocol groups onto piconets. Absent a group abstraction function, higher-level protocols would need to be exposed to the baseband protocol and Link Manager (LM) to manage the address groups efficiently.

## Basic Operation

The ACL link between two Bluetooth units is set up using the Link Manager Protocol (LMP). The baseband provides orderly delivery of data packets, despite occasional corrupt and duplicate packets. The baseband also provides full-duplex communication channels. However, all L2CAP communications need not be bidirectional. Multicast and unidirectional traffic—video, for example—require not duplex, but rather simplex channels.

L2CAP provides a reliable channel using the mechanisms available at the baseband layer. Reliability is achieved through data integrity checks performed when requested and by the retransmission of data until they have been successfully acknowledged, or a timeout occurs. Because even acknowledgments may be lost, however, timeouts can occur after the data have been successfully sent. The baseband protocol uses a 1-bit sequence number that removes duplicate packets. The use of broadcast packets is disallowed if reliability is required. This is enforced by having all broadcasts start the first segment of an L2CAP packet with the same sequence bit.

There are some capabilities that fall outside the scope of L2CAP's responsibilities:

- It does not transport audio designated for SCO links.
- It does not enforce a reliable channel or ensure data integrity; that is, L2CAP does not perform retransmissions or checksum calculations.
- It does not support a reliable multicast channel.
- It does not support the concept of a global group name.

## Channel Identifiers

L2CAP is based on the concept of *channels*. Each endpoint of an L2CAP channel is referred to by a channel identifier (CID). Some identifiers are reserved for specific L2CAP functions, such as signaling, which is used to create and establish connection-oriented data channels and negotiate changes in the characteristics of these channels. Another CID is reserved for all incoming connectionless data traffic.

Except for the reserved channels, the remaining CIDs can be managed in a manner best suited for a particular implementation, with the condition that the same CID not be reused as a local L2CAP channel endpoint for multiple simultaneous L2CAP channels between a local and remote device.

Figure 5.3 illustrates the use of CIDs between L2CAP peer entities in separate devices. The connection-oriented data channels represent a connection between two devices, where a CID identifies each endpoint of the channel. The connectionless channels limit data flow to a single direction. These channels are used to support a channel "group" where the CID on the source represents one or more remote devices.

Support for a signaling channel within an L2CAP entity is mandatory. Another CID is reserved for all incoming connectionless data traffic. In the figure, a CID is used to represent a group consisting of devices C and D. Traffic sent from this channel ID is directed to the remote channel reserved for connectionless data traffic.

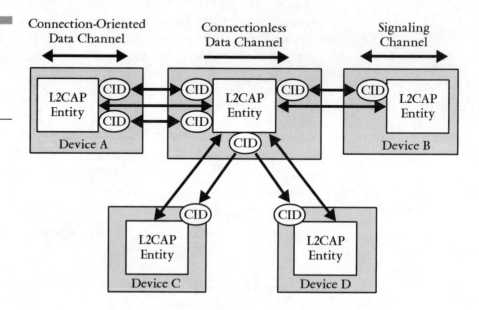

**Figure 5.3**
Channels between devices set up by L2CAP.
Source: Bluetooth Specification 1.0.

## Segmentation and Reassembly

Segmentation and reassembly (SAR) operations are used to improve efficiency by supporting a maximum transmission unit (MTU) size larger than the largest baseband packet. This reduces overhead by spreading the network and transport packets used by higher-layer protocols over several baseband packets. All L2CAP packets may be segmented for transfer over baseband packets. The protocol does not perform any segmentation and reassembly operations, but the packet format supports adaptation to smaller physical frame sizes. An L2CAP implementation exposes the outgoing MTU and segments higher-layer packets into smaller pieces that can be passed to the Link Manager via the Host Controller Interface (HCI), whenever one exists. On the receiving side, an L2CAP implementation receives segmented packets from the HCI and reassembles them into L2CAP packets using information provided through the HCI and the packet header.

### SEGMENTATION

The L2CAP MTU is exported using an implementation-specific service interface. It is the responsibility of the higher-layer protocol to limit the size of packets sent to the L2CAP layer below the MTU limit. An L2CAP implementation segments the packet into protocol data units (PDUs) to send to the lower layer. If L2CAP runs directly over the baseband protocol, an implementation may segment the packet into baseband packets for wireless transmission. If L2CAP runs above the HCI, as is typical, an implementation may send block-sized packets to the host controller, where they will be converted into baseband packets. All L2CAP segments associated with an L2CAP packet must be passed through to the baseband before any other L2CAP packet destined to the same Bluetooth unit may be sent.

### REASSEMBLY

The baseband protocol delivers ACL packets in sequence and protects the integrity of the data using a 16-bit Cyclic Redundancy Check (CRC). The baseband also supports reliable connections using an automatic repeat request (ARQ) mechanism. As the baseband controller receives ACL packets, it either signals the L2CAP layer on the arrival of each baseband packet, or accumulates a number of packets before the receive buffer fills up or a timer expires before signaling the L2CAP layer.

L2CAP implementations use the length field in the header of L2CAP packets as a consistency check and discard any that fail to match the length field. If channel reliability is not required, packets with improper lengths may be discarded. If channel reliability is required, L2CAP implementations must notify the upper layer that the channel has become unreliable. Reliable channels are defined in the Bluetooth specification as having a flush timeout value. This option is used to inform the recipient of the amount of time the originator's link controller/link manager will attempt to transmit an L2CAP segment before giving up and "flushing" the packet.

## State Machine

The L2CAP connection-oriented channel assumes various states during interactions between different layers. According to the Bluetooth specification, this "state machine" is only applicable to bidirectional

CIDs and is not representative of the signaling channel or the unidirectional channel. Figure 5.4 illustrates the events and actions performed by an implementation of the L2CAP layer.

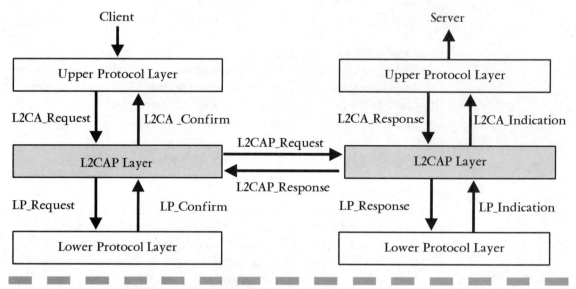

**Figure 5.4**  L2CAP interactions. Source: Bluetooth Specification 1.0.

Clients initiate requests, while servers accept the requests. An application-level client can both initiate and accept requests. The naming convention entails the interface between two layers (vertical interface) using the prefix of the lower layer that offers the service to the higher layer, such as L2CA. The interface between two entities of the same layer (horizontal interface) uses the prefix of the protocol, adding the letter "P" to the layer identification, so that L2CA becomes L2CAP. Events coming from above are called Requests (Req) and the corresponding replies are called Confirms (Cfm). Events coming from below are called Indications (Ind) and the corresponding replies are called Responses (Rsp).

Responses requiring further processing are said to be Pending (Pnd). The notation for Confirms and Responses assumes positive replies. Negative replies have a "Neg" suffix, as in L2CAP_ConnectCfmNeg. While Requests for an action always result in a corresponding Confirmation, Indications do not always have corresponding Responses. The latter is especially true if the Indications only provide information about locally triggered events.

# Logical Link Control

The message sequence chart in Figure 5.5 depicts a normal sequence of events. The two outer vertical lines represent the L2CA interface of the initiator (the device issuing a request) and the acceptor (the device responding to the initiator's request). Request commands at the L2CA interface result in Requests defined by the protocol. When the protocol communicates the request to the acceptor, the remote L2CA entity presents the upper protocol with an Indication. When the acceptor's upper protocol responds, the response is packaged by the protocol and communicated back to the initiator. The result is passed back to the initiator's upper protocol using a Confirm message.

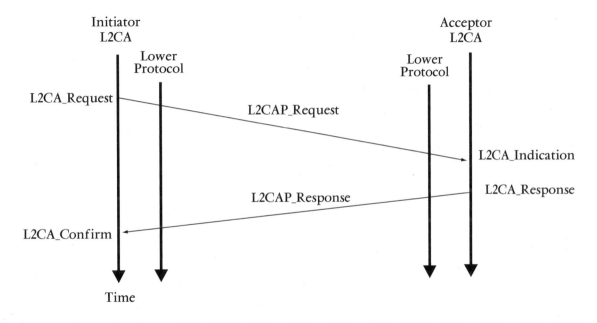

**Figure 5.5** Message sequence chart showing the interactions of different layers. Source: Bluetooth Specification 1.0.

## Events

Events are incoming messages to the L2CA layer along with any timeouts. Events are categorized as Indications and Confirms from lower layers, Requests and Responses from higher layers, data from peers, signal Requests and Responses from peers, and events caused by timer expirations.

## LOWER-LAYER TO L2CAP EVENTS

The Bluetooth specification describes the following messages that are exchanged between the lower-layer protocol (i.e., the baseband) and L2CAP during the connection establishment process:

- **LP_ConnectCfm**—Confirms the request to establish a lower-layer connection. This includes passing the authentication challenge if authentication is required to establish the physical link.

- **LP_ConnectCfmNeg**—Confirms the failure of the request to establish a lower-layer connection. This may happen when the device cannot be contacted, refused the request, or the LMP authentication challenge failed.

- **LP_ConnectInd**—Indicates the lower protocol has successfully established the connection. In the case of the baseband, this will be an ACL link. An L2CAP entity may use this information to keep track of physical links.

- **LP_DisconnectInd**—Indicates the lower protocol has been shut down by LMP commands or a timeout event.

- **LP_QoSCfm**—Confirms the request for a given quality of service.

- **LP_QoSCfmNeg**—Confirms the failure of the request for a given quality of service.

- **LP_QoSViolationInd**—Indicates the lower protocol has detected a violation of the QoS agreement specified in the previous `LP_QoSReq`.

## L2CAP-TO-L2CAP SIGNALING EVENTS

L2CAP-to-L2CAP signaling events are generated by each L2CAP entity following the exchange of the corresponding L2CAP signaling PDUs. L2CAP signaling PDUs, like any other L2CAP PDUs, are received from a lower layer via a lower protocol indication event. Per the Bluetooth specification, the following signaling messages are exchanged between L2CAP entities:

- **L2CAP_ConnectReq**—A connection request packet has been received.

- **L2CAP_ConnectRsp**—A connection response packet has been received with a positive result, indicating that the connection has been established.

- **L2CAP_ConnectRspPnd**—A connection response packet has been received, indicating the remote endpoint has received the request and is processing it.

# Logical Link Control

- **L2CAP_ConnectRspNeg**—A connection response packet has been received, indicating that the connection could not be established.
- **L2CAP_ConfigReq**—A configuration request packet has been received, indicating the remote endpoint wants to engage in negotiations concerning channel parameters.
- **L2CAP_ConfigRsp**—A configuration response packet has been received, indicating the remote endpoint agrees with all the parameters being negotiated.
- **L2CAP_ConfigRspNeg**—A configuration response packet has been received, indicating the remote endpoint does not agree to the parameters received in the response packet.
- **L2CAP_DisconnectReq**—A disconnection request packet has been received, indicating the channel must initiate the disconnection process. Following the completion of an L2CAP channel disconnection process, an L2CAP entity returns the corresponding local CID to the pool of unassigned CIDs.
- **L2CAP_DisconnectRsp**—A disconnection response packet has been received. Following the receipt of this signal, the receiving L2CAP entity returns the corresponding local CID to the pool of unassigned CIDs. There is no corresponding negative response because the disconnect request must succeed.

**L2CAP-TO-L2CAP DATA EVENTS**
There is only one message that is exchanged between L2CAP entities to indicate that a data packet has been received: `L2CAP_Data`.

**UPPER-LAYER TO L2CAP EVENTS**
The Bluetooth specification provdes for the following messages which are exchanged between the upper-layer protocol and L2CAP during the connection establishment process:

- **L2CA_ConnectReq**—A request from an upper layer for the creation of a channel to a remote device.
- **L2CA_ConnectRsp**—A response from an upper layer to the indication of a connection request from a remote device.
- **L2CA_ConnectRspNeg**—A negative response (rejection) from an upper layer to the indication of a connection request from a remote device.
- **L2CA_ConfigReq**—A request from upper layer to (re)configure the channel.

- **L2CA_ConfigRsp**—A response from an upper layer to the indication of a (re)configuration request.
- **L2CA_ConfigRspNeg**—A negative response from an upper layer to the indication of a (re)configuration request.
- **L2CA_DisconnectReq**—A request from an upper layer for the immediate disconnection of a channel.
- **L2CA_DisconnectRsp**—A response from an upper layer to the indication of a disconnection request; there is no corresponding negative response, the disconnect indication must always be accepted.
- **L2CA_DataRead**—A request from an upper layer for the transfer of received data from the L2CAP entity to the upper layer.
- **L2CA_DataWrite**—A request from an upper layer for the transfer of data from the upper layer to the L2CAP entity for transmission over an open channel.

### TIMER EVENTS

The Bluetooth specification provides for the following timer messages, which are exchanged between the endpoints during the channel termination process:

- **Response Timeout eXpired (RTX)**—Used to terminate the channel when the remote endpoint is unresponsive to signaling requests. This timer is started when a signaling request is sent to the remote device and disabled when the response is received. If the initial timer expires, a duplicate request message is sent or the channel identified in the request is disconnected. If a duplicate request message is sent, the RTX timeout value is reset to a new value of at least double the previous value. Implementations have the responsibility to decide on the maximum number of retransmission requests that will be performed at the L2CAP level before disconnecting the channel. The decision is based on the flush timeout of the signaling link. The longer the flush timeout, the more retransmissions may be performed at the physical layer and the greater the reliability of the channel, so that fewer retransmissions at the L2CAP level are required.
- **Extended Response Timeout eXpired (ERTX)**—Used in place of the RTX timer when it is suspected that the remote endpoint is performing additional processing of a request signal. This timer is started when the remote endpoint responds that a request is pending, as when an `L2CAP_ConnectRspPnd` event is received. This timer is disabled when the formal response is received or the physical link

is lost. If the initial timer expires, a duplicate request may be sent or the channel may be disconnected. If a duplicate request is sent, the particular ERTX timer disappears, replaced by a new RTX timer, and the whole timing procedure restarts as described previously for the RTX timer.

## Actions

Actions are categorized as Confirms and Indications to higher layers, Requests and Responses to lower layers, Requests and Responses to peers, data transmission to peers, and setting timers.

### L2CAP-TO-LOWER-LAYER ACTIONS
Per the Bluetooth specification, the following messages are exchanged between the lower-layer protocol (i.e., the baseband) and L2CAP during the connection establishment process:

- **LP_ConnectReq**—L2CAP requests the lower protocol to create a connection. If a physical link to the remote device does not exist, this message is sent to the lower protocol to establish the physical connection. Since no more than a single ACL link between two devices is assumed, additional L2CAP channels between these two devices must share the same baseband ACL link. Following the processing of the request, the lower layer returns with `LP_ConnectCfm` or `LP_ConnectCfmNeg` to indicate whether the request has been satisfied.

- **LP_QoSReq**—L2CAP requests the lower protocol to accommodate a particular QoS parameter. After processing the request, the lower layer returns with `LP_QoSCfm` or `LP_QoSCfmNeg` to indicate whether the request has been satisfied.

- **LP_ConnectRsp**—A positive response accepting the previous connection indication request.

- **LP_ConnectRspNeg**—A negative response denying the previous connection indication request.

### L2CAP-TO-L2CAP SIGNALING ACTIONS
These actions are the same as those discussed above under "L2CAP to lower-layer actions," except that they refer to the transmission, rather than reception, of the messages.

## L2CAP-TO-L2CAP DATA ACTIONS

This is the counterpart to "L2CAP to L2CAP data events," but in this case the action refers to data transmission rather than data reception.

## L2CAP-TO-UPPER-LAYER ACTIONS

The Bluetooth specification describes the following messages, which are exchanged between L2CAP and the upper-layer protocol during the connection establishment process:

- **L2CA_ConnectInd**—Indicates a connection request has been received from a remote device.
- **L2CA_ConnectCfm**—Confirms that a connection request has been accepted following the receipt of a connection message from the remote device.
- **L2CA_ConnectCfmNeg**—Negative confirmation (failure) of a connection request. (An RTX timer expiration for an outstanding connect request can substitute for a negative connect response and result for this action.)
- **L2CA_ConnectPnd**—Confirms that a connection response (pending) has been received from the remote device.
- **L2CA_ConfigInd**—Indicates a configuration request has been received from a remote device.
- **L2CA_ConfigCfm**—Confirms that a configuration request has been accepted following the receipt of a configuration response from the remote device.
- **L2CA_ConfigCfmNeg**—Negative confirmation (failure) of a configuration request; an RTX timer expiration for an outstanding connect request can substitute for a negative connect response and result for this action.
- **L2CA_DisconnectInd**—Indicates a disconnection request has been received from a remote device or the remote device has been disconnected because it failed to respond to a signaling request.
- **L2CA_DisconnectCfm**—Confirms that a disconnect request has been processed by the remote device following the receipt of a disconnection response from the remote device; an RTX timer expiration for an outstanding Disconnect Request can substitute for a Disconnect Response and result for this action. Upon receiving this event, the upper layer knows the L2CAP channel has been terminated. Consequently, there is no corresponding negative confirm.

Logical Link Control

- **L2CA_TimeOutInd**—Indicates that an RTX or ERTX timer has expired. This indication will occur an implementation-dependant number of times before the L2CAP implementation will give up and send the `L2CA_DisconnectInd` message.
- **L2CA_QoSViolationInd**—Indicates that the quality of service agreement has been violated.

## Channel Operational States

While the channel is being set up, it enters into various operational states, which are described in the Bluetooth specification as:

- **CLOSED**—In this state, there is no channel associated with the channel identifier (CID). This is the only state when a link-level connection (i.e., baseband) may not exist. Link disconnection forces all other states into the CLOSED state.
- **W4_L2CAP_CONNECT_RSP**—In this state, the CID represents a local endpoint and an `L2CAP_ConnectReq` message has been sent referencing this endpoint, which is now waiting for the corresponding `L2CAP_ConnectRsp` message.
- **W4_L2CA_CONNECT_RSP**—In this state, the remote endpoint exists and an `L2CAP_ConnectReq` has been received by the local L2CAP entity. An `L2CA_ConnectInd` has been sent to the upper layer and the part of the local L2CAP entity processing the received `L2CAP_ConnectReq` waits for the corresponding response. The response may require a security check to be performed.
- **CONFIG**—In this state, the connection has been established but both sides are still negotiating the channel parameters. This state may also be entered into when the channel parameters are being renegotiated. Prior to entering the CONFIG state, all outgoing data traffic is suspended since the traffic parameters of the data traffic are to be renegotiated. Incoming data traffic must be accepted until the remote channel endpoint has entered the CONFIG state. In the CONFIG state, both sides must issue `L2CAP_ConfigReq` messages. If a large amount of parameters need to be negotiated, the process can proceed incrementally with multiple messages sent to avoid any MTU limitations. Moving from the CONFIG state to the OPEN state requires both sides to be ready. An L2CAP entity is ready when it has received a positive response to its final request and has positively responded to the final request from the remote device.

- **OPEN**—In this state, the connection has been established and configured, and data flow may proceed.
- **W4_L2CAP_DISCONNECT_RSP**—In this state, the connection is shutting down and an `L2CAP_DisconnectReq` message has been sent. This state is now waiting for the corresponding response.
- **W4_L2CA_DISCONNECT_RSP**—In this state, the connection on the remote endpoint is shutting down and an `L2CAP_DisconnectReq` message has been received. An `L2CA_DisconnectInd` has been sent to the upper layer to notify the owner of the CID that the remote endpoint is being closed. This state is now waiting for the corresponding response from the upper layer before responding to the remote endpoint.

## Mapping Events to Actions

Table 5.1 summarizes the previous discussions and shows what events cause state transitions and what actions are taken during the transitions. For example, when the initiator is creating the first L2CAP channel between two devices, both sides start in the CLOSED state. After receiving the request from the upper layer, the entity requests the lower layer to establish a physical link. If no physical link exists, LMP commands are used to create the physical link between the devices. Once the physical link is established, L2CAP signals are sent over it.

Events that are not listed in Table 5.1, or have actions marked N/C (for no change), are assumed to be errors and are discarded. Data input and output events are only defined for the open and configuration states. Data may not be received during the initial configuration state, but may be received when the configuration state is re-entered due to a reconfiguration process. Data received during any other state are discarded.

# Logical Link Control

**TABLE 5.1**

Summary of Events to Actions that Occur in a Particular Channel State Source: Bluetooth Specification 1.0

| Event | Current State | Action | New State |
|---|---|---|---|
| `LP_ConnectCfm` | CLOSED | Flag physical link as active and initiate L2CAP connection. | CLOSED |
| `LP_ConnectCfmNeg` | CLOSED | Flag physical link as inactive and fail any outstanding service connection requests by sending `L2CA_ConnectCfmNeg` to the upper layer. | CLOSED |
| `LP_ConnectInd` | CLOSED | Flag link as active. | CLOSED |
| `LP_DisconnectInd` | CLOSED | Flag link as inactive. | CLOSED |
| `LP_DisconnectInd` | Any but CLOSED | Send upper layer `L2CA_DisconnectInd` message. | CLOSED |
| `LP_QoSViolationInd` | Any but OPEN | Discard | N/C |
| `LP_QoSViolationInd` | OPEN | Send upper layer `L2CA_QoSViolationInd` message. If service level is guaranteed, terminate the channel. | OPEN or W4_L2CA_DISCONNECT_RSP |
| `L2CAP_ConnectReq` | CLOSED (CID dynamically allocated from free pool) | Send upper layer `L2CA_ConnectInd`. Optionally, send peer `L2CAP_ConnectRspPnd`. | W4_L2CA_CONNECT_RSP |
| `L2CAP_ConnectRsp` | W4_L2CAP_CONNECT_RSP | Send upper layer `L2CA_ConnectCfm` message. Disable RTX timer. | CONFIG |
| `L2CAP_ConnectRspPnd` | W4_L2CAP_CONNECT_RSP | Send upper layer `L2CA_ConnectPnd` message. Disable RTX timer and start ERTX timer. | N/C |
| `L2CAP_ConnectRspNeg` | W4_L2CAP_CONNECT_RSP | Send upper layer `L2CA_ConnectCfmNeg` message. Return CID to free pool. Disable RTX and ERTX timers. | CLOSED |

*continued on next page*

**TABLE 5.1**

Summary of Events to Actions that Occur in a Particular Channel State Source: Bluetooth Specification 1.0 (continued)

| Event | Current State | Action | New State |
|---|---|---|---|
| L2CAP_ConfigReq | CLOSED | Send peer L2CAP_ConfigRspNeg message. | N/C |
| L2CAP_ConfigReq | CONFIG | Send upper layer L2CA_ConfigInd message. | N/C |
| L2CAP_ConfigReq | OPEN | Suspend data transmission at a convenient point. Send upper layer L2CA_ConfigInd message. | CONFIG |
| L2CAP_ConfigRsp | CONFIG | Send upper layer L2CA_ConfigCfm message. Disable RTX timer. If an L2CAP_ConfigReq message has been received and responded to, enter OPEN state; otherwise remain in CONFIG state. | N/C or OPEN |
| L2CAP_ConfigRspNeg | CONFIG | Send upper layer L2CA_ConfigCfmNeg message. Disable RTX timer. | N/C |
| L2CAP_DisconnectReq | CLOSED | Send peer L2CAP_DisconnectRsp message. | N/C |
| L2CAP_DisconnectReq | Any but CLOSED | Send upper layer L2CA_DisconnectInd message. | W4_L2CA_DISCONNECT_RSP |
| L2CAP_DisconnectRsp | W4_L2CAP_DISCONNECT_RSP | Send upper layer L2CA_DisconnectCfm message. Disable RTX timer. | CLOSED |
| L2CAP_Data | OPEN or CONFIG | If complete L2CAP packet received, send upper layer L2CA_Read message to confirm. | N/C |
| L2CA_ConnectReq | CLOSED (CID dynamically allocated from free pool) | Send peer L2CAP_ConnectReq message. Start RTX timer. | W4_L2CAP_CONNECT_RSP |
| L2CA_ConnectRsp | W4_L2CA_CONNECT_RSP | Send peer L2CAP_ConnectRsp message. | CONFIG |

*continued on next page*

# Logical Link Control

**TABLE 5.1**

Summary of Events to Actions that Occur in a Particular Channel State Source: Bluetooth Specification 1.0 (continued)

| Event | Current State | Action | New State |
|---|---|---|---|
| L2CA_ConnectRspNeg | W4_L2CA_CONNECT_RSP | Send peer L2CAP_ConnectRspNeg message. Return CID to free pool. | CLOSED |
| L2CA_ConfigReq | CLOSED | Send upper layer L2CA_ConfigCfmNeg message. | N/C |
| L2CA_ConfigReq | CONFIG | Send peer L2CAP_ConfigReq message. Start RTX timer. | N/C |
| L2CA_ConfigReq | OPEN | Suspend data transmission at a convenient point. Send peer L2CAP_ConfigReq message. Start RTX timer. | CONFIG |
| L2CA_ConfigRsp | CONFIG | Send peer L2CAP_ConfigRsp message. If all outstanding L2CAP_ConfigReq messages have received positive responses, enter OPEN state; otherwise remain in CONFIG state. | N/C or OPEN |
| L2CA_ConfigRspNeg | CONFIG | Send peer L2CAP_ConfigRspNeg message. | N/C |
| L2CA_DisconnectReq | OPEN or CONFIG | Send peer L2CAP_DisconnectReq message. Start RTX timer. | W4_L2CAP_DSCONNECT_RSP |
| L2CA_DisconnectRsp | W4_L2CA_DSCONNECT_RSP | Send peer L2CAP_DisconnectRsp message. Return CID to free pool. | CLOSED |
| L2CA_DataRead | OPEN | If payload complete, transfer payload to InBuffer. | OPEN |
| L2CA_DataWrite | OPEN | Send peer L2CAP_Data message. | OPEN |

*continued on next page*

**TABLE 5.1**

Summary of Events to Actions that Occur in a Particular Channel State (continued)
Source: Bluetooth Specification 1.0

| Event | Current State | Action | New State |
|---|---|---|---|
| `Timer_RTX` | Any | Send upper layer `L2CA_TimeOutInd` message. If final expiration, return CID to free pool or re-send Request. | CLOSED |
| `Timer_ERTX` | Any | Send upper layer `L2CA_TimeOutInd` message. If final expiration, return CID to free pool or re-send Request. | CLOSED |

*Source: Bluetooth Specification 1.0*

## Data Packet Format

L2CAP is packet based, but follows a communication model based on channels that represents a data flow between L2CAP entities in remote devices. Channels may be connection oriented or connectionless.

### Connection-Oriented Channel

Figure 5.6 illustrates the format of the L2CAP packet, also referred to as the L2CAP PDU, within a connection-oriented channel.

**Figure 5.6**
L2CAP packet format.
Source: Bluetooth Specification 1.0.

```
←─────── L2CAP Header ───────→
┌──────────┬────────────┬──────────────────────┐
│  Length  │ Channel ID │ Information (payload)│
├──────────┼────────────┼──────────────────────┤
│    16    │     16     │       Variable    Bits│
└──────────┴────────────┴──────────────────────┘
```

The fields shown are:

- **Length** (2 bytes or 16 bits)—Indicates the size of the information payload, excluding the length of the L2CAP header. This field provides a simple integrity check of the reassembled L2CAP packet on the receiving end.

- **Channel ID** (2 bytes or 16 bits)—Identifies the destination channel endpoint of the packet.

# Logical Link Control

- **Information** (up to 65,535 bytes or 524,280 bits)—Contains the payload received from the upper-layer protocol (outgoing packet), or delivered to the upper-layer protocol (incoming packet).

## Connectionless Data Channel

In addition to connection-oriented channels, L2CAP also supports the concept of a group-oriented channel in which data are sent to all members of the group in a best-effort fashion. Since this type of channel has no quality of service, it is inherently unreliable. Consequently, L2CAP makes no guarantee that data sent to the group will ultimately reach all members. If reliability is required, group transmission must be implemented at a higher layer. Since it is possible that non-group members may receive group transmissions, higher-level (or link-level) encryption should be used if privacy is a concern. Figure 5.7 shows the fields of a connectionless packet.

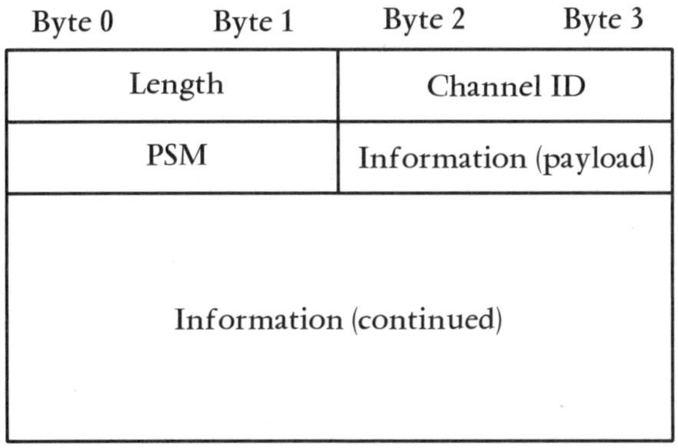

**Figure 5.7** Connectionless packet used for group-oriented data communication. Source: Bluetooth Specification 1.0.

| Byte 0 | Byte 1 | Byte 2 | Byte 3 |
|---|---|---|---|
| Length | | Channel ID | |
| PSM | | Information (payload) | |
| Information (continued) | | | |

The Bluetooth specification describes the fields used in the connectionless packet as:

- **Length** (2 bytes or 16 bits)—Indicates the size of information payload plus the PSM field in bytes, excluding the length of the L2CAP header.
- **Channel ID** (2 bytes or 16 bits)—Identifies the group destination of the packet.
- **Protocol/Service Multiplexer** (2 bytes minimum or 16 bits)—PSM values have two ranges. The values in the first range are assigned by the Bluetooth SIG and indicate protocols. The values in

the second range are dynamically allocated and used in conjunction with the Service Discovery Protocol (SDP).

- **Information** (up to 65,533 bytes or 524,264 bits)—Contains payload information to be distributed to all members of the group.

Basic group management mechanisms provided by the L2CAP group service interface include creating a group, adding members to a group, and removing members from a group. However, there are no predefined groups such as "all radios within range."

## Signaling

Bluetooth wireless technology uses signaling commands that are passed between two L2CAP entities on remote devices. The L2CAP implementation determines the Bluetooth address (BD_ADDR) of the device that sent the commands. Figure 5.8 shows the general format of all signaling commands.

| Byte 0 | Byte 1 | Byte 2 | Byte 3 |
|--------|------------|--------|--------|
| Code   | Identifier | Length          ||
| Data                                 ||||

**Figure 5.8**
General format of signaling commands passed between L2CAP entities on remote devices.
Source: Bluetooth Specification 1.0.

## Packet Structure

Per the Bluetooth specification, the fields used in the packets of signaling commands are:

- **Code** (1 byte or 8 bits)—Identifies the type of command. When a packet is received with an unknown Code field, a Command Reject packet is sent in response.
- **Identifier** (1 byte or 8 bits)—Used for matching a request with the reply. The requesting device sets this field and the responding device uses the same value in its response.

- **Length** (2 bytes or 16 bits)—Indicates the size in bytes of the data field of the signaling command only. It does not include the number of bytes in the Code, Identifier, and Length fields.
- **Data** (0 or more bytes)—This field is variable in length and discovered using the Length field. The Code field determines the format of the Data field.

## Signaling Commands

The Bluetooth specification describes 11 signaling commands, each with its own code, which is passed between L2CAP entities on remote Bluetooth devices.

### COMMAND REJECT

This packet is sent in response to a command packet with an unknown command code or invalid channel. For example, if a command refers to an invalid channel—typically because it does not exist—Command Reject is sent back as the response. Also, when multiple commands are included in an L2CAP packet and the packet exceeds the MTU of the receiver, Command Reject is sent back as the response.

### CONNECTION REQUEST

This packet is sent to create a channel between two Bluetooth devices. The channel connection must be established before configuration can begin.

### CONNECTION RESPONSE

When a Bluetooth unit receives this packet, it must send a Connection Response packet. On the receipt of a successful result, a logical channel is established.

### CONFIGURATION REQUEST

This packet is sent to establish an initial logical link transmission contract between two L2CAP entities and also to renegotiate this contract when required. During a renegotiation session, all data traffic on the channel is suspended pending the outcome. The decision on the amount of time or messages spent arbitrating the channel parameters before terminating the negotiation is left to the implementation, but in no case will last more than 120 seconds.

### CONFIGURATION RESPONSE

This packet is sent in reply to Configuration Request packets. Each configuration parameter value in a Configuration Response reflects an adjustment to a configuration parameter value that has been sent or implied in the corresponding Configuration Request. If a configuration parameter in a Configuration Request relates to traffic flowing from device A to device B, for example, the sender of the Configuration Response will only adjust this value again for the same traffic flowing from device A to device B.

### DISCONNECTION REQUEST

Terminating an L2CAP channel requires that a Disconnection Request packet be sent and acknowledged by a Disconnection Response packet. Disconnection is requested using the signaling channel, since all other L2CAP packets sent to the destination channel automatically get passed up to the next protocol layer. The receiver ensures that both source and destination CIDs match before initiating a connection disconnection. Once a Disconnection Request is issued, all incoming data in transit on the L2CAP channel are discarded and any new additional outgoing data are disallowed. Once a Disconnection Request for a channel has been received, all data queued to be sent out on that channel are discarded.

### DISCONNECTION RESPONSE

This packet is sent in response to each Disconnection Request.

### ECHO REQUEST

These packets are used to solicit a response from a remote L2CAP entity. Echo Request packets are typically used for testing the link or passing vendor-specific information using the optional data field. L2CAP entities must respond to Echo Request packets with Echo Response packets. The data field is optional and implementation dependent. L2CAP entities ignore the contents of this field.

### ECHO RESPONSE

Echo Response packets are sent upon receiving Echo Request packets. The identifier in the response must match the identifier sent in the Request. The optional and implementation-dependent data field may contain the contents of the data field in the Request, different data, or no data.

## Logical Link Control

**INFORMATION REQUEST**

Information Request packets are used to solicit implementation-specific information from a remote L2CAP entity. L2CAP entities must respond to Information Request packets with Information Response packets.

**INFORMATION RESPONSE**

Information Response packets are sent upon receiving Information Request packets. The identifier in the response must match the identifier sent in the Request. The optional data field may contain the contents of the data field in the Request, different data, or no data.

## Configuration Parameter Options

Bluetooth wireless technology uses option packets which are the mechanism for negotiating different connection requirements. This type of packet is comprised of an option type, an option length, and one or more option data fields. Figure 5.9 illustrates the basic format of an option packet.

| Byte 0 | Byte 1 | Byte 2 | Byte 3 |
|--------|--------|--------|--------|
| Type   | Length | Option Data ||

**Figure 5.9**
Format of a configuration option packet.
Source: Bluetooth Specification 1.0.

### Packet Structure

According to the Bluetooth specification, the fields used in configuration option packets are:

- **Type** (1 byte or 8 bits)—This field defines the parameters being configured.

- **Length** (1 byte or 8 bits)—This field defines the number of bytes in the option payload. An option type with no payload has a length of 0.

- **Option data** (2 bytes or 16 bits)—The contents of this field are dependent on the option type.

## Options

The Bluetooth specification allows several options to be negotiated between L2CAP entities:

### MAXIMUM TRANSMISSION UNIT

This option specifies the payload size, in bytes, that the sender is capable of accepting for that channel. L2CAP implementations support a minimum MTU size of 48 bytes. The default value is 672 bytes.

### FLUSH TIMEOUT

This option is used to inform the recipient of the amount of time the originator's link controller/link manager will attempt to successfully transmit an L2CAP segment before giving up and flushing the packet. This value represents units of time measured in milliseconds.

### QUALITY OF SERVICE

L2CAP implementations are only required to support best-effort service; support for any other service type is optional. Best effort does not require any guarantees, so if no QoS option is placed in the Configuration Request, best-effort service is assumed. If any QoS guarantees are required, then a QoS configuration request must be sent, specifying the following parameters:

- **Token rate**—This is the rate at which traffic credits are granted in bytes per second. An application may send data at this rate continuously. Burst data may be sent up to the token bucket size. Until a data burst has been drained, an application must limit itself to the token rate. For best-effort service, the application gets as much bandwidth as possible. For guaranteed service, the application gets the maximum bandwidth available at the time of the request.

- **Token bucket size**—This is the size of the token bucket (i.e., buffer) in bytes. If the bucket is full, then applications must either wait or discard data. For best-effort service, the application gets a bucket as big as possible. For guaranteed service, the maximum buffer space will be available to the application at the time of the request.

- **Peak bandwidth**—Expressed in bytes per second, this limits how fast packets may be sent back-to-back from applications. Some intermediate systems can take advantage of this information, to produce more-efficient resource allocation.

- **Latency**—This is the maximum acceptable delay between transmission of a bit by the sender and its initial transmission over the air, expressed in microseconds.
- **Delay variation**—This is the difference, in microseconds, between the maximum and minimum possible delay that a packet will experience as it goes out over a channel. This value is used by applications to determine the amount of buffer space needed at the receiving side in order to restore the original data transmission pattern.

## Configuration Process

The Bluetooth specification describes a three-step process for channel parameter negotiation that involves:

1. Informing the remote side of the non-default parameters that the local side will accept
2. Having the remote side agree or disagree to these values, including the defaults
3. Repeating steps 1 and 2 for the reverse direction from the (previous) remote side to the (previous) local side.

This process can be abstracted into a Request negotiation path and a Response negotiation path.

### REQUEST PATH
The Request Path negotiates the incoming MTU, flush timeout, and outgoing flow specification.

### RESPONSE PATH
The Response Path negotiates the outgoing MTU (remote side's incoming MTU), the remote side's flush timeout, and incoming flow specification (remote side's outgoing flow specification).

Before leaving the CONFIG state and moving into the OPEN state, both paths must reach closure. The request path requires the local device to receive a positive response to reach closure, while the response path requires the local device to send a positive response to reach closure.

# Service Primitives

Per the Bluetooth specification, the services offered by L2CAP are described in terms of service primitives and parameters.

## Event Indication

This primitive is used to request a callback when the selected indication Event occurs. The following callback functions are used:

### L2CA_CONNECTIND CALLBACK

This function includes the parameters for the address of the remote device that issued the connection request, the local CID representing the channel being requested, the Identifier contained in the request, and the PSM value the request is targeting.

### L2CA_CONFIGIND CALLBACK

This function includes the parameters indicating the local CID of the channel the request has been sent to, the outgoing MTU size (maximum packet that can be sent across the channel), and the flow specification describing the characteristics of the incoming data.

### L2CA_DISCONNECTIND CALLBACK

This function includes the parameter indicating the local CID to which the request has been sent.

### L2CA_QOSVIOLATIONIND CALLBACK

This function includes the parameter indicating the address of the remote Bluetooth device where the Quality of Service contract has been violated.

## Connect

The Connect primitive initiates an `L2CA_ConnectReq` message and blocks service until a corresponding `L2CA_ConnectCfm(Neg)` or `L2CA_TimeOutInd` message is received. This primitive is used to request the creation of a channel representing a logical connection to a physical address.

### Connect Response

This primitive represents the L2CA_ConnectRsp. This primitive is used to issue a response to a connection request event indication and is called no more than once after receiving the callback indication. This primitive returns once the local L2CAP entity has validated the request. A successful return indicates the response has been sent over the air interface.

### Configure

This primitive initiates the sending of an L2CA_ConfigReq message and blocks service until a corresponding L2CA_ConfigCfm(Neg) or L2CA_TimeOutInd message is received. This primitive is used to request the initial configuration (or reconfiguration) of a channel to a new set of channel parameters.

### Configuration Response

This primitive represents the L2CAP_ConfigRsp. This primitive is used to issue a response to a configuration request event indication.

### Disconnect

This primitive represents the L2CAP_DisconnectReq and the returned output parameters represent the corresponding L2CAP_DisconnectRsp or the RTX timer expiration. It is used to request the disconnection of the channel. Once the request is issued, no process will be able successfully to read or write from the CID. However, writes in progress will continue to be processed.

### Write

This primitive is used to request the transfer of data across the channel and may be used for both connection-oriented and connectionless traffic.

### Read

This primitive is used to request the reception of data. This request returns when data are available or the link is terminated. The data returned represent a single L2CAP payload. If not enough data are available, the command will block until the data arrive or the link is terminated. If the payload is bigger than the buffer, only the portion of the payload that fits into the buffer will be returned, and the remainder of the payload will be discarded. Like the Write command, the Read command may be used for both connection-oriented and connectionless traffic.

### Group Create

This primitive is used to request the creation of a CID to represent a logical connection to multiple devices.

### Group Close

This primitive is used to close down a Group.

### Group Add Member

This primitive is used to request the addition of a member to a group. The input parameter includes the CID representing the group and the BD_ADDR of the group member to be added. The output parameter Result confirms the success or failure of the request.

### Group Remove Member

This primitive is used to request the removal of a member from a group. The input parameters include the CID representing the group and BD_ADDR of the group member to be removed. The output parameter Result confirms the success or failure of the request.

### Get Group Membership

This primitive is used to request a report of the members of a group. The input parameter CID represents the group being queried. The output parameter Result confirms the success or failure of the operation. If the Result is successful, a list of Bluetooth addresses is returned showing members of the group.

### Ping

This primitive represents the initiation of an `L2CA_EchoReq` command and the reception of the corresponding `L2CA_EchoRsp` command.

### Get Info

This primitive represents the initiation of an `L2CA_InfoReq` command and the reception of the corresponding `L2CA_InfoRsp` command.

### Disable Connectionless Traffic

This primitive is used as a general request to disable the reception of connectionless packets. The input parameter is the PSM value indicating that service should be blocked.

### Enable Connectionless Traffic

This primitive is used as a general request to enable the reception of connectionless packets. The input parameter is the PSM value indicating that service should be unblocked.

## Summary

Like the Link Management Protocol (LMP), the Logical Link Control and Adaptation Protocol (L2CAP) is a link-level protocol that runs over the Bluetooth baseband. L2CAP is responsible for higher-level protocol multiplexing, which is supported by defining channels, each bound to a single protocol in many-to-one fashion. Although multiple channels can be bound to the same protocol, a single channel cannot be bound to multiple protocols. Each L2CAP packet received on a channel is directed to the appropriate higher-level protocol.

L2CAP supports large packet sizes up to 64 kilobytes using the low-overhead segmentation-and-reassembly mechanism. Group management provides the abstraction of a group of units, allowing for more efficient mapping between groups and members of a piconet. Group communication is connectionless and unreliable. When composed of only a pair of units, groups provide a connectionless channel alternative to L2CAP's connection-oriented channel. Finally, L2CAP conveys QoS information across channels and provides some admission control to prevent additional channels from violating existing QoS contracts.

# CHAPTER 6

# Bluetooth General Profiles

The Bluetooth SIG has identified various usage models, each of which is accompanied by a "profile." The profiles define the protocols and features that support a particular usage model. If devices from different manufacturers conform to the same Bluetooth SIG profile specification, they can be expected to interoperate when used for that particular service and usage case.

A profile defines the specific messages and procedures used to implement a feature. Some features are mandatory, others are optional, and some may be conditional. All defined features are process-mandatory, meaning that if a feature is implemented, it must be implemented in a specified manner. This ensures that the same feature works the same way for each device, regardless of manufacturer.

Four general profiles are used in the various usage models: the Generic Access Profile (GAP), the Serial Port Profile (SPP), the Service Discovery Application Profile (SDAP), and the Generic Object Exchange Profile (GOEP).[1] This chapter discusses the four general profiles, while the next chapter (Chapter 7) discusses the profiles that relate more closely to the specific usage models.

## Generic Access Profile

The Generic Access Profile defines the general procedures for discovering Bluetooth devices as well as the link management procedures for connecting them together. Thus, the main purpose of this profile is to describe the use of the lower layers of the Bluetooth protocol stack—Link Control (LC) and Link Manager Protocol (LMP). Also defined in this profile are security related procedures, in which case higher layers—L2CAP, RFCOMM, and OBEX—come into play.

In addition, the profile addresses the common format requirements for parameters accessible from the user interface. In other words, the Generic Access Profile describes how devices should behave while in standby and connecting states. This, in turn, guarantees that links and channels can always be established between Bluetooth devices. If the

---

[1] The Bluetooth SIG reserves the right to define a process for adding new Bluetooth profiles after the release of the Specification 1.0, in December 1999. The discussions in this book are based on Specification 1.0. As of October 2000 when this book was reprinted, the Bluetooth SIG had not issued a subsequent version.

# Bluetooth General Profiles

devices operate according to several profiles simultaneously, the GAP describes the mechanisms for handling all of them.

The GAP defines the general procedures for discovering the identities, names, and basic capabilities of other Bluetooth devices that are in "discoverable mode." A device in discoverable mode is ready to accept connections and service requests from other devices. Even if two Bluetooth devices do not share a common application, they must be able to communicate with each other to determine this. When two devices share the same application, but are from different manufacturers, the ability to establish a connection will not be impeded just because manufacturers choose to call basic Bluetooth capabilities by different names at the user interface level or because their products implement basic procedures in a different sequence.

Bluetooth devices that do not conform to any other Bluetooth profile must at least conform to the GAP. This ensures basic interoperability and coexistence between all Bluetooth devices, regardless of the type of applications they support. Devices that conform to another Bluetooth profile may use adaptations of the generic procedures as specified by that profile. However, they must still be compatible with the GAP at the generic procedures level (Figure 6.1).

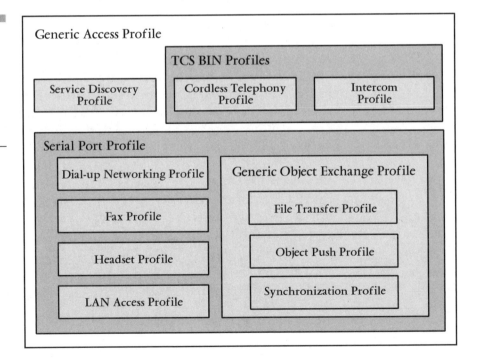

**Figure 6.1**
The relationship of the Generic Access Profile to other Bluetooth profiles and usage models. Source: Bluetooth Specification 1.0.

## Common Parameters

If a vendor claims that its product conforms to the GAP profile, all mandatory capabilities of the profile must be supported as described in the Bluetooth specification. This also applies for all optional and conditional capabilities for which support is claimed by the vendor. All capabilities—mandatory, optional, and conditional—for which support is indicated by the vendor are subject to verification as part of the Bluetooth certification program. The common parameters that Bluetooth devices must support in order to interoperate at the generic procedures level are discussed below.

### DEVICE NAMES

The GAP permits Bluetooth devices to have user-friendly names that can be up to 248 bytes long. However, it cannot be assumed that a remote device will be able to handle the maximum number of characters for a Bluetooth device name. If the remote device has limited display capabilities, as is typical with most cell phones, for example, it may allow only the first 20 characters.

### BLUETOOTH PIN

The Bluetooth PIN (personal identification number) is used as the first step in authenticating two Bluetooth devices that have not previously exchanged link keys to establish a trusted relationship. The PIN is used in the pairing procedure (discussed later) to generate the initial link key used for further authentication. The PIN may be entered at the user interface level, but may also be stored in the device if it does not have a sufficient user interface for entering and displaying digits.

### CLASS OF DEVICE

During the device discovery procedure, a parameter called "Class of Device" is conveyed from the remote device, indicating the type of device it is and the types of services it supports. Either the user initiates the bonding procedure by entering the PIN, or the user is requested to enter the PIN during the establishment procedure since the devices did not share a common link key previously. In the first case, the user is said to perform "bonding" and in the second case the user is said to perform "authentication."

### DISCOVERY MODES

Bluetooth devices are in either discoverable mode or non-discoverable mode. There are two discoverable modes: limited and general. When a device is in limited discoverable mode, it makes itself available only for a limited period of time, during temporary conditions, or for a specific event. The device makes itself available by going into a "scan state" periodically to check for certain types of inquiry codes. On such occasions, it can respond to another Bluetooth device that makes a limited inquiry using a special code called the Limited Inquiry Access Code (LIAC).

When a device is in general discoverable mode, it makes itself available continuously and can respond to another Bluetooth device that makes a general inquiry using another special code called the General Inquiry Access Code (GIAC). However, a device can be in only one discoverable mode at a time, either limited or general. Even when a Bluetooth device is discoverable it may be unable to respond to an inquiry due to other baseband activity. When a Bluetooth device is in non-discoverable mode it does not respond to any type of inquiry.

### CONNECTIVITY MODES

With respect to paging, a Bluetooth device must be in either the connectable or the non-connectable mode. Paging is a procedure that involves transmitting a series of messages with the objective of setting up a communication link to a unit active within the coverage area. When a Bluetooth device is in connectable mode it puts itself into the "page scan" state. Upon receipt of a page, it can respond. When a Bluetooth device is in non-connectable mode it does not put itself into the page-scan state. Since it cannot receive pages, it does not respond to paging.

### PAIRING MODES

Pairing is an initialization procedure by which two devices that are communicating for the first time create a common link key that will be used for subsequent authentication. For first-time connection, pairing requires the user to enter a Bluetooth security code or PIN. Bluetooth devices must be in either the pairable mode or the non-pairable mode. When a Bluetooth device is in the pairable mode it accepts pairing (i.e., the creation of bonds) initiated by the remote device, and in non-pairable mode it does not.

## SECURITY MODES

There are three security modes for Bluetooth devices. When a Bluetooth device is in security mode 1 it never initiates a security procedure. When a Bluetooth device is in security mode 2 it does not initiate any security procedure before a channel establishment request has been received or it has initiated a channel establishment procedure.

Whether a security procedure is initiated or not depends on the security requirements of the requested channel or service. At the least, a Bluetooth device in security mode 2 must classify the security requirements of its services in terms of authorization, authentication, and encryption. (From the perspective of a remote device, security mode 1 is a special case of security mode 2 where no service has registered any security requirement.)

When a Bluetooth device is in security mode 3 it initiates security procedures before it sends a message indicating that link setup is complete. A Bluetooth device in security mode 3 may reject the connection request based on internal settings (e.g., only communication with pre-paired devices is allowed).

# Idle Mode Procedures

There are several idle mode procedures that can be initiated by a Bluetooth device and directed at a remote Bluetooth device.

## GENERAL INQUIRY

The general inquiry procedure provides the initiating device with the Bluetooth device address, clock, Class of Device, and page scan mode of general discoverable devices. Those within range of the initiating device can be set up to scan for inquiry messages with the General Inquiry Access Code. All devices in limited discoverable mode are discovered using the general inquiry procedure.

## LIMITED INQUIRY

The limited inquiry procedure provides the initiating device with the Bluetooth device address, clock, Class of Device, and page scan mode of limited discoverable devices. Those within range of the initiating device can be set up to scan for inquiry messages with the Limited Inquiry Access Code (LIAC) in addition to the General Inquiry Access Code (GIAC). Since it is not guaranteed that the discoverable device scans for the LIAC, the initiating device may choose either the general or limited inquiry procedure.

### NAME DISCOVERY

The purpose of name discovery is to provide the initiating device with the device name of other connectable devices—specifically, Bluetooth devices within range that will respond to paging. Name discovery is the procedure for retrieving the device name from connectable Bluetooth devices by targeting the request toward known Bluetooth devices for which the device addresses are available.

A related procedure is "name request," which is also used for retrieving the device names from connectable Bluetooth devices. It is not necessary that full link establishment be performed just to get the name of another device. In the name request procedure, the initiating device uses the device access code of the remote unit that was retrieved beforehand through an inquiry procedure.

### DEVICE DISCOVERY

The purpose of device discovery is to provide the initiating device with the address, clock, class of device, page scan mode, and device name of discoverable Bluetooth units. During the device discovery procedure, either a general or limited inquiry is performed; this is followed by name discovery toward some or all of the devices that responded to the inquiry.

## Bonding

The purpose of bonding is to create a relationship between two Bluetooth devices based on a common link key (a bond). The link key is created and exchanged (pairing) during the bonding procedure and is typically stored by both Bluetooth devices for future authentication. In addition to pairing, the bonding procedure can involve higher-layer initialization procedures.

There are two types of bonding: dedicated and general. Dedicated bonding occurs when the devices only create and exchange a common link key, whereas general bonding is included as part of the normal channel and connection establishment procedures. In fact, bonding is in principle the same as link establishment. This means that pairing may be performed successfully if a device has initiated bonding while the target device is in its normal connectable and security modes.

Before bonding can be started, the initiating device must know the device access code of the unit to pair with. This is normally done by first performing device discovery. A Bluetooth device that can initiate bonding uses limited inquiry, and a Bluetooth device that accepts bonding supports the limited discoverable mode.

## Establishment Procedures

There are three establishment procedures defined in the Bluetooth specification: link, channel, and connection. But before any establishment procedure can be initiated, certain information has to be available to the initiating device, which is obtained during device discovery:

- The device address from which the device access code is generated.
- The system clock of the remote device.
- The page scan mode used by the remote device.

Additional information obtained during device discovery that is useful for making the decision to initiate an establishment procedure is the class of device and the device name.

### LINK ESTABLISHMENT
The link establishment procedure is used to set up a physical link—specifically, an Asynchronous Connectionless (ACL) link—between two Bluetooth devices using procedures from the Bluetooth IrDA Interoperability Specification and Generic Object Exchange Profile.

Within Bluetooth, the paging procedure conforms to the Bluetooth IrDA Interoperability Specification. The paging unit uses the device access code and page mode received through a previous inquiry. After paging is completed, a physical link between the two Bluetooth devices is established. If a reversal of master-slave roles is required—typically, it is the paged device that has an interest in changing roles—it is done immediately after the physical link is established. If the paging device does not accept master-slave role reversal, the paged device determines whether or not to maintain the physical link. Both devices can perform link setup using Link Manager Protocol (LMP) procedures that require no interaction with the host on the remote side. Optional LMP features can be used, but only upon confirmation that the other device supports the requested feature.

When the paging device needs to go beyond the link setup phase, it issues a request to be connected to the host of the remote device. If the paged device is in security mode 3, this is the trigger for initiating authentication. The paging device then sends a host connection request during link establishment, but before channel establishment, and may initiate authentication after having sent the host connection request. After authentication has been performed, any of the devices can initiate encryption.

Further link configuration may take place after the host connection request. When the requirements of both devices are satisfied, each sends a message to the other indicating that setup is complete.

**CHANNEL ESTABLISHMENT**

The channel establishment procedure is used to set up a channel (logical link) between two devices using the Bluetooth File Transfer Profile Specification. Channel establishment starts after link establishment is completed. At that point, the initiating device sends a channel establishment request. Security procedures may take place after channel establishment has been initiated. Channel establishment is completed when the remote device answers the channel establishment request with a positive response.

**CONNECTION ESTABLISHMENT**

The connection establishment procedure is used to set up a connection between applications on two Bluetooth devices. Connection establishment starts after channel establishment is completed. At that point, the initiating device sends a connection establishment request. The specific request used depends upon the application. It may be a TCS SETUP message in the case of a Bluetooth telephony application, for example, or initialization of RFCOMM and establishment of a data-link connection in the case of a serial port-based application. Whatever the application, connection establishment is completed when the remote device accepts the connection establishment request.

When one Bluetooth device has established a connection with another Bluetooth device, it may be available for a second connection on the same channel, a second channel on the same link, or a second physical link—or any combination of these. If the new establishment procedure is directed toward the same remote device, the security part of the establishment depends on the security modes already in use. If the new establishment procedure is directed toward a different remote device, the initiating device behaves according to active modes, apart from the fact that a physical link is already established.

# Serial Port Profile

When Bluetooth wireless technology is used for cable replacement, the Serial Port Profile (SPP) is employed for the resulting connection-oriented channel. This profile is built upon the Generic Access Profile and defines how Bluetooth devices can be set up to emulate a serial cable connection using RFCOMM, a simple transport protocol that emulates RS-232 serial ports between two peer devices (Figure 6.2). RFCOMM is used to transport the user data, modem control signals and configuration commands. The RFCOMM session runs on an L2CAP channel. The applications on both devices are typically legacy applications that expect communication to occur over a serial cable, which this profile emulates.

Any legacy application can be run on either device, using the virtual serial port as if a physical serial cable, with RS-232 control signaling, were connecting the two devices. However, since the legacy applications are not aware of Bluetooth procedures for setting up emulated serial cables, they may need the assistance of a helper application utilizing the Bluetooth specification on both sides of the link. The Bluetooth specification does not address the requirements of helper applications because the overriding concern of the Bluetooth SIG is with interoperability. It is assumed that some kind of helper application will be offered by the solution provider to bridge the two environments.

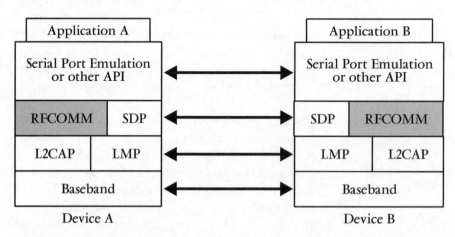

**Figure 6.2**
The protocol model for an emulated serial cable connection.
Source: Bluetooth Specification 1.0.

In a simple serial port configuration in which there are two computers connected via the emulated serial cable (Figure 6.3), one device takes the initiative to form a connection to the other device. This device is called the Initiator, while the device targeted for the connection is called the Acceptor. When the Initiator starts link establishment, service discovery procedures are performed to set up the emulated serial cable connection.

**Figure 6.3**
Two computers, with one taking the role of Initiator and the other taking the role of Acceptor for setting up an emulated serial cable connection.
Source: Bluetooth Specificaton 1.0.

Under this profile, data rates of up to 128 Kbps are supported. Although the Bluetooth specification describes a single emulated serial port connection between two devices in a point-to-point configuration, nothing prevents multiple executions of the SPP from running concurrently in the same device in support of multiple connections. In such cases, the devices may even take on the two different roles of Initiator and Acceptor concurrently. No fixed master-slave roles are specified for this profile, since both devices are assumed to be peers.

Support for security features—authorization, authentication, and encryption—is optional. However, a device must support the corresponding security procedures if requested to do so from a peer device. If the use of security features is desired, the two devices are paired during the connection establishment phase. Bonding is not explicitly used in the Serial Port Profile, so support for bonding is optional.

## Application-Level Procedures

Three application-level procedures are described in the Bluetooth specification, and are required for establishing an emulated serial cable connection between two devices.

**ESTABLISH LINK/SET UP VIRTUAL SERIAL CONNECTION**
This procedure describes the steps required to establish a connection to an emulated serial port of a remote device.

1. Submit a query using the Service Discovery Protocol (SDP) to determine the RFCOMM server channel number of the application in the remote device. If a browsing capability is included, the user can select from among the available ports (or services) in the peer device. If the user knows exactly which service to contact, it is only necessary to look up the parameters using the Service Class ID associated with that service.
2. As an option, the remote device may be required to authenticate itself; as another option, encryption can be enabled.
3. Request a new L2CAP channel to the remote RFCOMM entity.
4. Initiate an RFCOMM session on the L2CAP channel.
5. Start a new data-link connection on the RFCOMM session using the server channel number (mentioned in step 1).

When this procedure is completed, the virtual serial cable connection is ready for use by the applications on both devices. If an RFCOMM session already exists between the devices when setting up a new data-link connection, the new connection is established on the existing RFCOMM session, in which case steps 3 and 4 become unnecessary.

## ACCEPT LINK/ESTABLISH VIRTUAL SERIAL CONNECTION

This procedure requires that the Acceptor take part in the following steps:

1. Provide authentication if requested, and upon further request, turn on encryption.
2. Accept a new channel establishment indication from L2CAP.
3. Accept an RFCOMM session establishment on that channel.
4. Accept a new data-link connection on the RFCOMM session.

If the user requesting the emulated serial port wants security, and the procedures have not already been carried out, the last step may trigger a local request to authenticate the remote device and turn on encryption.

## REGISTER SERVICE RECORD IN LOCAL SDP DATABASE

All services/applications reachable through RFCOMM must have an SDP service record that includes the parameters necessary to reach the corresponding service/application. This calls for a Service Database and the ability to respond to SDP queries. To support legacy applica-

# Bluetooth General Profiles

tions running on virtual serial ports, service registration is done by a helper application, which assists the user in setting up the port.

## Power Mode and Link Loss Handling

The power requirements of units connected via emulated serial ports may be quite different, so there is no requirement under SPP that power-saving modes be used. However, requests to use a low-power mode, if possible, will not be denied. If sniff, park, or hold mode is used, neither the RFCOMM data-link connection nor the L2CAP channel is released. If a unit detects link loss, RFCOMM is considered as shut down. But before communication at higher layers can resume, the RFCOMM session initialization procedure must be performed.

## RS-232 Control Signals

Wire-based serial port connections implement flow control between devices either through software control using characters such as Transmitter On/Transmitter Off (XON/XOFF) or signals such as Request to Send/Clear to Send (RTS/CTS) or Data Terminal Ready/Data Set Ready (DTR/DSR). These methods may be used by both sides of a wired link, or may be used in only one direction.

The RFCOMM protocol provides two flow control mechanisms. One is the Modem Status Command (MSC), which conveys the RS-232 control signals and the break signal for all emulated serial ports. This flow control mechanism operates on individual data link connections. The RFCOMM protocol also uses its own flow control commands—Flow Control On/Flow Control Off (FCON/FCOFF)—that operate on the aggregate data flow between two devices.

If the local device relays information from a physical serial port where overrun, parity, or framing errors may occur, the remote device must be informed of any changes in RS-232 line status with the Remote Line Status Indication command.

RS-232 port settings are relayed between devices with the Remote Port Negotiation command directly before data-link connection establishment. This is required if the API to the RFCOMM adaptation layer exposes settings like baud rate and parity. Information conveyed in the remote port negotiation procedure is useful for devices with a physical serial port, or when data pacing is done at an emulated serial port interface for any reason. RFCOMM does not impose throughput limits based on baud rate settings.

## L2CAP Interoperability Requirements

As noted earlier, the Serial Port Profile (SPP) is used for connection-oriented channels only. Although connectionless channels are not used within the execution of this profile, concurrent use of both channel types by other profiles and applications is allowed.

**SIGNALING**
Only the Initiator may issue an L2CAP connection request within the execution of this profile. Other than that, the Serial Port Profile does not impose any additional restrictions or requirements on L2CAP signaling.

**CONFIGURATION OPTIONS**
Three configuration options are available for L2CAP in the Serial Port Profile. One is the size of the Maximum Transmission Unit (MTU), which refers to the largest possible unit of data that can be sent over the data link. L2CAP implementations must support a minimum MTU size of 48 bytes. The default value is 672 bytes. For efficient use of the communication resources, however, the MTU is set as large as possible, while respecting any physical constraints imposed by the devices involved, and the need of these devices to continue honoring any previously agreed upon Quality of Service (QoS) commitments with other devices and/or applications.

Another configuration option is Flush Timeout. This option is used to inform the recipient of the amount of time the originator's link controller/link manager will attempt to successfully transmit an L2CAP segment before giving up and "flushing" the packet.

Finally, there is the Quality of Service option. Since L2CAP implementations are only required to support "best-effort" service, support for any other service type is optional. Best effort does not require any performance guarantee. If any QoS guarantees are required then a QoS configuration request must be sent, specifying such performance parameters as delay variation (microseconds), peak bandwidth (bytes/second), and latency (microseconds). If no QoS option is placed in the request, best-effort service is assumed.

## SDP Interoperability Requirements

There are no SDP Service Records related to the Serial Port Profile in the local device. However, there are Serial Port related entries in the SDP database of the remote device; these include the attributes for such items as `ServiceClassIDList`, `ProtocolDescriptorList`, and `ServiceName`. To retrieve the service records in support of this profile, the SDP client entity in the local device connects and interacts with the SDP server entity in the remote device via the SDP and L2CAP procedures.

## Link Manager Interoperability Requirements

In addition to the requirements on supported procedures stated in the Link Manager specification itself, the Serial Port Profile also requires support for encryption in both the connected devices.

### ERROR BEHAVIOR
If a Bluetooth unit attempts to use a mandatory feature, and the other Bluetooth unit answers that it does not support that feature, the initiating unit sends a detach message indicating as the reason, "unsupported LMP feature." A unit will always be able to handle the rejection of a request for an optional feature.

### LINK POLICY
As noted, there are no fixed master-slave roles for the implementation of the Serial Port Profile. The profile also does not impose any requirements concerning which low-power modes to use, or when to use them. The Link Manager of each device decides these matters and issues requests when appropriate, subject to the latency limitation specified in the configurable QoS commitment.

### LINK CONTROL INTEROPERABILITY
The Link Control (LC) level defines several capabilities, including inquiry, inquiry scan, paging, and error behavior.

### INQUIRY
When inquiry is invoked in the local device, it uses the General Inquiry procedure specified under the Generic Access Profile (GAP)

discussed earlier. Only the local device can invoke the inquiry procedure within the execution of the Serial Port Profile.

### INQUIRY SCAN

For inquiry scan, the General Inquiry Access Code (GIAC) is used, according to the limited discoverable mode defined in GAP, if appropriate for the application residing in the remote device. The remote device will answer the inquiry scan with inquiry response messages.

### PAGING

Only the local device can page within the Serial Port Profile implementation. However, the paging step is skipped if there is already a baseband connection between the local and remote devices. In this case, the connection may have been set up as a result of a previous page.

### ERROR BEHAVIOR

Since most features at the LC level have to be activated by LMP procedures, errors will usually be caught at that layer. However, there are some LC procedures that are independent of the LMP layer, such as inquiry and paging. Improper use of such features is often difficult to detect, and there is no mechanism defined in the Serial Port Profile to detect or guard against their improper use.

## Service Discovery Application Profile

The Service Discovery Application Profile (SDAP) describes the features and procedures used to discover services registered in other Bluetooth devices and retrieve information about these services. With the number of services that can be provided over Bluetooth links expected to increase, standardized procedures aid users in locating and identifying them.

In this profile, as in the Serial Port Profile discussed earlier, only connection-oriented channels are used. In addition, no L2CAP broadcasts are used. Before any two devices equipped with Bluetooth wireless technology can communicate with each other they must be powered on and be initialized. Initialization may require providing a PIN for the creation of a link key for device authorization and data encryption. Then a link has to be created; this may require the discovery of the other unit's Bluetooth device address via an inquiry process, and the paging of the other device.

Included in the Bluetooth protocol stack is the Service Discovery Protocol (SDP), which is used to locate services available on or via devices within range of a Bluetooth device. Once the link is created, services can be located, and one or more of them can be selected through the user interface. Although SDP is not directly involved in accessing a particular service, it facilitates access by prompting the local Bluetooth stack to access the desired service. Unlike other profiles, where service discovery interactions result from the need to enable a transport service like RFCOMM, or a usage scenario like file transfer, this profile requires that service discovery be specifically invoked by the user.

SDP supports the following service inquiries:

- Search by service class
- Search service attributes
- Service browsing

The first two types of inquiries are used when searching for specific services and provide the user with the answers to such questions as: "Is service X available, or is service X with characteristics 1 and 2 available?" Service browsing is used for a general service search and provides the user with answers to such questions as: "What services are available?" or "What services of type X are available?" Implementing any of these service inquiries requires that devices first be discovered and linked with, then queried about the services they support.

## Client and Server Roles

The Service Discovery Profile defines the roles of Bluetooth devices in terms of local device (LocDev) and remote device (RemDev). A LocDev is the device that initiates the service discovery procedure. To accomplish this, it must contain at least the client portion of the Bluetooth SDP architecture. A LocDev contains the service discovery application (SrvDscApp), which enables the user to initiate discoveries and display the results of these discoveries.

A RemDev is any device that participates in the service discovery process by responding to the service inquiries generated by a local device. To accomplish this, it must contain at least the server portion of the Bluetooth SDP architecture. A RemDev contains a service records database, which the server portion of SDP consults to create responses to service discovery requests.

The roles of LocDev or RemDev for Bluetooth devices are not permanent or exclusive to one device or another. For example, a RemDev may have the SrvDscApp installed as well as an SDP client, and a LocDev may have an SDP server. Thus, a device could be a LocDev for a particular SDP transaction, while acting as a RemDev for another SDP transaction.

The relationship of "master" and "slave" is not imposed on the devices participating in the service discovery profile—in other words, a local device is not considered as a Bluetooth master and the remote devices as Bluetooth slaves. Service discovery can be initiated by either a master or slave device while they are members of the same piconet. At the same time, a slave in a piconet can initiate service discovery in a new piconet, provided that it notifies the master of the original piconet that it will be unavailable, possibly entering the "hold" operational mode until it returns.

Figure 6.4 shows the Bluetooth protocols and supporting entities involved in the service discovery profile.

The service discovery application interfaces with the baseband via the Bluetooth module control function, which prompts the Bluetooth module when to enter various search modes of operation. The service discovery user application in a local device interfaces with the Bluetooth SDP client to send service inquiries and receive service inquiry responses from the SDP servers of one or more remote devices (Figure 6.5). SDP uses the connection-oriented (CO) transport service in L2CAP, which in turn uses the baseband asynchronous connectionless (ACL) links to carry the SDP protocol data units (PDUs) over the air.

The service discovery profile does not require the use of authentication and/or encryption for SDP transactions. If these security measures are used by any of the devices involved, service discovery is performed only on the subset of devices that are able to pass the authentication and encryption security "roadblocks." In other words, any security measures that may be in place for SDP transactions are really determined by those already in place for the Bluetooth link.

# Bluetooth General Profiles

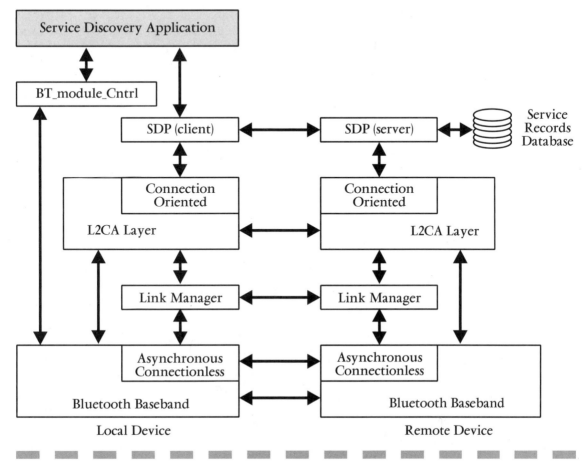

**Figure 6.4** The Bluetooth protocol stack for the service discovery profile. Source: Bluetooth Specification 1.0.

## Pairing

No particular requirements regarding pairing are imposed by the service delivery profile—pairing may be performed or not performed. As discussed earlier, pairing is an initialization procedure whereby two devices that are communicating for the first time create a common link key that will be used for subsequent authentication. Whenever a local device tries to perform service discovery against unconnected remote devices, it is the responsibility of the service discovery application to allow pairing prior to connection, or to bypass any devices that may require pairing first. The service delivery profile is concerned only with performing service discovery whenever the local device can establish a baseband link with one or more remote devices.

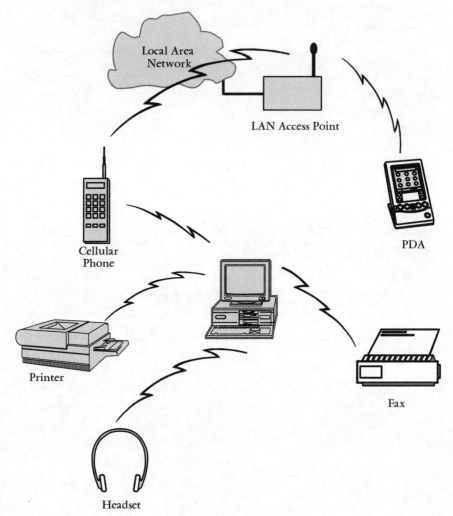

**Figure 6.5**
A typical service discovery scenario in which the computer (center) sends out service inquiries to various remote devices. The computer will receive back service inquiry responses from the SDP servers of one or more of the other devices.
Source: Bluetooth Specification 1.0.

## Service Discovery Application

With the SrvDscApp, the local device is assumed to be able to enter the inquiry and/or page states. Also, a remote device that has services it wants to make available to other devices (e.g. printer, a LAN access point, a PSTN gateway, etc.) is assumed to be able to enter the inquiry scan and/or page scan states. As noted, since the SrvDscApp may also perform service inquiries against already connected remote devices, it is not required that a local device always be the master of a connection with a remote device, or that a remote device always be the slave of a connection with a local device.

There are several alternatives for implementing service discovery. For example, the user of the local device can provide information for the desired service search through the service discovery application. The SrvDscApp then searches for devices via the Bluetooth inquiry procedure. For each device found, the local device will connect to it, perform any necessary link setup, per the Generic Access Profile, and then look for the desired services. Alternatively, device inquiry may be done before the collection of user input for the service search.

In either case, page, link creation, and service discovery are all done on one remote device at a time; the local device does not page another remote device before completing the service search with the previous remote device and, if necessary, disconnecting from it. However, service discovery can also be implemented across a maximum of seven devices at the same time. In this case, the local device under the control of the SrvDscApp pages all remote devices, creates links with all of them, and then queries all the connected devices for the desired services at the same time.

The features of the service discovery application are summarized as follows:

- Bluetooth inquiries are activated following a user request for a service search.
- For any new remote device found following an inquiry, the SrvDscApp finishes service discovery and terminates its link with the first device prior to attempting a connection with the next remote device.
- For any remote device already connected, the local device does not disconnect following service discovery.

Another feature of the service discovery application is that the user may choose a trusted or untrusted mode of operation. This means the user can decide whether the SrvDscApp will permit connections only with trusted devices, or with any newly discovered remote devices that require only pairing.

When a local device performs a service discovery search, it can do so against three types of remote devices:

- **Trusted devices**—These are devices not currently connected with the local device, but which already have an established trusted relationship with it.
- **Unknown (new) devices**—These are untrusted devices that are not currently connected with the local device.
- **Connected devices**—These are devices that are already connected to the local device.

To discover trusted or unknown (new) remote devices, the SrvDscApp activates the Bluetooth inquiry and/or page processes. For connected remote devices, only paging is needed. To perform its task, the SrvDscApp must have access to the Bluetooth device address of the units in the vicinity of the local device, regardless of whether these devices have been located via a Bluetooth inquiry process or are already connected to the local device. Thus, the Bluetooth module controller in a local device maintains the list of devices in the vicinity and makes it available to the SrvDscApp.

## Message Sequence

The Service Discovery Profile addresses three Bluetooth procedures: device discovery, device name discovery, and service discovery. The first two procedures do not require host intervention, while the third does. Figure 6.6 summarizes the key message exchange phases encountered during the execution of this profile. Not all procedures are present at all times, and not all devices need to go through them all the time. For example, if authentication is not required, the authentication phase will not be executed. If the SrvDsvApp needs to inquire for services on a specific remote device to which the local device is currently connected, inquiries and pages may not be executed. The figure also notes the conditions under which particular phases are executed or not executed.

## Service Discovery

The service discovery application does not make use of SDP as a means of accessing a service; instead, SDP is used as a means of informing the local device user about the services available. Bluetooth applications running in a local device can also retrieve any pertinent information that will facilitate the application's access to a desired service in a remote device.

# Bluetooth General Profiles

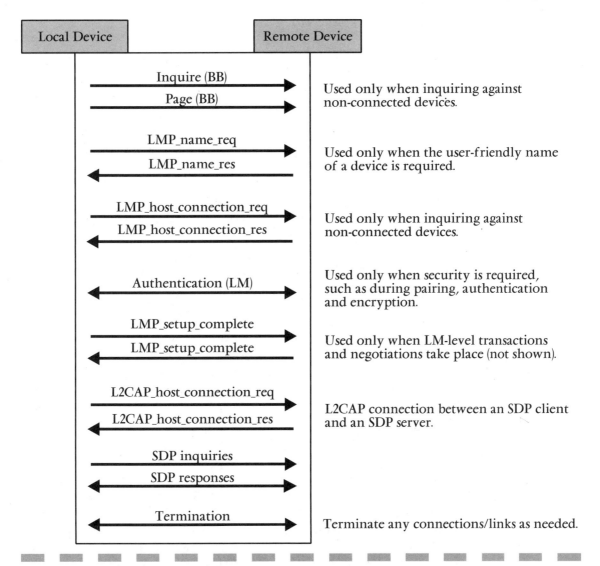

**Figure 6.6** A summary of Bluetooth processes that support the Service Discovery Profile. Source: Bluetooth Specification 1.0.

## Signaling

For the purpose of retrieving SDP-related information, only a local device can initiate an L2CAP connection. Other than that, SDAP does not impose any additional restrictions or requirements on L2CAP signaling.

## Configuration Options

As with the Serial Port Profile, there are three configuration options for L2CAP in the Serial Discovery Profile. One is the size of the Maximum Transmission Unit (MTU), which refers to the largest possible unit of data that can be sent over the data link. As noted, L2CAP implementations must support a minimum MTU size of 48 bytes. The default value is 672 bytes. For greater efficiency of communication resource usage, however, the MTU is set as large as possible, while respecting any physical constraints imposed by the devices involved, and the need of these devices to continue honoring any previously agreed upon Quality of Service (QoS) commitments with other devices and/or applications.

During the lifetime of an L2CAP connection for SDP transactions between two devices—also referred to as the SDP session—either or both devices may become engaged in an L2CAP connection with still other devices or applications. If any new connection has "non-default" QoS requirements, the MTU for the original SDP session can be re-negotiated while the session is in progress to accommodate the QoS parameters of the new L2CAP connection.

Another configuration option is Flush Timeout. As with the Serial Port Profile, this option is used to inform the recipient of the amount of time the originator's link controller/link manager will attempt to transmit successfully an L2CAP segment before giving up and "flushing" the packet.

Since L2CAP implementations are only required to support "best effort" service, support for any other QoS is optional. Best effort does not require any performance guarantee. If any performance guarantee is required, then a QoS configuration request must be sent, specifying such performance parameters as delay variation (microseconds), peak bandwidth (bytes/second), and latency (microseconds). If no QoS option is placed in the request, SDP traffic is assumed to be best effort and is treated as such.

## SDP Transactions and L2CAP Connections

SDP transactions consist of a sequence of request-and-response PDU exchanges, but SDP itself is actually a connectionless datagram service. The reason is that no SDP-level connections are formed prior to the exchange of PDUs. Instead, SDP delegates the creation of connections to the L2CAP layer, which acts on its behalf. The SDP layer also delegates the tear-down of connections to the L2CAP layer.

Normally, in a true connectionless service, the connection would be torn down after a service request PDU is sent. Since SDP servers are considered stateless, having L2CAP tear down the connection after a service request PDU is sent would disrupt the SDP transaction. In addition, there would be a significant performance penalty if a new L2CAP connection had to be created for each subsequent SDP PDU transmission. To avoid these problems, the L2CAP connections for SDP transactions stay in place longer than the transmission of a single SDP PDU. In fact, an SDP session is maintained for as long as there is a need to interact with a specific device.

At a minimum, an SDP transaction represents a single exchange of a request PDU transmission from an SDP client to an SDP server, and the transmission of a response PDU from the SDP server back to the SDP client. However, an SDP session may last less than the minimum in the event of unrecoverable errors that occur in layers below SDP in either the local device and remote device, or in the SDP layer and the service records database in the remote device. Also, the user may terminate an SDP session prior to transaction completion.

Specifying a minimum duration of an SDP session makes for a smooth execution of SDP transactions. For improved performance, implementers of Bluetooth wireless technology may allow SDP sessions to last longer than the minimum duration. In general, an SDP session can be maintained for as long as there is a need for interactivity with a specific device. Since this introduces an element of unpredictability, SDP implementers may use timers to track periods of transaction inactivity over a specific SDP session. When the specified time period expires, the L2CAP connection is torn down.

SDP implementations may also rely on explicit input received from a higher layer, perhaps initiated from the SrvDscApp itself, to open and close an SDP session. These inputs come from the use of low-level primitives[2] indicating "open search" and "close search." Finally, an implementation may permit users to interrupt an SDP session at any time through a service primitive indicating "terminate."

---

[2] As used here, primitives are abstract descriptions of the services offered by a protocol. The format of the primitives discussed above is `openSearch(.)`, `closeSearch(.)`, and `terminatePrimitive(.)`. Further discussion of primitives is beyond the scope of this book.

Usually a remote device does not terminate an SDP session, but such an occurrence is possible. For example, the remote device can terminate the SDP session by using the L2CAP connection termination PDU, or terminate the SDP session by no longer responding to SDP requests or L2CAP signaling commands. The latter occurrence may indicate a problem with the vendor's implementation.

## Link Manager

As discussed earlier, Link Manager carries out link setup, authentication, link configuration, and other management functions. It provides services like encryption control, power control, and QoS capabilities. It also manages devices in different modes (park, hold, sniff, and active).

Table 6.1 lists all the LMP features, noting those that are mandatory (M) to support with respect to the Service Discovery Profile, those that are optional (O), and those that are excluded (X). The reason for excluding features from being activated in this profile is that they may degrade the operation of devices. However, such features may be used by other applications running concurrently with this profile. The table also notes conditional (C) features that are not specifically required in this profile, but when they are encountered no attempt will be made to change them. Any deviation from these rules, per Table 6.1, may cause the Bluetooth units to behave as discussed in the next section, Error Behavior.

**TABLE 6.1**

Summary of LMP Procedures Used with the Service Discovery Application Profile (SDAP)
Source: Bluetooth Specification 1.0

| | LM Procedure | Support in LMP | Support in LocDev | Support in RemDev |
|---|---|---|---|---|
| 1 | Authentication | M | C | C |
| 2 | Pairing | M | | |
| 3 | Change link key | M | | |
| 4 | Change current link key | M | | |
| 5 | Encryption | O | C | C |
| 6 | Clock offset request | M | | |
| 7 | Timing accuracy information request | O | | |
| 8 | LMP version | M | | |

*continued on next page*

# Bluetooth General Profiles

**TABLE 6.1**

Summary of LMP Procedures Used with the Service Discovery Application Profile (SDAP) Source: Bluetooth Specification 1.0 (continued)

| | LM Procedure | Support in LMP | Support in LocDev | Support in RemDev |
|---|---|---|---|---|
| 9 | Supported features | M | | |
| 10 | Switch to master slave role | O | | |
| 11 | Name request | M | | |
| 12 | Detach | M | | |
| 13 | Hold mode | M | | |
| 14 | Sniff mode | O | | |
| 15 | Park mode | O | | |
| 16 | Power control | O | | |
| 17 | Channel quality-driven DM/DH (Data-Medium rate/Data-High rate) | O | | |
| 18 | Quality of service | M | | |
| 19 | SCO links | O | X | X |
| 20 | Control of multi-slot packets | M | | |
| 21 | Concluding parameter negotiation | M | | |
| 22 | Host connection | M | | |

**ERROR BEHAVIOR**

If a Bluetooth unit tries to use a mandatory feature, and the other unit replies that it does not support it, the initiating unit sends a "detach" PDU with the reason "unsupported LMP feature." A unit will always be able to handle the rejection of the request for an optional feature.

**LINK POLICY**

No fixed master-slave roles are specified for Bluetooth devices executing the Service Discovery Profile. In addition, this profile does not specify which low-power modes to use, or when to use them. It is the responsibility of the Link Manager of each device to decide and request special link features as appropriate for the situation.

## Link Control

In Table 6.2, the features of the Link Control (LC) level are listed, noting those that are mandatory (M) to support with respect to the Service Discovery Profile, those that are optional (O), and those that are excluded (X) but may be used by other applications running concurrently with this profile. The table also notes those features that are not specifically required in this profile; when they are encountered, no attempt will be made to change them. These are referred to as conditional (C) features.

**TABLE 6.2** Summary of LC Procedures Used with the Service Discovery Application Profile (SDAP)
Source: Bluetooth Specification 1.0

| Procedure | Support in Baseband | Support in LocDev | Support in RemDev |
|---|---|---|---|
| 1  Inquiry | M | C | |
| 2  Inquiry scan | M | | C |
| 3  Paging | M | C | |
| 4  Page scan | | | |
|    a  Type R0 | M | | C |
|    b  Type R1 | M | | C |
|    c  Type R2 | M | | C |
| 5  Packet types | | | |
|    a  ID packet | M | | |
|    b  NULL packet | M | | |
|    c  POLL packet | M | | |
|    d  FHS packet | M | | |
|    e  DM1 packet | M | | |
|    f  DH1 packet | M | | |
|    g  DM3 packet | O | | |
|    h  DH3 packet | O | | |
|    i  DM5 packet | O | | |
|    j  DH5 packet | O | | |

*continued on next page*

# Bluetooth General Profiles

**TABLE 6.2**

Summary of LC Procedures Used with the Service Discovery Application Profile (SDAP)
Source: Bluetooth Specification 1.0 (continued)

| Procedure | Support in Baseband | Support in LocDev | Support in RemDev |
|---|---|---|---|
| k  AUX packet | M | X | X |
| l  HV1 packet | M | X | X |
| m  HV2 packet | O | X | X |
| n  HV3 packet | O | X | X |
| o  DV packet | M | X | X |
| 6  Inter-piconet capabilities | O | | |
| 7  Voice codec | | | |
| a  PCM (A-law) | O | X | X |
| b  PCM (µ-law) | O | X | X |
| c  CVSD | O | X | X |

When a local device starts a service search with unconnected remote devices, it needs to inquire about them or page them.

**INQUIRY**

When prompted by the SrvDscApp, the local device advises its baseband to enter the inquiry state. If there are already existing and ongoing connections with QoS requirements that must be met, entry into this state may not be immediate. To compensate for this possibility, the user of the SrvDscApp can set the criteria for the duration of an inquiry. When inquiry is invoked from the local device, the General Inquiry Access Code (GIAC) procedure is typically used.

**INQUIRY SCAN**

Inquiry scans are device-dependent policies that fall outside the scope of this profile. Devices that operate in the discoverable mode could be discovered by inquiries sent by other devices. To be discovered by an inquiry resulting from SrvDscApp, a remote device must scan for inquiries using the GIAC.

### PAGING

When the SrvDscApp connects to a specific remote device for the purpose of inquiring about its service records, the local device prompts its baseband to enter the page state, ceding to the QoS requirements of any existing and ongoing connections. Depending on the paging class (R0, R1, or R2) indicated by a remote device, the local device pages accordingly. For the pages, the 48-bit Bluetooth device address of the remote device is used.

### PAGE SCAN

Like inquiry scans, page scans are device-dependent policies that fall outside the scope of this profile. Devices that operate in a connectable mode of operation, as defined in the Generic Access Profile (GAP), could establish Bluetooth links with other devices from pages sent by them. To establish a link with a remote device, a local device sends a page that results from a SrvDscApp action using the remote device's 48-bit Bluetooth device address.

### ERROR BEHAVIOR

With most features at the LC level activated by LMP procedures, errors typically are caught at that layer. However, the misuse of inquiry or paging features may be difficult or impossible to detect. There is no mechanism defined in this profile to detect or prevent the improper use of such features.

## Generic Object Exchange Profile (GOEP)

GOEP defines how Bluetooth devices support the object exchange usage models, including the File Transfer Profile, Object Push Profile and Synchronization Profile, which are discussed in the next chapter along with other usage models. The most common devices that make use of these models include notebook PCs, PDAs, smart phones, and mobile phones using Bluetooth wireless components.

The GOEP specification provides generic interoperability for the application profiles using Object Exchange (OBEX) capabilities and defines the interoperability requirements of the lower protocol layers (e.g., Baseband and LMP) for the application profiles.

OBEX provides object exchange services similar to the HTTP used on the World Wide Web. However, OBEX works for the many devices that cannot afford the substantial resources required by an HTTP server and it also targets devices with usage models different from the Web. The major use of OBEX is in supporting "Push" or "Pull" applications, allowing timely and efficient communications among portable devices in dynamic environments.

OBEX is not limited to quick connect-transfer-disconnect scenarios. It also allows sessions in which transfers take place over a period of time, maintaining the connection even when it is idle. This means OBEX can be used to perform complex tasks such as database transactions and synchronization. It was designed to be application friendly and provide cross-platform interoperability. It is compact, flexible, extensible, minimizes strain on resources of small devices, and maps easily into Internet data transfer protocols.

## Profile Stack

As shown in Figure 6.7, both the client and server are equipped with the GOEP profile stack, which consists of the Baseband, LMP, and L2CAP, which are the OSI layer 1 and 2 Bluetooth protocols. RFCOMM is the Bluetooth adaptation of GSM TS 07.10, which provides a transport protocol for serial port emulation. SDP is the Bluetooth Service Discovery Protocol. OBEX is the Bluetooth adaptation of IrOBEX, developed by the Infrared Data Association. The Application Client layer is the entity sending and retrieving data objects from the Application Server using the OBEX operations. The Application Server is the data storage element to and from which data objects can be pushed and pulled, respectively.

## Server and Client

Under the Generic Object Exchange Profile, a client pushes (sends) data objects to the server or pulls (retrieves) data objects from the server.

It is assumed that the user may have to put the device containing the server into the discoverable and connectable modes when the inquiry and link establishment procedures, respectively, are processed in the client, per the requirements of the Generic Access Profile (GAP).

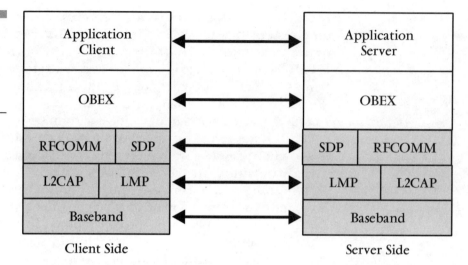

**Figure 6.7**
Protocol model for the Generic Object Exchange Profile. Source: Bluetooth Specification 1.0.

The Generic Object Exchange Profile defines procedures for use in point-to-point configurations. Consequently, the server is assumed to offer services for only one client at a time, although this is not a requirement. It is possible for an implementation to support multiple clients simultaneously over separate point-to-point links.

## Profile Basics

Before a server is used with a client for the first time, a bonding procedure that includes pairing may be performed. Although this procedure must be supported, its use is dependent on the application profiles. Bonding usually requires the user to manually activate the process and enter a Bluetooth PIN code on the keyboards of both the client and server devices.

In addition to the link-level bonding, an OBEX initialization procedure may be performed before the client can use the server for the first time. The application profiles using GOEP must specify whether this procedure is supported to provide the required security level.

Link and channel establishment must be done according to the procedures defined in GAP. Security can be provided by authenticating the other party upon connection establishment, and by encrypting all user data on the link level. Whether authentication and encryption are supported in the devices will depend on the application profile using GOEP.

The user interface is not defined in GOEP, but instead is defined in the application profiles, where necessary. In addition, GOEP does not specify fixed master-slave roles.

## Features

The Generic Object Exchange Profile provides three features. The use of other features, such as setting the current directory, must be defined by the application profiles that require them.

### ESTABLISHING AN OBJECT EXCHANGE (OBEX) SESSION
This feature is used to establish the object exchange session between the client and server. Before a session is established, payload data cannot be exchanged between the client and the server. Use of the OBEX operations for establishing an OBEX session is described later.

### PUSHING A DATA OBJECT
This feature is used if data need to be transferred from the client to the server. The data object is pushed to the server using the PUT operation of the OBEX protocol. The data can be sent in one or more OBEX packets.

### PULLING A DATA OBJECT
This feature is used if data need to be transferred from the server to the client. The data object is pulled from the server using the GET operation of the OBEX protocol. The data can be sent in one or more OBEX packets.

## OBEX Operations

The operations specified by the OBEX protocol are Connect, Disconnect, Put, Get, Abort, and SetPath. The application profiles using GOEP must specify which of these operations is supported to provide the functionality defined in the application profiles.

The IrOBEX specification does not define how long a client should wait for a response to an OBEX request, except that it should be a reasonable period between a request and response before automatically canceling the operation. The Bluetooth specification suggests a reasonable time period is 30 seconds or more.

### INITIALIZATION OF OBEX

If authentication is used by the server and client, its initialization must be accomplished before the first OBEX connection can be established. This can be done at any time before the first OBEX connection, but requires user intervention on both the client and server devices.

Authentication is done using an OBEX password, which may be the same as a Bluetooth PIN code. Even if the user enters the same code for link authentication and OBEX authentication, they must be entered separately. After the OBEX password is entered in both the client and server, it is stored in each for future use.

### OBEX SESSION ESTABLISHMENT

The OBEX connection can be made with or without authentication, but at a minimum all application profiles using GOEP must support an OBEX session without authentication. Figure 6.8 illustrates how an OBEX session is established using the CONNECT operation. The CONNECT request indicates a need for connection and may also indicate which service is used.

Although the OBEX authentication scheme used on the client and server is based on HTTP, it does not support all its features and options. In GOEP, OBEX authentication is used to authenticate the client and server. Figure 6.9 depicts the establishment of an OBEX session with authentication.

## Summary

The Generic Object Exchange Profile (GOEP) defines the requirements for Bluetooth devices necessary for the support of the object exchange (OBEX) usage models, which include the File Transfer Profile, Object Push Profile, and the Synchronization Profile; these are expected to be used primarily in devices that use Bluetooth wireless technology, such as notebook PCs, PDAs, smart phones, and mobile phones. However, digital cameras, printers, auto-tellers, information kiosks, calculators, data collection devices, home electronics, medical instruments, automobiles, office equipment, toys, and even watches are all candidates for using OBEX as well.

In addition to GOEP, the Bluetooth specification includes the following other specifications for OBEX and the applications using it.

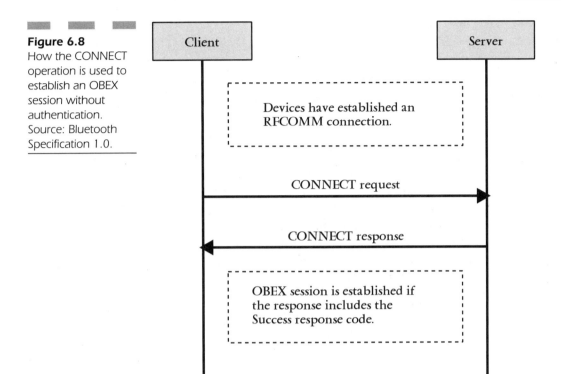

**Figure 6.8**
How the CONNECT operation is used to establish an OBEX session without authentication.
Source: Bluetooth Specification 1.0.

- **Bluetooth Synchronization Profile Specification**—This defines the interoperability requirements for the applications within the Synchronization application profile. However, it does not define the requirements for the Baseband, LMP, L2CAP, or RFCOMM.

- **Bluetooth File Transfer Profile Specification**—This defines the interoperability requirements for the applications within the File Transfer application profile, but does not define the requirements for the Baseband, LMP, L2CAP, or RFCOMM.

- **Bluetooth Object Push Profile Specification**—This defines the interoperability requirements for the applications within the Object Push application profile, but not the requirements for the Baseband, LMP, L2CAP, or RFCOMM.

- **Bluetooth IrDA Interoperability Specification**—This defines how the applications can function over both Bluetooth and IrDA wireless devices, specifies how OBEX is mapped over RFCOMM and TCP, and defines the application profiles using OBEX over the Bluetooth specification.

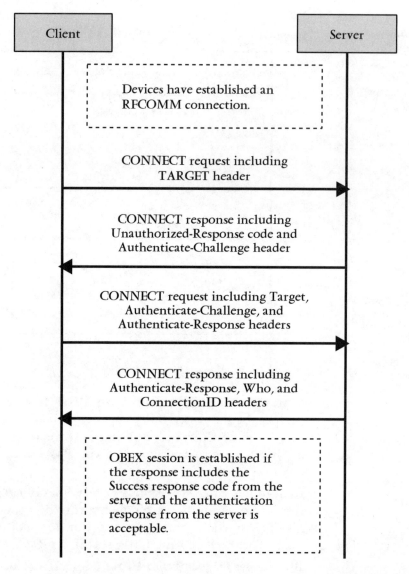

**Figure 6.9**
Establishment of an OBEX session with authentication.
Source: Bluetooth Specification 1.0.

The last specification is noteworthy because it represents the first session layer common to both infrared and Bluetooth environments. In adopting common usage models and then exploiting the unique advantages of each technology, the combination of Bluetooth wireless technology and infrared creates the only short-range wireless standards that can meet user needs, varying from wireless voice transmission to high-speed (16 Mbps) robust data transfer.

# CHAPTER 7

# Bluetooth Profiles for Usage Models

As noted in the previous chapter, the Bluetooth SIG has identified various usage models, each of which has a supporting "profile" that defines the protocols and features the usage models draw upon for implementation. This dependency is shown in Figure 7.1 where, for example, the Intercom Profile defines the protocols and procedures that would be used by devices implementing the intercom part of the usage model called "3-in-1 phone," also known as the "walkie-talkie" application of Bluetooth wireless technology. As indicated in the figure, the Intercom Profile is dependent on the Generic Access Profile (GAP), which was discussed in the previous chapter.

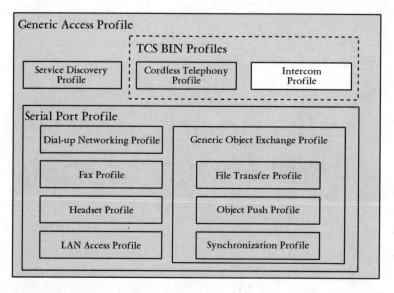

**Figure 7.1**
The Intercom Profile is dependent on the Generic Access Profile.
Source: Bluetooth Specification 1.0.

## Intercom Profile

The Intercom Profile supports the usage scenarios that require a direct speech link between two Bluetooth devices, such as two cellular phone users engaged in a speech call on a Bluetooth connection. Even though the call is a direct phone-to-phone connection using Bluetooth wireless technology only, the link is established using telephony-based signaling. The voice coder/decoder (codec) used may be either Pulse Code Modulation (PCM)[1] or Continuously Variable Slope Delta (CVSD) modulation. Negotiation of Quality of Service (QoS) is optional.

---

[1] Either A-law or μ-law PCM is supported.

# Bluetooth Profiles for Usage Models

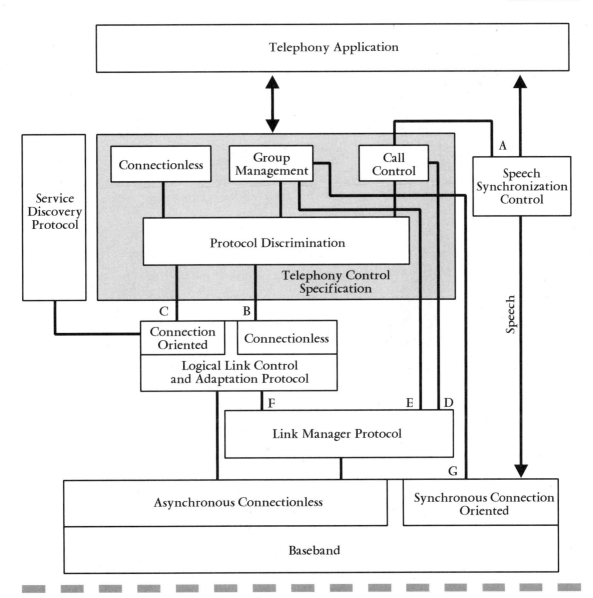

**Figure 7.2** Block diagram of the Intercom Protocol model. Source: Bluetooth Specification 1.0.

In the Intercom Protocol model, the interfaces labeled A, B, and C in Figure 7.2 are used for the following purposes:

The Call Control (CC) element uses interface A to the speech synchronization control to connect and disconnect the speech paths. Interface B delivers TCS messages on the connection-oriented (point-to-point) L2CAP channel. Interface C is used by the Call Control element directly to control the Link Manager for the purpose of establishing and releasing Synchronous Connection Oriented (SCO) links. For initialization purposes, it also directly controls the Link Control/Baseband elements to enable inquiry, paging, inquiry scan, and page scan.

Figure 7.3 shows a typical configuration of devices that make use of the Intercom Profile. Because the intercom usage model is completely symmetrical, there are no specific roles defined for each device. A device supporting the Intercom Profile is generally referred to as a Terminal (TL).

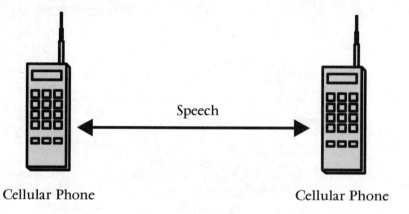

**Figure 7.3**
System configuration of two devices making use of the Intercom Profile.
Source: Bluetooth Specification 1.0.

Several interactions take place when a terminal wants to establish an intercom call with another terminal. If the initiator of the intercom call does not have the Bluetooth address of the acceptor, it has to obtain it using the device discovery procedure described in the Generic Access Profile. Since the Intercom Profile does not mandate a particular security mode, if the users of either device (initiator/acceptor) want to enforce security in the execution of this profile, the authentication procedure of the Generic Access Profile must be performed to create a secure connection.

It is the responsibility of the initiator to establish the link and channel, per the Generic Access Profile. Based on the security requirements enforced by users of either device, authentication may be performed and encryption may be enabled. When the intercom call is established, two-way speech between the terminal users can occur. When either user goes "on hook," the intercom call is cleared, and the channel and link are released.

# Call Procedures

Before a call request can be made, a connection-oriented L2CAP channel must be set up between the initiator and acceptor devices. When the L2CAP channel is established, the acceptor will start a timer. If the acceptor has not received a SETUP message initiating the call request before the timer expires, it terminates the L2CAP channel. If the SETUP message is received within the time limit, the timer is canceled and the call request goes through.

The call request procedure used in the Intercom Profile is defined in TCS Binary, as are the call confirmation and call connection procedures. For call connection, SCO link establishment is initiated with the appropriate Link Management Protocol procedure before sending a CONNECT message. The speech path is set up by a unit when it receives a CONNECT or CONNECT ACKNOWLEDGE message.

Figure 7.4 summarizes the signaling flow between Bluetooth units for connection establishment and SCO link establishment, while Figure 7.5 summarizes the signaling flow for call clearing. Other signaling flows are possible within the Intercom Profile.

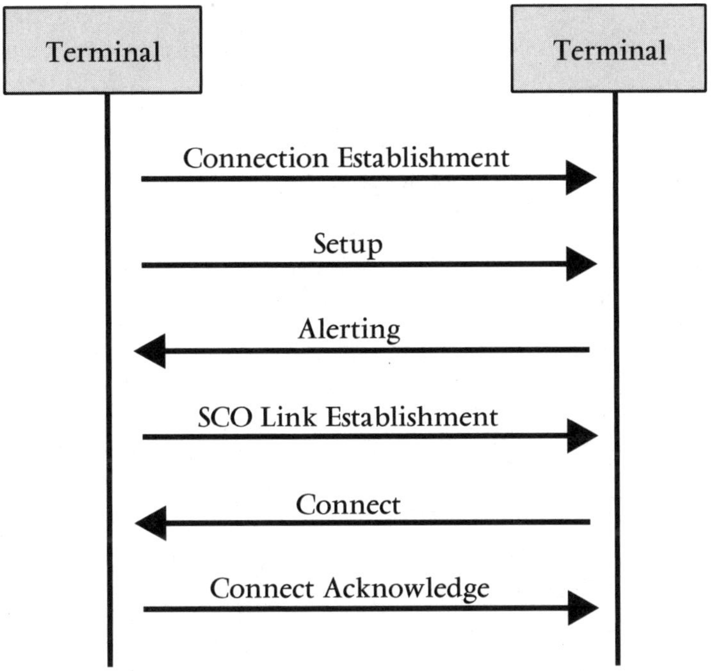

**Figure 7.4** Signal flow for successful call establishment. Source: Bluetooth Specification 1.0.

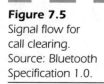

**Figure 7.5**
Signal flow for call clearing.
Source: Bluetooth Specification 1.0.

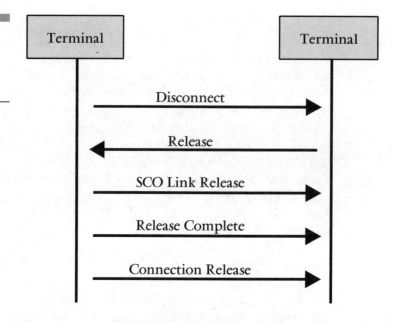

Table 7.1 lists all LMP capabilities, along with what features are mandatory (M) and optional (O) with respect to the Intercom Profile.

All call-clearing and call-collision procedures are defined in TCS Binary and are supported by both terminals. After the last call-clearing message has been sent, a unit releases the SCO link by invoking the appropriate Link Management Protocol procedure. In addition, the L2CAP channel used for TCS call-control signaling is terminated. When the SCO link and L2CAP channel are torn down, the two units are detached from each other and are free to make calls to other Bluetooth devices.

**TABLE 7.1**

Summary of LMP Procedures
Source: Bluetooth Specification 1.0

| Procedure | Support |
|---|---|
| Authentication | M |
| Pairing | M |
| Change link key | M |
| Change the current link key | M |
| Encryption | O |
| Clock offset request | M |

*continued on next page*

# Bluetooth Profiles for Usage Models

**TABLE 7.1**

Summary of LMP Procedures
Source: Bluetooth Specification 1.0
(continued)

| Procedure | Support |
|---|---|
| Slot offset information | O |
| Timing accuracy information request | O |
| LMP version | M |
| Supported features | M |
| Switch master-slave role | O |
| Name request | M |
| Detach | M |
| Hold mode | O |
| Sniff mode | O |
| Park mode | O |
| Power control | O |
| Channel quality driven DM/DH | O |
| Quality of service | O |
| SCO links | M |
| Control of multi-slot packets | O |
| Paging scheme | O |
| Link supervision | M |
| Connection establishment | M |

## Message Summary

Table 7.2 summarizes the TCS Binary messages used in the Intercom Profile, indicating whether they are mandatory (M) or optional (O).

## Call Failure

According to the Bluetooth specification, there are circumstances when a call request fails to result in a connection between devices under the Intercom Profile. The reasons for call failure include:

**TABLE 7.2**

Summary of TCS Binary Messages Used in the Intercom Profile Source: Bluetooth Specification 1.0

| Message | Support |
| --- | --- |
| ALERTING | M |
| CONNECT | M |
| CONNECT ACKNOWLEDGE | M |
| DISCONNECT | M |
| INFORMATION | O |
| RELEASE | M |
| RELEASE COMPLETE | M |
| SETUP | M |

- Normal call clearing
- User busy
- No user responding
- No answer from user (user alerted)
- Call rejected by user
- No circuit/channel available
- Temporary failure
- Requested circuit/channel not available
- Bearer capability not presently available
- Bearer capability not implemented
- Requested facility not implemented
- Timer expired

Table 7.3 summarizes the support status for discoverability, connectability, and pairing modes used in the Intercom Profile, indicating whether they are mandatory (M), optional (O), or conditional (C).

In terms of idle mode procedures supported within the Intercom Profile, only general inquiry is mandatory. Limited inquiry, name discovery, device discovery, and bonding are supported as options.

**TABLE 7.3**

Support Status for Various Modes Used in the Intercom Profile Source: Bluetooth Specification 1.0

| Procedure | Support |
| --- | --- |
| 1. Discoverability modes | |
| Non-discoverable mode | M |
| Limited discoverable mode | O |
| General discoverable mode | M |
| 2. Connectability modes | |
| Non-connectable mode | N/A |
| Connectable mode | M |
| 3. Pairing modes | |
| Non-pairable mode | O |
| Pairable mode | C* |

*If the bonding procedure is supported, pairable mode is mandatory; otherwise it is optional.

## Cordless Telephony Profile

In addition to the intercom application, the 3-in-1 phone usage model can support cordless-only telephony or cordless telephony services available through a multimedia PC, further extending the usefulness of Bluetooth wireless technology in a residential or small office environment. The Cordless Telephony Profile defines the procedures for and the features associated with making calls via the base station and placing direct intercom calls between two terminals. It is also good for accessing supplementary services provided by the external Public Switched Telephone Network (PSTN). This mode of operation allows cellular phones to use Bluetooth wireless technology as a short-range bearer to access PSTN services via a cordless telephone base station, which is among several devices that can act as a "gateway" to the PSTN.

To accomplish all this, the Cordless Telephony Profile makes use of the Bluetooth baseband, Link Manager Protocol, L2CAP, Service Discovery Protocol, and the Telephony Control Protocol Specification (TCS-Binary). As indicated in Figure 7.6, the Cordless Telephony Profile is dependent on the Generic Access Profile.

**Figure 7.6**
Dependency relationship of the Cordless Telephony Profile to the Generic Access Profile. Source: Bluetooth Specification 1.0.

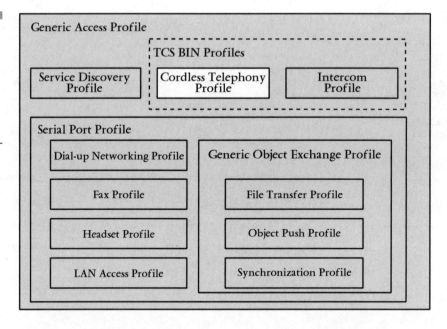

In the Cordless Telephony Profile, the interfaces labeled A through G (refer to Figure 7.2) are used for the following purposes:

As in the Intercom Profile, the Call Control (CC) element uses interface A to the speech synchronization control to connect and disconnect internal speech paths. Interface B is used by the gateway to send and by the terminal to receive broadcast TCS-Binary messages. Interface B is used to deliver all TCS messages that are sent on a point-to-point synchronous connection-oriented (SCO) L2CAP channel. Interface D is used by the Call Control element to directly control the Link Manager (LM) for the purpose of establishing and releasing SCO links. Interface E is used by the Group Management procedures to control Link Manager functions during initialization and for key handling purposes. Interface F is not used within the Cordless Telephony Profile. Interface G is used by the Group Management procedures to directly control the Link Control/baseband to enable inquiry, paging, inquiry scan, and page scan.

## Device Roles

As noted above, two roles are defined for devices that interact under the procedures described in the Cordless Telephony Profile: gateways and terminals. In this profile, both connection-oriented channels and connectionless channels are used. Connectionless channels are used to broadcast information from the gateway to the terminals. Only the gateway can use connectionless channels for sending.

The gateway (GW) acts as a terminal endpoint from the perspective of the external network and handles all communication towards that network. The gateway is the central point with respect to external calls, which means that it handles all call setup requests to/from the external network. Among the types of devices that can act as a gateway include a PSTN or ISDN home base station, GSM gateway, satellite gateway, or H.323 gateway.[2]

Gateways fall into two types: those that support multiple active terminals simultaneously and those that support only one active terminal. The latter type of gateway does not support multiple ringing terminals, multiple active calls, or services involving more than one terminal simultaneously.

The terminal (TL) is the wireless user terminal, which may be a cordless telephone, a dual-mode cellular/cordless phone, or a multimedia PC. The Cordless Telephony profile supports a topology of one gateway and a small number of terminals, as shown in Figure 7.7.

## Typical Call Scenarios

The following typical call scenarios are described in the Cordless Telephony Profile:

- Connecting to the gateway so that incoming calls can be routed to the terminal and outgoing calls can be originated.
- Placing a call from a terminal to a user on the network to which the gateway is connected.
- Receiving a call from the network to which the gateway is connected.
- Making direct calls between two terminals, in which case, it is an intercom call.

---

[2] H.323 is an International Telecommunication Union (ITU) umbrella standard that covers a number of audio and video encoding standards for communication over networks with non-guaranteed bandwidth. Such networks include Ethernet and Token Ring LANs, and can include the Internet as well.

**Figure 7.7**
Typical system configuration of gateway and terminal devices under the Cordless Telephony Profile. Source: Bluetooth Specification 1.0.

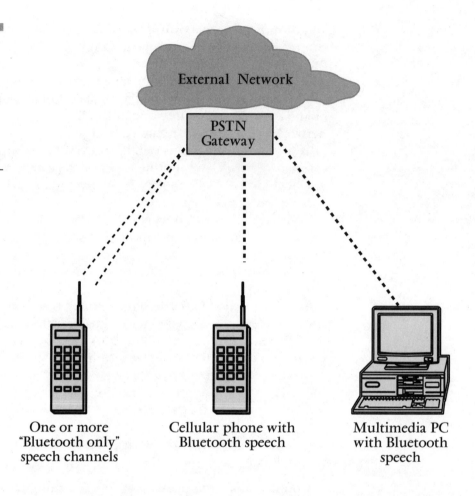

- Using supplementary services provided by the external network by means of Dual Tone Multi-Frequency (DTMF) signaling and register recall (i.e., flash hook).[3]

The gateway is normally the master of the piconet in the Cordless Telephony Profile. As such, it controls the power mode of the terminals and may broadcast information to them. The gateway is as conservative as possible when deciding what power mode to put the ter-

---

[3] The specific set of external supplementary services is not defined in the Cordless Telephony Profile and is dependent on the network to which the gateway is connected. The Cordless Telephony Profile provides the means for accessing them through the use of DTMF signaling and register recall.

minals in. When a terminal is not engaged in signaling, the gateway puts it in a low-power mode. Per the Bluetooth specification, the recommended low-power mode is referred to as "park" mode. This mode is power efficient, allows for reasonable call setup times, and allows broadcasting to the attached terminals. The low-power mode parameters chosen by the gateway are such that the terminal can always return to the active state within 300 milliseconds (ms.). If the gateway can save power during a call, it may use the sniff mode. A terminal may also request to be put in sniff mode.

A terminal out of range of a gateway searches for it by periodically trying to page it. A gateway devotes much of its free capacity to page scanning in order to allow roaming terminals that come within range to find it as quickly as possible. When a terminal has successfully paged a gateway, the gateway requests a master-slave switch when the terminal connects. If the terminal rejects the request, the gateway may detach it. A terminal that does not accept master-slave switch requests will not be guaranteed service.

The gateway enters active mode whenever an incoming call arrives or when a terminal wants to make an outgoing call. The L2CAP channels are used for all TCS control signaling. Voice is transported using SCO links. Quality of Service negotiation is optional. For security purposes, the terminals and gateway employ authentication procedures, and all user data are encrypted for transmission. To facilitate secure communication between cordless units, communication is restricted to members of the Wireless User Group (WUG). The gateway always acts as the WUG master.

## Features

The Bluetooth specification describes the following features available to devices that implement the Cordless Telephony Profile:

- **Calling Line Identification Presentation (CLIP)**—This refers to the ability to view the number of the calling party before deciding whether to accept the call.

- **Call Information**—This refers to the ability to provide additional information during the active phase of a call.

- **Connection Management**—This refers to the ability to accept and the terminals to request connections for the purposes of implementing TCS-BIN procedures.

- **DTMF Signaling**—This refers to the ability, when placing external calls, to send a DTMF signal over the external network to call another party.
- **Incoming External Call**—This refers to a call originating from the external network connected to the local gateway.
- **Initialization**—This refers to the process by which a terminal is granted access rights to a gateway.
- **Intercom Call**—This refers to a call originating from a terminal toward another terminal, in which case the Intercom Profile is used.
- **Multi-Terminal Support**—This refers to the ability of a gateway to handle multiple active terminals that are registered at the same time. The terminals comprise a Wireless User Group (WUG).
- **On Hook**—This refers to the ability of a terminal to terminate a call (i.e., hang up), thereby releasing all radio resources that supported the call.
- **Outgoing External Call**—This refers to a call originated by a terminal towards the external network connected to the local gateway.
- **Post-dialing**—This refers to the ability of a terminal to send dialing information after the outgoing call request setup message is sent.
- **Register recall**—This refers to the ability of the terminal to request "register recall," and of the gateway to transmit the request to the local network. Register recall means to seize a register (with dial tone) to permit input of further digits or perform other actions. This is also known as "flash hook."

Table 7.4 summarizes the features supported by the Cordless Telephony Profile, indicating whether support for a particular feature is mandatory (M) or optional (O) in both the terminal and gateway devices.

## Terminal-to-Gateway Connection

Only trusted terminals are allowed to connect to a gateway. When a terminal connects to a gateway, the link is configured and the L2CAP connection used for further signaling during the TCS-BIN session is set up and configured. The connecting terminal is responsible for setting up the connection-oriented L2CAP channel.

**TABLE 7.4**

Summary of Application-layer Features Available Through the Cordless Telephony Profile
Source: Bluetooth Specification 1.0

| Feature | Support in Terminal | Support in Gateway |
|---|---|---|
| Connection Management | M | M |
| Outgoing External Call | M | M |
| Incoming External Call | M | M |
| Intercom Call | M | N/A |
| On-Hook | M | M |
| Post-Dialing | O | O |
| Multi-Terminal Support | O | O |
| Call Information | O | O |
| Calling Line Identification Presentation (CLIP) | M | O |
| DTMF Signaling | M | M |
| Register Recall | M | M |

To avoid the paging delay at call setup and to facilitate message broadcasts, the terminal establishes an L2CAP connection to the gateway when it comes into range, and not before every call. This L2CAP connection remains in place until the radio link is lost or the terminal ceases activity. Consequently, the L2CAP connection may be idle for long periods of time.

A gateway supporting multiple terminals uses a connectionless L2CAP channel for TCS-BIN broadcasted messages. A terminal is added to the connectionless group when it connects to the gateway.

## Terminal-to-Terminal Connection

As noted earlier, a terminal-to-terminal connection is actually an intercom call. When a terminal initiates a call and establishes a direct link to another terminal, the procedures used are those from the Intercom Profile.

If the terminal has the capability to participate in two piconets simultaneously, it can remain a member of the gateway piconet and participate in signaling toward another gateway during the intercom call. However, if the terminal is not able to participate in two piconets simultaneously, it must detach itself from the first gateway while the intercom call is active. After the intercom call with another gateway is terminated, it can re-establish the connection to the first gateway.

## Call Control

In an outgoing external call, the terminal is referred to as the "outgoing side" and the gateway is referred to as the "incoming side." In an incoming external call, the terminal that terminates the call is the incoming side and the gateway is the outgoing side. The call control procedures used in the Cordless Telephony Profile are performed in accordance with those defined in TCS Binary. As described in the Bluetooth specification, these call control procedures are:

- **Call request**—This initiates call establishment with a SETUP message.
- **Overlap sending**—When incomplete called-number information is received, or it cannot be determined that the called-number information is complete, timers are reset in an effort to respond appropriately to the SETUP message with such messages as CALL PROCEEDING or CONNECT.
- **Call proceeding**—This acknowledges the SETUP message and indicates that the call is being processed.
- **Call confirmation**—Upon receiving an indication that user alerting has been initiated at the called address, the incoming side sends an ALERTING message, and enters the call received state. When the outgoing side receives the ALERTING message, it begins an internally generated alerting indication and enters the call delivered state.
- **Call connection**—An incoming side indicates acceptance of an incoming call by sending a CONNECT message to the outgoing side, and stopping the user alerting. Upon sending the CONNECT message, the incoming side starts a timer. On receipt of the CONNECT message, the outgoing side stops any internally generated alerting indications, stops any running timers, completes the requested channel to the incoming side, sends a CONNECT ACKNOWLEDGE message, and enters the active state. The CONNECT ACKNOWLEDGE message indicates completion of the

# Bluetooth Profiles for Usage Models

requested channel. On receiving this message, the incoming side connects to the channel, stops the timer, and enters the active state. When the timer expires before receiving a CONNECT ACKNOWLEDGE message, the incoming side initiates call clearing.

- **Non-selected user clearing**—When a call has been delivered over a connectionless channel, as in the case of a multi-point configuration, in addition to sending a CONNECT ACKNOWLEDGE message to the incoming side selected for the call, the outgoing side sends a RELEASE message to all other incoming sides that have sent SETUP ACKNOWLEDGE, CALL PROCEEDING, ALERTING, or CONNECT messages in response to the SETUP message. These RELEASE messages are used to notify the incoming sides that the call is no longer being offered to them.

- **In-band tones and announcements**—If the requested channel carries a speech call, a progress indicator is sent simultaneously with the application of the in-band tone and/or announcement. This progress indicator may be included in any call control message. Upon receipt of this message, the outgoing side may connect, if it is not already connected, to the channel to receive the in-band tone and/or announcement.

- **Failure of call establishment**—The incoming side may initiate call clearing for a number of reasons:
    — Unassigned (unallocated) number
    — No route to destination
    — User busy
    — No user responding
    — Number changed
    — Invalid number format (incomplete number)
    — No circuit/channel available
    — Requested circuit/channel not available
    — Bearer capability not presently available
    — Bearer capability not implemented
    — No answer from user (user alerted)
    — Call rejected by user

- **Call clearing**—Clearing may be initiated from the outgoing or the incoming side. In the case of the outgoing side, a DISCONNECT message is sent, disconnection from the channel begins, and the device enters the Disconnect request state. The incoming side enters the Disconnect indication state upon receipt of a DISCONNECT message which prompts the incoming side to disconnect from the bearer channel. Once the channel used for the call has been disconnected, the incoming side sends a RELEASE message to the outgoing side and enters the Release request state. On receipt of the RELEASE message the outgoing side releases the channel, and sends the RELEASE COMPLETE message.

- **Call information**—While in the active state, both the incoming and outgoing sides may exchange other information related to the ongoing call by using INFORMATION messages.
- **Calling line identity**—To inform the incoming side of the identity of the caller, the outgoing side may include the calling party number information element in the SETUP message transferred as part of the call request. The calling party number may consist of up to 24 digits.

## Group Management

Group Management is another set of procedures performed as defined in TCS Binary. Terminals that want to become members of a Wireless User Group may initiate a request towards the gateway, which has responsibility for Group Management. The gateway may accept or reject the request based on such things as configuration, or if the user already has physical access to the base station. A gateway that accepts the access rights request adds the terminal to the WUG and initiates the Configuration Distribution procedure, which is also performed in accordance with TCS Binary.

**CONFIGURATION DISTRIBUTION**
Because of the security implications of the Configuration Distribution procedure, a terminal is not forced to store the key information received during this procedure. In addition, the gateway can always reject the ACCESS RIGHTS REQUEST from a terminal for implementation-dependent reasons. A user may be required to press a button on the gateway, for example, before being granted access to the group.

A terminal that stores link keys during the Configuration Distribution procedure does not overwrite existing link keys to other WUG members. Only if there was no previous link key to a specific device is the key obtained during the Configuration Distribution procedure. Link-loss handling is applicable to this procedure as well.

### LINK LOSS DETECTION BY THE GATEWAY

If the gateway detects loss of link before receiving the INFO ACCEPT message, it considers the WUG update to be terminated unsuccessfully and the terminal as detached. If the gateway detects loss of link after receiving the INFO ACCEPT message, it considers the WUG update to be terminated successfully.

### FAST INTER-MEMBER ACCESS

The fast inter-member access procedure is used when two terminals that are members of the same WUG need to establish their own piconet. This may be needed when an intercom call is established, for example. Terminal X may detach from the gateway after having sent the LISTEN ACCEPT message by terminating the L2CAP channel to the gateway and sending an LMP detach message. Terminal Y may detach from the gateway after having received the LISTEN ACCEPT message by terminating the L2CAP channel to the gateway and sending an LMP detach message.

## Periodic Key Update

A master key is used during a gateway-terminal connection. This key is issued to the terminal when it connects to the gateway and is valid for only a single session. Since the gateway piconet is operational all the time, this means that the same master key would always be used. To increase the security level, the master key is changed periodically under control of a timer.

The timer determines the interval between key changes. When the timer expires, the gateway tries to do a key update on all terminals. However, some terminals may be out of range or powered off. In such cases, the new key is given to the terminal the next time it attaches. After there has been an attempt to update all terminals, the timer is reset.

The key-update process is performed by the gateway on each terminal in turn. If a terminal is parked, it is unparked. The new link key is then issued. Activating the link key is accomplished by turning off encryption, and then turning it back on. Finally, the terminal may be returned to park. If any of these sub-procedures fails, further sub-procedures are not performed on that terminal. The gateway then attempts to update the next terminal.

## Inter-Piconet Capability

Inter-piconet capability refers to the ability of a master device to keep the synchronization of a piconet while page scanning in free slots and allowing for new members to join the piconet. While a new unit is in the process of joining the piconet, and until the master-slave switch is performed, operation may be temporarily degraded for the other members. A gateway that supports multiple terminals has this inter-piconet capability. The terminals also may have the inter-piconet capability.

## Service Discovery Procedures

Table 7.5 lists all entries in the SDP database of the gateway, as defined by the Cordless Telephony Profile, indicating whether the presence of the field is mandatory (M) or optional (O).

**TABLE 7.5**

Entries in the Gateway's Service Discovery Database Source: Bluetooth Specification 1.0

| Item | Value | Status |
| --- | --- | --- |
| Service Class ID | | M |
| Service Class #0 | Generic Telephony | O |
| Service Class #1 | Cordless Telephony | M |
| Protocol Descriptor List | | M |
| Protocol #0 | L2CAP | M |
| Protocol #1 | TCS-BIN Cordless | M |
| Service Name (displayable text name) | Defined by service-provider (default, "Cordless Telephony") | O |
| External Network | PSTN, ISDN, GSM, CDMA, analog cellular, packet switched, other | O |
| Bluetooth Profile Descriptors | | M |
| Profile #0 | Cordless Telephony | M |
| Parameter for Profile #0 (version) | | O |

# Bluetooth Profiles for Usage Models

## LMP Procedures

Table 7.6 lists all LMP features, indicating which are mandatory (M) to support with respect to the Cordless Telephony profile, which are optional (O), and which are conditional (C).

**TABLE 7.6**

LMP Features Supported by the Cordless Telephony Profile
Source: Bluetooth Specification 1.0

| Procedure | Support in LMP | Support in Terminal | Support in Gateway |
|---|---|---|---|
| Authentication | M | | |
| Pairing | M | | |
| Change link key | M | | |
| Change current link key | M | | |
| Encryption | O | M | M |
| Clock offset request | M | | |
| Slot offset information | O | | |
| Timing accuracy information request | O | | |
| LMP version | M | | |
| Supported features | M | | |
| Switch of master-slave role | O | M | C |
| Name request | M | | |
| Detach | M | | |
| Hold mode | O | | |
| Sniff mode | O | | |
| Park mode | O | M | M |
| Power control | O | | |
| Channel-quality driven DM/DH* | O | | |
| Quality of service | M | | |
| SCO links | O | M | M |
| Control of multi-slot packets | O | | |
| Paging scheme | O | | |
| Link supervision | M | | |
| Connection establishment | M | | |

* DM/DH = Data, Medium-rate/Data, High-rate

## Link Control Features

Table 7.7 lists all features available on the Link Control (LC) level, including which are mandatory (M), which are optional (O), which are conditional (C), and which are specifically excluded (X) with respect to the Cordless Telephony Profile.

**TABLE 7.7**
Summary of Link Control (LC) Features
Source: Bluetooth Specification 1.0

| | Procedure | | Support in Terminal | Support in Gateway |
|---|---|---|---|---|
| 1 | Inquiry | | | X |
| 2 | Inquiry scan | | X | |
| 3 | Paging | | | X |
| 4 | Page scan | | | |
| | a | Type R0 | | |
| | b | Type R1 | | |
| | c | Type R2 | | |
| 5 | Packet types | | | |
| | a | ID packet | | |
| | b | NULL packet | | |
| | c | POLL packet | | |
| | d | FHS packet | | |
| | e | DM1 packet | | |
| | f | DH1 packet | | |
| | g | DM3 packet | | |
| | h | DH3 packet | | |
| | i | DM5 packet | | |
| | j | DH5 packet | | |
| | k | AUX packet | X | X |
| | l | HV1 packet | | |
| | m | HV2 packet | | |
| | n | HV3 packet | M | M |

*continued on next page*

# Bluetooth Profiles for Usage Models

**TABLE 7.7**

Summary of Link Control (LC) Features
Source: Bluetooth Specification 1.0 (continued)

| | Procedure | Support in Terminal | Support in Gateway |
|---|---|---|---|
| | o    DV packet | X | X |
| 6 | Inter-piconet capabilities | O | C |
| 7 | Voice codec | | |
| | a    PCM (A-law) | | |
| | b    PCM (µ-law) | | |
| | c    CVSD | M | M |

## GAP Compliance

The Cordless Telephony Profile requires compliance with the Generic Access Profile (GAP) with regard to modes of operation, security, idle mode procedures, and bonding.

### MODES

Table 7.8 shows the support status for modes within the Cordless Telephony Profile.

**TABLE 7.8**

Support Status for Operational Modes Within the Cordless Telephony Profile
Source: Bluetooth Specification 1.0

| | Procedure | Support in Terminal | Support in Gateway |
|---|---|---|---|
| 1 | Discoverability modes | | |
| | Non-discoverable mode | N/A | M |
| | Limited discoverable mode | N/A | O |
| | General discoverable mode | N/A | M |
| 2 | Connectability modes | | |
| | Non-connectable mode | N/A | X |
| | Connectable mode | N/A | M |
| 3 | Pairing modes | | |
| | Non-pairable mode | M | M |
| | Pairable mode | O | M |

## SECURITY

The Generic Access Profile specifies authentication and three security modes for a device. Support for authentication is mandatory for both the gateway and terminals operating within the Cordless Telephony Profile. The security modes are:

- **Security mode 1** (non-secure)—A device will not initiate any security procedure.
- **Security mode 2** (service-level enforced security)—A device does not initiate security procedures before channel establishment at the L2CAP level. This mode allows different and flexible access policies for applications, especially those running in parallel and requiring different levels of security.
- **Security mode 3** (link-level enforced security)—A device initiates security procedures before link setup at the LMP level is completed.

Table 7.9 summarizes the support status for the security aspects of the Cordless Telephony Profile.

**TABLE 7.9** Security Aspects of the Cordless Telephony Profile Source: Bluetooth Specification 1.0

| | Procedure | Support in Terminal | Support in Gateway |
|---|---|---|---|
| 1 | Authentication | M | M |
| 2 | Security modes | | |
| | Security mode 1 | X | X |
| | Security mode 2 | C | C |
| | Security mode 3 | C | C |

*Note:* Support for at least one security mode (2 or 3) is mandatory.

## IDLE MODE PROCEDURES

The Generic Access Profile specifies the Idle mode procedures used within the Cordless Telephony Profile. It is mandatory for the terminal to support initiation of bonding, and for the gateway to accept bonding. Table 7.10 summarizes the support status for the Idle mode procedures used within the Cordless Telephony Profile.

**TABLE 7.10**

Summary of Idle Mode Procedures Used in the Cordless Telephony Profile
Source: Bluetooth Specification 1.0

| Procedure | Support in Terminal | Support in Gateway |
|---|---|---|
| General inquiry | M | N/A |
| Limited inquiry | O | N/A |
| Name discovery | O | N/A |
| Device discovery | O | N/A |
| Bonding | M | M |

## Headset Profile

The Headset Profile defines the protocols and procedures for the usage model known as "Ultimate Headset," which can be implemented by such devices as cellular phones and personal computers. The headset can act as a device's audio input and output interface for the purpose of increasing the user's freedom of movement while maintaining call privacy. The headset must be able to send AT-commands and receive result codes. This ability allows the headset to answer incoming calls and then terminate them without the user having to physically manipulate the telephone handset.

Figure 7.8 shows the dependence of the Headset Profile on both the Serial Port Profile and the Generic Access Profile. Figure 7.9 shows the protocols and entities used in the Headset Profile.

With regard to Figure 7.9, baseband corresponds to OSI Layer 1, while LMP and L2CAP correspond to OSI Layer 2. RFCOMM is the Bluetooth specification's adaptation of GSM TS 07.10 for serial port emulation and SDP is the Bluetooth Service Discovery Protocol. For all of these protocols/entities, the Serial Port Profile is the base standard and all requirements stated in the Serial Port Profile apply, except where the Headset Profile explicitly states deviations.

Headset Control is the entity responsible for headset-specific control signaling and is based on AT commands. It is assumed in this profile that Headset Control has access to some lower-layer procedures, such as Synchronous Connection-Oriented (SCO) link establishment. The audio port emulation layer is the entity that emulates the audio port on the cellular phone or PC, and the audio driver is the driver software in the headset.

**Figure 7.8**
The Headset Profile is dependent on both the Serial Port Profile and the Generic Access Profile.
Source: Bluetooth Specification 1.0.

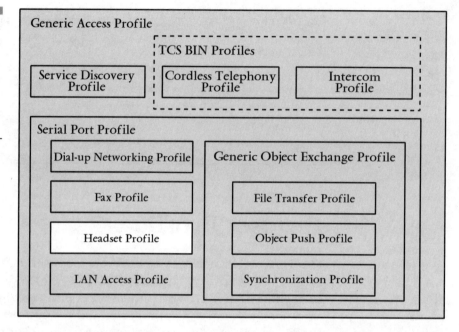

There are two device roles defined for the Headset Profile. The Audio Gateway (AG) is the gateway of the audio, both for input and output. Typical devices acting as Audio Gateways are cellular phones and PCs. The Headset (HS) acts as the Audio Gateway's remote audio input and output mechanism. The Headset Profile requires that both devices support SCO links.

**Figure 7.9**
Protocols and entities used in the Headset Profile.
Source: Bluetooth Specification 1.0.

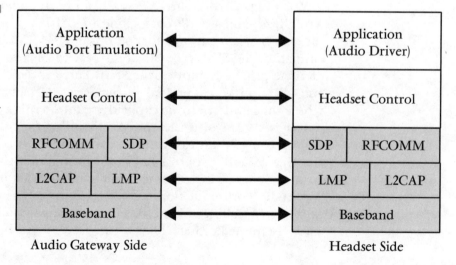

## Profile Restrictions

The Bluetooth specification describes the following restrictions that apply to the Headset Profile:

- The ultimate headset is assumed to be the only use case active between the two devices.
- The transmission of audio is based on Continuously Variable Slope Delta (CVSD) modulation. The result is monophonic audio of a quality that normally will not have perceived audio degradation.
- Only one audio connection at a time is supported between the headset and audio gateway.
- The audio gateway controls SCO link establishment and release. The headset directly connects and disconnects the internal audio streams upon SCO link establishment and release. Once the link is established, valid speech exists on the SCO link in both directions.
- The Headset Profile offers only basic interoperability such that the handling of multiple calls at the audio gateway is not supported.
- It is assumed that the headset's user interface can detect a user-initiated action, such as the pressing of a button.

## Basic Operation

A headset may be able to use the services of an audio gateway without a secure connection; it is left to the user to enforce security on devices that support authentication and/or encryption. If baseband authentication and/or encryption are to be used, the two devices have to create a secure connection using the GAP authentication procedure. This procedure may include entering a PIN code and the creation of link keys. Since the headset will usually be a device with a limited user interface, the fixed PIN of the headset will most likely be used during the GAP authentication procedure.

A link has been established when a call is initiated or received. This requires paging of the other device but, as an option, it may require unparking. No fixed master-slave roles apply to the headset or audio gateway. The audio gateway and headset provide serial port emulation via RFCOMM. Serial port emulation is used to transport the user data, modem control signals, and AT commands from the headset to the audio gateway. AT commands are then parsed by the audio gateway and responses are sent back to the headset.

## Features

Table 7.11 summarizes the features available in Bluetooth units that comply with the Headset Profile, indicating which are mandatory (M) and which are optional (O). The terms "incoming" and "outgoing" are from the perspective of the headset (HS).

**TABLE 7.11**

Summary of Features Available in Bluetooth Units that are Compliant with the Headset Profile
Source: Bluetooth Specification 1.0

| Feature | Support in Headset | Support in Audio Gateway |
| --- | --- | --- |
| Incoming audio connection | M | M |
| Outgoing audio connection | M | O |
| Audio connection transfer | M | M |
| Remote audio volume control | O | O |

### INCOMING AUDIO CONNECTION

To initiate connection establishment, the audio gateway responds to an internal event or remote user request. When the connection is established, the audio gateway sends one or more AT-based RING indications to alert the local user. As an option, the audio gateway may provide an in-band ringing tone, in which case SCO link establishment takes place first. The tone is used to alert the user through the earpiece when the user is wearing the headset. This process is summarized in Figure 7.10.

To accept the incoming audio connection, the user presses a button on the headset, upon which the headset sends an AT-based keypad control (AT+CKPD) command to the audio gateway. The gateway then establishes the SCO link, if one is not already established.

### OUTGOING AUDIO CONNECTION

To initiate an outgoing audio connection, the user pushes a button on the headset. The headset initiates connection establishment, and sends an AT+CKPD command to the audio gateway. Various other internal processes may be needed at the audio gateway to establish and/or route an audio stream to the headset. With a cellular phone, for example, a cellular call may have to be established to the external network using the last dialed number or a preprogrammed number. With a personal computer, an audio file may have to be called or an audio

## Bluetooth Profiles for Usage Models

CD track selected for playback. The audio gateway is responsible for establishing the SCO link for the outgoing audio connection.

With regard to audio connection release, a call can be terminated either on the headset when a button is pushed or on the audio gateway when some internal action or user intervention occurs. Regardless of what side initiates connection release, the audio gateway has the responsibility for actually releasing the connection.

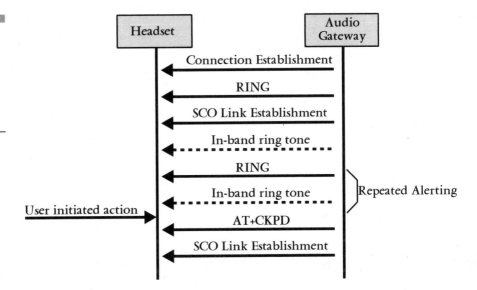

**Figure 7.10**
Sequence of events leading to an incoming audio connection.
Source: Bluetooth Specification 1.0.

### AUDIO CONNECTION TRANSFER

An audio connection can be transferred from the audio gateway to the headset or from the headset to the audio gateway. The connection is transferred to the device initiating the transfer. The audio connection transfer from the audio gateway to the headset is initiated by a user action on the headset, which results in an AT+CKPD command being sent to the audio gateway. The audio connection transfer from headset to audio gateway is initiated by a user action at the audio gateway.

### REMOTE AUDIO VOLUME CONTROL

As an option, the audio gateway can control the volume of the microphone and speaker of the headset by sending special AT-based codes during an active audio connection.

## CONNECTION HANDLING

The Bluetooth layers below the Headset Control entity (refer to Figure 7.9) are used to establish and release a connection. When park mode is not supported, either the headset or the audio gateway can initiate connection establishment. However, if there is no RFCOMM session between the two devices, the initiating device must first initialize RFCOMM. Connection establishment is performed per the Generic Access Profile (GAP) and Serial Port Profile (SPP). When the audio connection is released, the connection may also be released. The audio gateway always initiates connection release.

If park mode is supported, the connection is established on the first request for an audio connection. Later, when an audio connection is required, the parked device is unparked. (In this profile, connection establishment is referred to as initial connection establishment, while unparking is referred to as connection establishment.) Initial connection establishment is performed according to GAP and SPP. Both sides may initiate initial connection establishment, after which park mode is activated.

## Link Control Features

Table 7.12 lists all features available on the Link Control (LC) level, including which are mandatory (M), which are optional (O), and which are specifically excluded (X) with respect to the Headset Profile.

**TABLE 7.12** Summary of Link Control (LC) Features Supported in Headset Profile Source: Bluetooth Specification 1.0

| | Procedure | Support in Baseband | Support in Audio Gateway | Support in Headset |
|---|---|---|---|---|
| 1 | Inquiry | M | | X |
| 2 | Inquiry scan | M | X | |
| 3 | Paging | M | | |
| 4 | Page scan | | | |
| a | Type R0 | M | | |
| b | Type R1 | M | | |
| c | Type R2 | M | | |
| 7 | Voice codec | | | |
| c | CVSD | O | M | M |

# Bluetooth Profiles for Usage Models

## GAP Compliance

The Headset Profile requires compliance with the Generic Access Profile (GAP) with regard to modes of operation, idle mode procedures, and bonding.

### MODES

Table 7.13 shows the support status for modes within the Headset Profile, including which are mandatory (M) and which are optional (O) with respect to the Headset Profile.

**TABLE 7.13**

Support Status for Operational Modes Within the Headset Profile
Source: Bluetooth Specification 1.0

| | Procedure | Support in Headset | Support in Audio Gateway |
|---|---|---|---|
| 1 | Discoverability modes | | |
| | Non-discoverable mode | M | N/A |
| | Limited discoverable mode | O | N/A |
| | General discoverable mode | M | N/A |
| 2 | Connectability modes | | |
| | Non-connectable mode | N/A | N/A |
| | Connectable mode | M | M |
| 3 | Pairing modes | | |
| | Non-pairable mode | O | O |
| | Pairable mode | O | O |

### IDLE MODE PROCEDURES

The Generic Access Profile specifies the Idle mode procedures used within the Head Profile. It is mandatory for the audio gateway to support initiation of bonding, and for the headset to accept bonding. Table 7.14 summarizes the support status for the Idle mode procedures used within the Headset Profile.

**TABLE 7.14**

Summary of Idle Mode Procedures Used in the Headset Profile Source: Bluetooth Specification 1.0

| Procedure | Support in Headset | Support in Audio Gateway |
|---|---|---|
| General inquiry | N/A | M |
| Limited inquiry | N/A | O |
| Name discovery | N/A | O |
| Device Discovery | N/A | O |
| Bonding | M | M |

## Dialup Networking Profile

The Dialup Networking Profile defines the protocols and procedures used by devices such as modems and cellular phones for implementing the usage model called "Internet Bridge." Among the possible scenarios for this model are the use of a cellular phone as a wireless modem for connecting a computer to a dialup Internet access server, or the use of a cellular phone or modem by a computer to receive data.

Figure 7.11 shows the dependence of the Dialup Networking Profile on both the Serial Port Profile and the Generic Access Profile. Figure 7.12 shows the protocols and entities used in the Dialup Networking Profile.

With regard to Figure 7.12, baseband corresponds to OSI Layer 1, while LMP and L2CAP correspond to OSI Layer 2. RFCOMM is the Bluetooth specification's adaptation of GSM TS 07.10 and SDP is the Bluetooth Service Discovery Protocol. Dialing and control are the commands and procedures used for automatic dialing and control over the asynchronous serial link provided by the lower layers.

The modem emulation layer is the entity emulating the modem and the modem driver is the driver software in the data terminal. For all these protocols/entities, the Serial Port Profile is the base standard and all requirements stated in the Serial Port Profile apply, except where the Dialup Networking Profile explicitly states deviations. It is assumed by this profile that the application layer has access to some lower-layer procedures, such as Synchronous Connection-Oriented (SCO) link establishment.

There are two device roles defined for the Dialup Networking Profile: the Gateway (GW) is the device that provides access to the public

# Bluetooth Profiles for Usage Models

network. Typical devices that can act as gateways include cellular phones and modems. The Data Terminal (DT) is the device that uses the dialup services of the gateway. Typical devices that act as data terminals are laptops and desktop PCs.

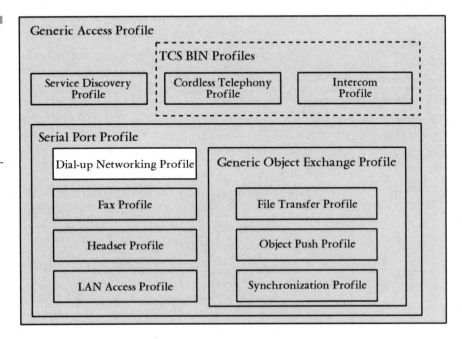

**Figure 7.11**
The Dialup Networking Profile is dependent on both the Serial Port Profile and the Generic Access Profile.
Source: Bluetooth Specification 1.0.

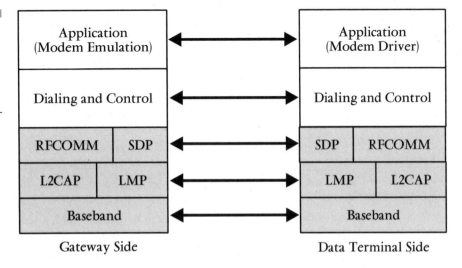

**Figure 7.12**
Protocols and entities used in the Dialup Networking Profile.
Source: Bluetooth Specification 1.0.

## Profile Restrictions

The Bluetooth specification describes the following restrictions that apply to the Dialup Networking Profile:

- For incoming calls, the modem is not required to have the capability to report and/or discriminate between different call types.
- Support is provided for one-slot packets only, ensuring that data rates of up to 128 Kbps can be used. Support for higher rates is optional.
- Only one call at a time is supported.
- Only point-to-point configurations are supported.
- There is no way to discriminate between two SCO channels originating from the same device. (The manufacturer decides how to deal with this situation.)
- Before a cell phone or modem can be used with a PC/laptop for the first time, an initialization procedure must be performed; this typically involves manually activating initialization support and entering a PIN code from the PC/laptop.
- There is no support for multiple instances of the implementation in the same device.

## Basic Operation

Before the services of a gateway can be used for the first time by a data terminal, the two devices have to initialize, which entails the exchange of a PIN code, creation of link keys, and implementation of service discovery. A link has to be established before calls can be initiated or received, and this requires paging of the other device. It is the responsibility of the data terminal to initiate link establishment. There are no fixed master-slave roles among the gateway and data terminals.

The gateway and data terminal provide serial port emulation using the Serial Port Profile. Serial port emulation is used to transport the user data, modem control signals, and AT commands between the gateway and the data terminal. AT commands are then parsed by the gateway and responses are sent back to the data terminal. An SCO link is used for data transport, as well as optional audio feedback during call establishment.

For authentication and encryption, baseband and LMP mechanisms are used, per the Generic Access Profile. Security is accomplished by authenticating the other party upon connection establishment and by encrypting all user data for transmission.

## Services

Table 7.15 summarizes the services available in Bluetooth units that comply with the Dialup Networking Profile, indicating which are mandatory (M) and which are optional (O).

**TABLE 7.15**

Summary of Services Available in Bluetooth Units that are Compliant with the Dialup Networking Profile Source: Bluetooth Specification 1.0

| Service | Support in Data Terminal | Support in Gateway |
|---|---|---|
| Data call without audio feedback | M | M |
| Data call with audio feedback | O | O |
| Fax services without audio feedback | N/A | N/A |
| Fax services with audio feedback | N/A | N/A |
| Voice call | N/A | N/A |

Support for data calls is mandatory from both gateways and data terminals. As an option, audio feedback may be provided as well. The gateway emulates a modem connected via a serial port. The Serial Port Profile is used for RS-232 emulation, and a modem emulation entity running on top of the Serial Port Profile provides the modem emulation. Support for fax transmissions is not covered in the Dialup Networking Profile, but is covered in the Fax Profile, which is discussed later. Support for voice calls is not covered in the Dialup Networking Profile, but is covered under the Cordless Telephony Profile.

## Gateway Commands

The gateway device must be able to support the commands listed in Table 7.16 to ensure that basic functionality can always be provided.

The gateway device must also be able to support the responses listed in Table 7.17 to ensure that basic functionality can always be provided.

**TABLE 7.16**

Required Commands Supported by the Gateway in the Dialup Networking Profile
Source: Bluetooth Specification 1.0

| Command | Function |
|---|---|
| &C | Receive line signal detection |
| &D | Data terminal ready |
| &F | Set to factory-defined configuration |
| +GCAP | Request complete capabilities list |
| +GMI | Request manufacturer identification |
| +GMM | Request model identification |
| +GMR | Request revision identification |
| A | Answer |
| D | Dial |
| E | Echo |
| H | Hook control |
| L | Monitor speaker loudness |
| M | Monitor speaker mode |
| O | Return to online data state |
| P | Select pulse dialing |
| Q | Result code suppression |
| S0 | Automatic answer |
| S10 | Automatic disconnect delay |
| S3 | Command line termination character |
| S4 | Response formatting character |
| S5 | Command line editing character |
| S6 | Pause before blind dialing |
| S7 | Connection completion timeout |
| S8 | Comma dial modifier time |
| T | Select tone dialing |
| V | DCE response format |
| X | Result code selection and call progress monitoring control |
| Z | Reset to default configuration |

# Bluetooth Profiles for Usage Models

**TABLE 7.17**
Required Responses Supported by the Gateway in the Dialup Networking Profile
Source: Bluetooth Specification 1.0

| Response | Description |
|---|---|
| OK | Acknowledges execution of a command |
| CONNECT | Connection has been established |
| RING | The DCE has detected an incoming call signal from the network |
| NO CARRIER | The connection has been terminated, or the attempt to establish a connection failed |
| ERROR | Error |
| NO DIALTONE | No dial tone detected |
| BUSY | Busy signal detected |

## Audio Feedback

As an option, the gateway or data terminal may provide audio feedback during call establishment. SCO links are used to transport the digitized audio over the Bluetooth link. The gateway takes the initiative for SCO link establishment. The setting of the M parameter (see Table 7.16) controls whether audio feedback is provided by the gateway. If a gateway provides audio feedback for a call, it initiates the SCO Link procedure per the Link Manager Protocol to establish the audio link when the DCE goes off hook.

The gateway releases the audio link when the DCE has detected a carrier or when the DCE goes on hook. The Remove SCO Link procedure is used for audio link release, per the Link Manager Protocol. If SCO link establishment fails, call establishment proceeds without the audio feedback.

The data terminal must not be active in any other profile that uses SCO links while it is operating in the Dialup Networking Profile. Consequently, the behavior in a situation where multiple SCO links are established simultaneously is undefined in the Bluetooth specification. As noted, the Dialup Networking Profile leaves it up to the device manufacturer how to deal with this situation.

## Service Discovery Procedures

Table 7.18 lists all entries in the SDP database of the gateway, as defined by the Dialup Networking Profile, indicating whether the presence of the field is mandatory (M) or optional (O).

**TABLE 7.18** Entries in the Gateway's Service Discovery Database per the Dialup Networking Profile Source: Bluetooth Specification 1.0

| Item | Value | Status |
|---|---|---|
| Service Class ID List | | M |
| Service Class #0 | Generic Networking | O |
| Service Class #1 | Dialup Networking | M |
| Protocol Descriptor List | | M |
| Protocol #0 | L2CAP | M |
| Protocol #1 | RFCOMM | M |

## Link Control Features

Table 7.19 lists all features available on the Link Control (LC) level, including which are optional (O) and which are conditional (C) with respect to the Dialup Networking Profile.

**TABLE 7.19** Summary of Link Control (LC) Features Supported in Dialup Networking Profile Source: Bluetooth Specification 1.0

| | Procedure | Support in Baseband | Support in Gateway | Support in Data Terminal |
|---|---|---|---|---|
| 5 | Packet types | | | |
| n | HV3 packet | O | C | C |
| 7 | Voice codec | | | |
| c | CVSD | O | C | C |

## GAP Compliance

The Dialup Networking Profile requires compliance with the Generic Access Profile (GAP) with regard to modes of operation, security, idle mode procedures, and bonding.

# Bluetooth Profiles for Usage Models

## MODES

Table 7.20 shows the support status for Modes within the Dialup Networking Profile, including which are mandatory (M), which are optional (O), and which are specifically excluded (X) with respect to the Dialup Networking Profile.

**TABLE 7.20**

Support Status for Operational Modes within the Dialup Networking Profile Source: Bluetooth Specification 1.0

| | Procedure | Support in Data Terminal | Support in Gateway |
|---|---|---|---|
| 1 | Discoverability modes | | |
| | Non-discoverable mode | N/A | M |
| | Limited discoverable mode | N/A | O |
| | General discoverable mode | N/A | M |
| 2 | Connectability modes | | |
| | Non-connectable mode | N/A | X |
| | Connectable mode | N/A | M |
| 3 | Pairing modes | | |
| | Non-pairable mode | M | M |
| | Pairable mode | O | M |

## SECURITY

The Generic Access Profile specifies authentication and three security modes for a device. Support for authentication is mandatory for both the gateway and terminals operating within the Dialup Networking Profile. Table 7.21 summarizes the support status for the security aspects of the Dialup Networking Profile. An explanation of the security modes is provided within the discussion of the Cordless Telephony Profile.

## IDLE MODE PROCEDURES

The Generic Access Profile specifies the Idle mode procedures used within the Dialup Networking Profile. It is mandatory for the terminal to support initiation of bonding, and for the gateway to accept bonding. The support status for the Idle mode procedures used within the Dialup Networking Profile is the same as that for the Cordless Telephony Profile.

**TABLE 7.21**

Security Aspects of the Dialup Networking Profile
Source: Bluetooth Specification 1.0

|   | Procedure       | Support in Terminal | Support in Gateway |
|---|-----------------|---------------------|--------------------|
| 1 | Authentication  | M                   | M                  |
| 2 | Security modes  |                     |                    |
|   | Security mode 1 | N/A                 | X                  |
|   | Security mode 2 | C                   | C                  |
|   | Security mode 3 | C                   | C                  |

*Note: Support for at least one security mode (2 or 3) is mandatory.*

## Fax Profile

The Fax Profile defines the protocols and procedures used by devices implementing the fax part of the usage model called "Data Access Points, Wide Area Networks." A cellular phone or modem using Bluetooth wireless technology may be used by a computer as a wireless fax-modem to send or receive fax messages. As indicated in Figure 7.13, the Fax Profile is dependent upon both the Serial Port Profile and the Generic Access Profile. Figure 7.14 shows the protocols and entities used in the Fax Profile.

**Figure 7.13**
The Fax Profile is dependent on both the Serial Port Profile and the Generic Access Profile.
Source: Bluetooth Specification 1.0.

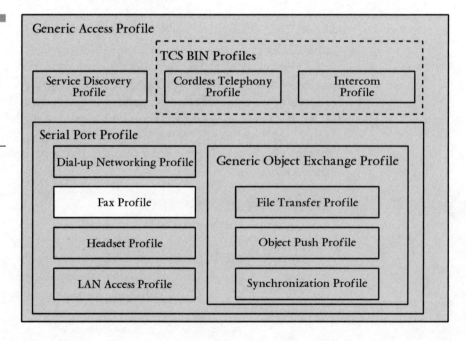

# Bluetooth Profiles for Usage Models

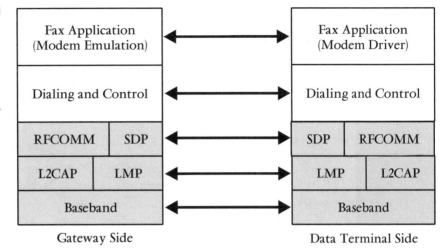

**Figure 7.14**
Protocols and entities used in the Fax Profile.
Source: Bluetooth Specification 1.0.

With regard to Figure 7.14, baseband corresponds to OSI Layer 1, while LMP and L2CAP correspond to OSI Layer 2. RFCOMM is the Bluetooth specification's adaptation of GSM TS 07.10 for providing serial port emulation and SDP is the Bluetooth Service Discovery Protocol. For all these protocols/entities, the Serial Port Profile is the base standard and all requirements stated in the Serial Port Profile apply, except where the Fax Profile explicitly states deviations.

The dialing and control layer defines the commands and procedures for automatic dialing and control for the asynchronous serial link provided by the lower layers. The modem emulation layer is the entity responsible for emulating the modem, and the modem driver is the driver software in the data terminal. It is assumed by this profile that the application layer has access to some lower-layer procedures, such as Synchronous Connection-Oriented (SCO) link establishment.

The two device roles defined for the Fax Profile are the same as those of the Dialup Networking Profile. The gateway (GW) is the device that provides facsimile services. Typical devices that can act as gateways are cellular phones and modems. The data terminal (DT) is the device that uses the facsimile services of the gateway. Typical devices acting as data terminals are laptops and desktop PCs.

## Profile Restrictions

As described in the Bluetooth specification, the following restrictions apply to the Fax Profile:

- For incoming calls, the gateway (cell phone or modem) is not required to have the capability to report and/or discriminate between different call types.
- Support is provided for one-slot packets only, ensuring that data rates of up to 128 Kbps can be used. Support for higher rates is optional.
- Only one call at a time is supported.
- Only point-to-point configurations are supported.
- There is no way to discriminate between two SCO channels originating from the same device. (The manufacturer must decide how to deal with this situation.)
- There is no support for multiple instances of the implementation in the same device.

The support of data calls is not covered by this profile; however, under the Dialup Networking Profile, support for data calls is mandatory from both the gateway and data terminals. The support of voice calls also is not covered by this profile; however, under the Cordless Telephony Profile, support for voice calls is mandatory from both gateway and terminals.

## Basic Operation

Although there are no fixed master-slave roles defined for devices in the Fax Profile, link establishment is always initiated by the data terminal. When the data terminal needs to use the facsimile services of a gateway and it does not have the Bluetooth address of the gateway, it must be obtained through the device discovery procedure described in the Generic Access Profile.

The Fax Profile specifies the use of a secure connection of all user data via authentication and encryption through the baseband/LMP encryption mechanisms described in the Generic Access Profile.

When the fax call is established, the gateway and data terminal provide serial port emulation via the Serial Port Profile. The gateway emulates a modem connected via a serial port. The Serial Port Profile is used for RS-232 emulation, and RFCOMM running on top of the serial port profile provides the modem emulation. The serial port emulation is used to transport the user data, modem control signals, and AT commands between the gateway and data terminal. AT commands are then parsed by the gateway and responses sent back to the

# Bluetooth Profiles for Usage Models

data terminal. An optional SCO link may be used to transport audio feedback. After the fax call has been cleared, the channel and link will be released as well.

## Services

Table 7.22 summarizes the services available in Bluetooth units that comply with the Fax Profile, indicating which are mandatory (M) and which are optional (O).

**TABLE 7.22** Summary of Services Available in Bluetooth Units that are Compliant with the Fax Profile Source: Bluetooth Specification 1.0

| Service | Support in Data Terminal | Support in Gateway |
|---|---|---|
| Data call without audio feedback | N/A | N/A |
| Data call with audio feedback | N/A | N/A |
| Fax services without audio feedback | M | M |
| Fax services with audio feedback | O | O |
| Voice call | N/A | N/A |

## Gateway Commands

Although the Fax Profile does not require a specific class of fax, the gateway device must be able to support the commands and responses as defined in the following supported fax classes:

- Fax Class 1 TIA-578-A and ITU T.31
- Fax Class 2.0 TIA-592 and ITU T.32
- Fax Service Class 2 (manufacturer specific)

Bluetooth devices implementing this profile must support a minimum of one fax service class, but may support any or all fax service classes. The data terminal will check the gateway via the Service Discovery Protocol (SDP) or perform an AT+FCLASS command to discover the fax class of service supported by the gateway. The required commands, protocols, and result codes used are the same as described in the TIA/ITU standards documents for each fax service class.

## Audio Feedback

The gateway or data terminal may optionally provide audio feedback during call establishment. The procedure for handling audio feedback under the Fax Profile is exactly the same as described for the Dialup Networking Profile, discussed previously.

## Service Discovery Procedures

Table 7.23 lists all entries in the SDP database of the gateway, as defined by the Fax Profile, indicating whether the presence of the field is mandatory (M) or optional (O).

**TABLE 7.23**

Entries in the Gateway's Service Discovery Database per the Fax Profile Source: Bluetooth Specification 1.0

| Item | Value | Status |
|---|---|---|
| Service Class ID List | | M |
| Service Class #0 | Generic telephony | O |
| Service Class #1 | Facsimile | M |
| Protocol Descriptor List | | M |
| Protocol #0 | L2CAP | M |
| Protocol #1 | RFCOMM | M |
| Parameter for Protocol #1 | Server channel number | M |

## Link Control Features

With respect to the Fax Profile, all features available on the Link Control (LC) level, including which are optional (O) and which are conditional (C) for the baseband, gateway, and data terminal, are the same as those described in the Dialup Networking Profile (refer to Table 7.19).

## GAP Compliance

The Fax Profile requires compliance with the Generic Access Profile (GAP) with regard to modes of operation, security, idle mode procedures, and bonding.

### MODES

Support status for modes within the Fax Profile is exactly the same as that within the Dialup Networking Profile (refer to Table 7.20), including which are mandatory (M), which are optional (O), and which are specifically excluded (X).

### SECURITY

Like the Dialup Networking Profile, the Fax Profile draws upon the Generic Access Profile for the authentication and three security modes used by both the gateway and terminals (refer to Table 7.21).

### IDLE MODE PROCEDURES

The Generic Access Profile specifies the Idle mode procedures used within the Fax Profile. It is mandatory for the data terminal to support initiation of bonding, and for the gateway to accept bonding. The support status for the Idle mode procedures used within the Fax Profile is the same as that for the Cordless Telephony Profile (refer to Table 7.10).

## LAN Access Profile

The LAN Access Profile defines how devices using Bluetooth wireless technology can access the services of a LAN using the Point-to-Point Protocol (PPP) over RFCOMM and use the same PPP mechanisms to network two devices using Bluetooth wireless technology. In this usage model, multiple data terminals (DTs) use a LAN access point (LAP) as a wireless connection to a local area network (LAN). Once connected, the data terminals operate as if it they are connected to the LAN via dialup networking and can access all the services provided by the LAN.

PPP is an Internet Engineering Task Force (IETF) standard widely deployed as a means of allowing access to networks. It provides authentication, encryption, data compression, and support for multiple protocols. Although PPP is capable of supporting various networking protocols (e.g. IP, IPX, etc.), the LAN Access Profile does not require the use of any particular protocol. The LAN Access Profile merely defines how PPP is supported to provide LAN access for a single Bluetooth device, LAN access for multiple Bluetooth devices, and PC-to-PC communication using PPP networking over serial cable emulation.

As shown in Figure 7.15, the LAN Access Profile is dependent upon both the Serial Port Profile and the Generic Access Profile. Figure 7.16 shows the protocols and entities used in the LAN Access Profile.

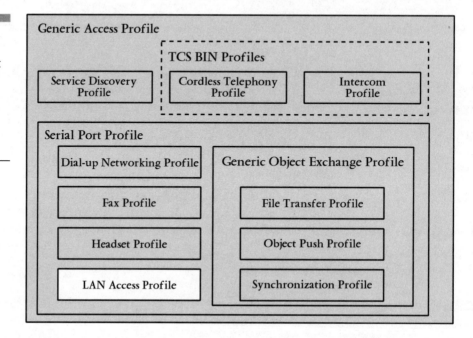

**Figure 7.15**
The LAN Access Profile is dependent on both the Serial Port Profile and the Generic Access Profile.
Source: Bluetooth Specification 1.0.

With regard to Figure 7.15, baseband corresponds to OSI Layer 1, while LMP and L2CAP correspond to OSI Layer 2. RFCOMM is the Bluetooth specification's adaptation of GSM TS 07.10 and SDP is the Bluetooth Service Discovery Protocol. In this profile there is a Management Entity (ME) that coordinates procedures during initialization, configuration, and connection management. PPP Networking is the means of taking IP packets to/from the PPP layer and putting them onto the LAN. The specific mechanism that does this is not defined in the LAN Access Profile, but is a well-understood feature of Remote Access Server (RAS) products.

The LAN Access Profile specifies two device roles: the LAN Access Point (LAP) and the data terminal (DT). The LAP provides access to such networks as Ethernet, Token Ring and Fibre Channel, as well as to Cable Modem, Firewire, USB, and Home Networking products. The LAP provides the services of a PPP Server. The PPP connection is carried over RFCOMM, which is used to transport the PPP packets and provide flow control of the PPP data stream.

# Bluetooth Profiles for Usage Models

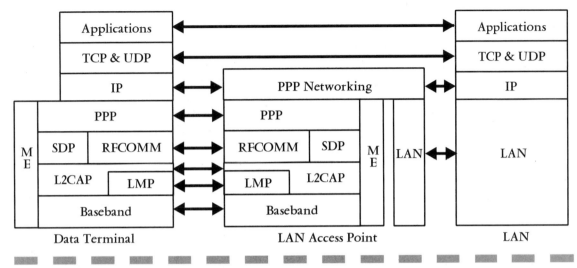

**Figure 7.16** Protocols and entities used in the LAN Access Profile. Source: Bluetooth Specification 1.0.

The data terminal is the device that uses the services of the LAP. Typical devices that can act as data terminals are laptops, notebooks, desktop PCs, and PDAs. The data terminal is a PPP client. As such, it establishes a PPP connection with a LAP for the purpose of gaining access to a LAN. This profile assumes that the LAP and the DT each have a single Bluetooth radio.

## Profile Restrictions

As described in the Bluetooth specification, the following restrictions apply to the LAN Access Profile:

- A single data terminal can use a LAP for a wireless connection to a LAN, and once connected, will operate as if it were connected to the LAN via dialup networking. The data terminal can then access all the services provided by the LAN.
- Multiple data terminals can use a LAP for a wireless connection to a LAN, and once connected, will operate as if they were connected to the LAN via dialup networking. The data terminals can access all the services provided by the LAN. The data terminals can also communicate with each other via the LAP.
- Two Bluetooth devices can communicate with each other via a PC-to-PC connection, similar to the way a direct cable connection is

often used to connect two PCs. In this scenario, one of the devices will assume the role of a LAP, while the other will assume the role of a DT.

Some LAP products may have an internal LAN or use the Public Switched Telephone Network (PSTN) to access the global Internet or a private corporate intranet. The dialup mechanisms to achieve these connections are specific to the LAP, and the DT users may be totally unaware of these activities except by experiencing longer connection times and traffic delays.

## Basic Operation

A data terminal must find a LAP within radio range that provides PPP/RFCOMM/L2CAP service.[4] The user could use an application to find and select a suitable LAP. If there is no existing baseband physical link, then the data terminal requests a baseband physical link with the selected LAP. At some point after physical link establishment, the devices perform mutual authentication, with each device insisting that encryption be used on the link.

When the data terminal establishes a PPP/RFCOMM/L2CAP connection, the LAP may as an option, use an appropriate PPP authentication mechanism such as the Challenge Handshake Authentication Protocol (CHAP). For example, the LAP may challenge the data terminal user to authenticate himself or herself, whereupon the data terminal must then supply a username and password. If these mechanisms are used and the data terminal fails to authenticate itself, then the PPP link will be dropped.

Using the appropriate PPP mechanisms, a suitable IP address is negotiated between the LAN access point and data terminal. IP traffic can then flow across the PPP connection. At any time either the data terminal or the LAN access point may terminate the PPP connection.

---

[4] The speed of RFCOMM connections is not configurable by the user. RFCOMM transfers data as fast as it can. The actual transfer rate varies, depending on the volume of Bluetooth traffic on the baseband link. The connection speed is not set at some typical serial port value.

# Bluetooth Profiles for Usage Models

## Security

In any wireless environment, Bluetooth included, security is of paramount concern. Both the LAN access point and the data terminal enforce encryption over the baseband physical link while PPP traffic is being sent or received. Both devices will refuse any request to disable encryption. Bluetooth pairing must occur as a means of authenticating the users. For this, a PIN or link key must be supplied. Failure to complete the pairing process prevents access to the LAN access service. Depending on the manufacturer, a product may require further authentication, encryption, and/or authorization.

## GAP Compliance

With regard to modes of operation, the LAN Access Profile requires compliance with the Generic Access Profile (GAP). Table 7.24 shows the support status for Modes within the LAN Access Profile, including which are mandatory (M), which are optional (O), and which are specifically excluded (X).

**TABLE 7.24** Support Status for Operational Modes Within the LAN Access Profile Source: Bluetooth Specification 1.0

| | Procedure | Support in LAN Access Point | Support in Data Terminal |
|---|---|---|---|
| 1 | Discoverability modes | | |
| | Non-discoverable mode | O | X |
| | Limited discoverable mode | X | X |
| | General discoverable mode | M | X |
| 2 | Connectability modes | | |
| | Non-connectable mode | O | X |
| | Connectable mode | M | X |
| 3 | Pairing modes | | |
| | Non-pairable mode | O | X |
| | Pairable mode | M | X |

One use for the non-discoverable mode is when the LAP is intended for personal use only. In this case, the data terminal would remember the identity of the LAP and not need to use the Bluetooth inquiry mechanism. One use for the general discoverable mode is when the LAP is intended for general use. The data terminal would not be expected to remember the identity of all the LAPs that it uses. In this case, the data terminal is expected to use the Bluetooth inquiry mechanism to discover the LAPs within range.

In addition, there is a parameter for "maximum number of users," which is mandatory for the LAP, but as an option is configurable by the LAP administrator.

Different products have different capabilities and resource limitations that will limit the number of simultaneous users they can support. The administrator of the LAP may choose further to limit the number of simultaneous users. The fewer simultaneous users there are using a Bluetooth radio, the more bandwidth will be available to each.

A LAP can also be restricted to a single user. In this single-user mode, either the data terminal or the LAN Access Point may be the master of the piconet. There are situations in which a data terminal may want to connect to a LAP and still remain the master of an existing piconet. For example, a PC is the master of a piconet with connections to a Bluetooth mouse and a Bluetooth video projector. The PC then requires a connection to the LAP, but must remain master of the existing piconet. If, for some reason, the PC can only be a member of one piconet, then the LAP must be a piconet slave. This situation is only possible if the LAP's "maximum number of users" parameter has been configured to single-user mode.

When the "maximum number of users" parameter is set to allow more than one user to access the LAP, the LAP is always the master of the piconet. If the data terminal refuses to allow the LAP to become master, then the data terminal will not be able to access the LAN.

## Service Discovery Procedures

A LAN Access Point is capable of providing one or more services for connecting to a LAN. For example, different services could provide access to different IP subnets on the LAN. The data terminal user can choose which of the LAN access services he or she requires. When the access point provides more than one PPP/RFCOMM service, selection is based on service attributes that are made public via the Service Discovery Protocol (SDP).

# Bluetooth Profiles for Usage Models

Table 7.25 lists all entries in the SDP database of the gateway, as defined by the LAN Access Profile, indicating whether the presence of the field is mandatory (M) or optional (O).

**TABLE 7.25**

Entries in the Gateway's Service Discovery Database per the LAN Access Profile
Source: Bluetooth Specification 1.0

| Item | Value | Status |
|---|---|---|
| Service Class ID List | | M |
| Service Class #0 | LAN Access using PPP | M |
| Protocol Descriptor List | | M |
| Protocol #0 | L2CAP | M |
| Protocol #1 | RFCOMM | M |
| Parameter for Protocol #0 | Server channel number | M |
| Profile Descriptor List | | O |
| Profile #0 | LAN Access using PPP | |
| Parameter 0 | Version | |
| Service Name | Displayable name | O |
| Service Description | Displayable information | O |
| Service Availability | Load factor | O |
| IP Subnet | Displayable information | O |

## Link Control

The LAN Access Profile is built on the Serial Port Profile, which is described in Chapter 6. To discover the LAPs that may be in range, a data terminal must use the General Inquiry procedure defined in the Generic Access Profile (GAP), which is also described in Chapter 6.

## Management Entity Procedures

Link Establishment is required for communication between a LAN Access Point and a data terminal. The DT first performs a General Inquiry to discover what LAPs are within radio range, per the Generic Access Profile. Having performed the inquiry, the data terminal will have gathered a list of responses from nearby LAPs. The DT sorts the list according to some product-specific criteria. It starts with the LAP at the top of the list and tries to establish a link with it. Any error or failure to establish a link causes the DT to skip this LAP. The DT will then attempt to establish a link with the next LAP in the list. If there are no more LAPs in the list, the DT does not proceed with further link establishment procedures. At this point, link establishment must be reinitiated.

## File Transfer Profile

The File Transfer Profile supports the file transfer usage model, which offers the ability to transfer data objects from one Bluetooth device to another. These devices would typically be PCs, smart phones, or PDAs. Object types include, but are not limited to, Excel spreadsheets (.xls), PowerPoint (.ppt) presentations, audio files (.wav), image files (.jpg or .gif), and Microsoft Word files (.doc). This usage model also offers users the ability to browse the contents of folders that reside on a remote device. New folders can be created and others deleted. Entire folders, directories, or streaming media formats can be transferred between devices.

As shown in Figure 7.17, the File Transfer Profile is dependent upon both the Serial Port Profile and the Generic Access Profile, but uses the Generic Object Exchange Profile (GOEP) as a base profile to define the interoperability requirements for the protocols needed by the applications. Figure 7.18 shows the protocols and entities used in the File Transfer Profile.

With regard to Figure 7.18, baseband corresponds to OSI Layer 1, while LMP and L2CAP correspond to OSI Layer 2. RFCOMM is the Bluetooth specification's adaptation of GSM TS 07.10 and SDP is the Bluetooth Service Discovery Protocol. OBEX is the Bluetooth adaptation of the Infrared Object Exchange (IrOBEX) protocol standardized by the Infrared Data Association (IrDA).

The File Transfer Profile specifies two device roles: client and server. The client device initiates the operation, which pushes and pulls objects to and from the server. The server device is the target remote Bluetooth device that provides an object exchange server and folder browsing capability using the OBEX Folder Listing format. Servers are allowed to support read-only folders and files, enabling them to restrict folder/file deletion and creation.

# Bluetooth Profiles for Usage Models

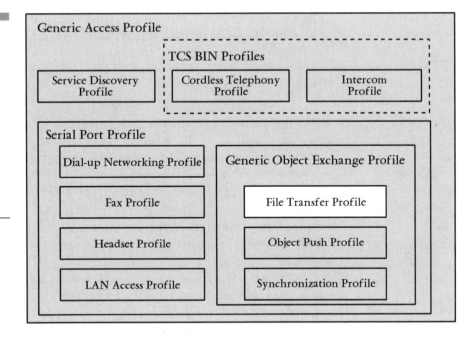

**Figure 7.17**
The File Transfer Profile is dependent on both the Serial Port Profile and the Generic Access Profile, but uses the Generic Object Exchange Profile (GOEP) as a base profile.
Source: Bluetooth Specification 1.0.

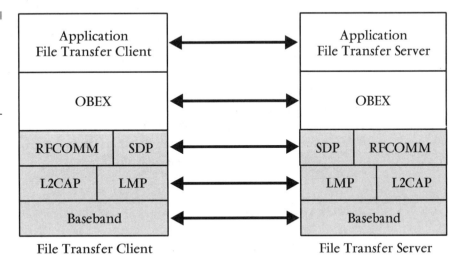

**Figure 7.18**
Protocols and entities used in the File Transfer Profile.
Source: Bluetooth Specification 1.0.

Support for link-level authentication and encryption is required, but their actual use is optional. Likewise, support for OBEX authentication is required but its use is optional. The File Transfer Profile does not mandate that the server or client enter any discoverable or connectable modes automatically, even if they can. On the client side, end-user intervention is always needed to initiate file transfer. Support of bonding is required, but its use is optional.

## Basic Operation

When the client wants to select a server the user puts the client device into File Transfer mode. A list of servers that may support the File Transfer service is displayed to the user for selection. The connection to a server may require the user to enter a password for authentication. If both link-level authentication and OBEX authentication are required, the user is prompted for two passwords. If the client requires authentication of the server, the server prompts the user for a password. After the connection is complete, including any authentication, the server's root folder is displayed.

The first presentation has the root folder selected as the current folder. The user chooses a folder to be the current folder. The contents of this folder are displayed. To push a file from the client to the server, the user selects a file on the client and activates the Push Object function. The object is transferred to the current folder on the server. To pull a file from the server, the user selects a file in the current folder of the server and activates the Pull Object function. The user is notified of the result of the operation. To delete a file on the server, the user selects the file in the server's current folder and activates the Delete Object function. The user is notified of the result of the operation. To create a new folder on the Server, the user activates the Create Folder function. This function requests a name from the user for the folder. When the operation is complete, a new folder is created in the server's current folder.

## Functions

Clients provide file transfer functions to the user via a user interface. An example of a file transfer user interface is the file-tree viewer in Microsoft Windows, which allows the user to browse through folders and files. Using such a viewer, the user can browse through and manipulate files on another PC, which appears in the network view. The following functions are available through the user interface:

### SELECT SERVER
This entails the users selecting the server from a list of possible servers, and setting up a connection to it.

### NAVIGATE FOLDERS
This provides a display of the server's folder hierarchy, including the files in the folders, and moving through the server's folder hierarchy to select the current folder. The current folder is where items are pulled and/or pushed.

### PULL OBJECT
This enables the user to copy a file or a folder from the server to the client.

### PUSH OBJECT
This enables the user to copy a file or folder from the client to the server.

### DELETE OBJECT
This enables the user to delete a file or folder on the server.

### CREATE FOLDER
This enables the user to create a new folder on the server.

## Features

The File Transfer application is divided into three main features: folder browsing, object transfer, and object manipulation. Browsing in an object store involves displaying folder contents and setting the current folder. Transferring folders requires transferring all the items stored in a folder, including other folders. The process of transferring a folder may require that new folders be created. Object manipulation includes deleting folders and files on a server, and creating new folders on a server. These application layer features are summarized in Table 7.26 which indicates whether support is mandatory (M) or optional (O) at the client and server.

**TABLE 7.26**

Application-Layer Features Available in the File Transfer Profile
Source: Bluetooth Specification 1.0

| Feature | Support in Client | Support in Server |
|---|---|---|
| Folder browsing | M | M |
| Object transfer: | | |
| File transfer | M | M |
| Folder transfer | O | O* |
| Object manipulation | O | O* |

*Although server support for folder transfer and object manipulation is optional, the server must have the ability to respond with an appropriate error code if it does not support the feature.

## OBEX Operations

There are various OBEX operations that both the client and server must support; these are implemented with their own command. The CONNECT command is used to establish the initial connection between client and server, while the DISCONNECT command takes down that connection when the file transfer application is no longer needed. The PUT command is used to transfer objects from the client to the server (i.e., push), while the GET command is used to transfer objects from the server to the client (i.e., pull). The ABORT command is used to terminate the request, while SETPATH is used to set the current folder or create a new folder. Table 7.27 summarizes the OBEX operations available in the File Transfer Profile.

**TABLE 7.27**

OBEX Operations Available in the File Transfer Profile
Source: Bluetooth Specification 1.0

| Operation | Client | Server |
|---|---|---|
| Connect | M | M |
| Disconnect | M | M |
| Put | M | M |
| Get | M | M |
| Abort | M | M |
| SetPath | M | M |

# Bluetooth Profiles for Usage Models

## Service Discovery Procedures

The service belonging to the File Transfer Profile is a server, which enables bidirectional generic file transfer. OBEX is used as a session protocol for this service. The following service records must be put into the Service Discovery database. Table 7.28 summarizes the information in the database, indicating whether the presence of the field is mandatory (M) or optional (O).

**TABLE 7.28**

Entries in the Service Discovery Database per the File Transfer Profile Source: Bluetooth Specification 1.0

| Item | Value | Status |
|---|---|---|
| Service Class ID List | | M |
|   Service Class #0 | OBEX File Transfer | M |
| Protocol Descriptor List | | M |
|   Protocol #0 | L2CAP | M |
|   Protocol #1 | RFCOMM | M |
|     Parameter for Protocol #0 | Server channel number (varies) | M |
|   Protocol #2 | OBEX | M |
| Service Name | Displayable text name | O |
| Bluetooth Profile Descriptor List | | O |
|   Profile #0 | OBEX File Transfer | |
|     Parameter for Protocol #0 | Profile version | |

## Object Push Profile

The Object Push Profile defines the application requirements for support of the Object Push usage model between Bluetooth devices. The profile makes use of the Generic Object Exchange Profile (GOEP) to define the interoperability requirements for the protocols needed by the applications. Among the most common devices that would use the Object Push usage model are notebook PCs, PDAs, and mobile phones.

The Object Push Profile allows a Bluetooth device to push an object to the inbox of another Bluetooth device. The object might be a business card or an appointment. The device can also pull a business card from another Bluetooth device. Two Bluetooth devices can exchange business cards with each other, in which case a push of a business card is followed by a pull of a business card.

As shown in Figure 7.19, the Object Push Profile is dependent on both the Generic Object Exchange Profile and the Serial Port Profile, but uses the Generic Object Exchange Profile (GOEP) as a base profile to define the interoperability requirements for the protocols needed by the applications. Figure 7.20 shows the protocols and entities used in the Object Push Profile.

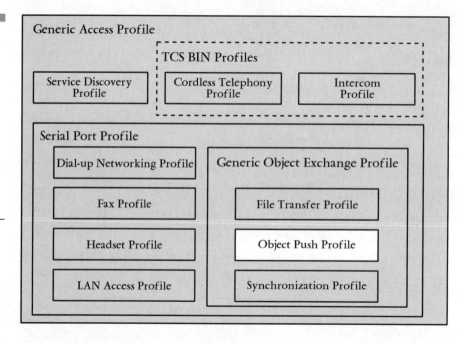

**Figure 7.19**
The Object Push Profile is dependent on both the Serial Port Profile and the Generic Access Profile, but uses the Generic Object Exchange Profile (GOEP) as a base profile.
Source: Bluetooth Specification 1.0.

With regard to Figure 7.20, baseband corresponds to OSI Layer 1, while LMP and L2CAP correspond to OSI Layer 2. RFCOMM is the Bluetooth specification's adaptation of GSM TS 07.10 and SDP is the Bluetooth Service Discovery Protocol. OBEX is the Bluetooth specification's adaptation of the Infrared Object Exchange (IrOBEX) protocol standardized by the Infrared Data Association (IrDA).

The Object Push Profile specifies two device roles: Push Server and Push Client. The Push Server is the device that provides an object exchange server. The Push Client is the client device that pushes and pulls objects to and from the Push Server.

# Bluetooth Profiles for Usage Models

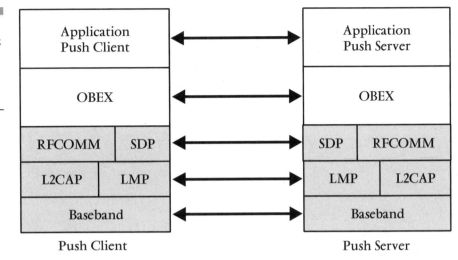

**Figure 7.20**
Protocols and entities used in the Object Push Profile.
Source: Bluetooth Specification 1.0.

In this profile, support for link-level authentication and encryption is required, but their actual use is optional. OBEX authentication is not used. The Object Push Profile does not mandate that the server or client enter any discoverable or connectable modes automatically, even if they can. On the Push Client side, end-user interaction is always needed to initiate object push, business card pull, or business card exchange. Support of bonding is required, but its use is optional.

## Functions

There are three different functions associated with the Object Push profile: Object Push, Business Card Pull, and Business Card Exchange. The Object Push function initiates the process that pushes one or more objects to a Push Server. The Business Card Pull function initiates the process that pulls the business card from a Push Server. The Business Card Exchange function initiates the process that exchanges business cards with a Push Server.

The three functions are activated by the user, and are not performed automatically without user interaction. When the user selects one of these functions, an inquiry procedure is performed to produce a list of available devices in the vicinity. The inquiry procedure used falls under the Generic Object Exchange Profile (GOEP).

## Basic Operation

When a Push Client wants to push an object to a Push Server, the user puts the device into Object Exchange mode.[5] The user selects the Object Push function on the device, and a list of Push Servers that may support the Object Push service is displayed on the client device. (If authentication is requested, the user might have to enter a Bluetooth PIN at some point.) The user then selects a Push Server to which to push the object. If the selected device does not support the Object Push service, the user is prompted to select another device. When an object is received in the Push Server, the user typically will be asked to accept or reject the object.

When a Push Client wants to pull the business card from a Push Server, the user puts the device into Object Exchange mode. The user selects the Business Card Pull function on the device, and a list of Push Servers that may support the Object Push service is displayed on the client device. (As in the previous case, if authentication is requested, the user might have to enter a Bluetooth PIN at some point.) The user then selects a Push Server from which to pull the business card. If the selected device does not support the Object Push service, the user is prompted to select another device. Some devices might ask the user whether or not to accept the request to pull the business card from the user's device.

When a Push Client wants to exchange business cards with a Push Server, the user puts the device into Object Exchange mode. The user then selects the Business Card Exchange function on the device and a list of Push Servers that may support the Object Push service is displayed to the user. (As in the previous two cases, if authentication is requested, the user might have to enter a Bluetooth PIN at some point.) The user then selects a Push Server with which to exchange business cards. If the selected device does not support the Object Push service, the user is prompted to select another device. When a Push Client tries to exchange business cards with the Push Server, the user of the Push Server may be asked to accept or reject the business card offered by the Push Client. Some devices might also ask the user

---

[5] When entering the Object Exchange mode, the Push Server sets the device in Limited Discoverable Mode, per the Generic Access Profile. Public devices that want to be visible at all times, or devices that cannot supply a user interface to enable Object Exchange mode, use General Discoverable Mode, instead of Limited Discoverable Mode. The mode will usually be set and unset by the user.

whether or not to accept the request to pull the business card from his or her device.

## Features

The Object Push Profile offers three main features: Object Push, Business Card Pull, and Business Card Exchange. These application-layer features are summarized in Table 7.29, which indicates whether support is mandatory (M) or optional (O) at the Push Client and Push Server.

**TABLE 7.29**

Application-layer Features of the Object Push Profile
Source: Bluetooth Specification 1.0

| Features | Support in Push Client | Support in Push Server |
|---|---|---|
| Object Push | M | M |
| Business Card Pull | O | O* |
| Business Card Exchange | O | O* |

*Optional, but the server must be able to respond with an error code on a pull request, even if it does not support this feature.

## Content Formats

To achieve application-level interoperability within the Object Push Profile, standardized content formats are defined for phone book, calendar, messaging, and notes applications:

- **Phone book applications**—These support data exchange using the vCard content format. Even if a phone book application supports another content format, it must still support vCard. However, if a device does not have a phone book application, it need not support vCard.
- **Calendar applications**—These support data exchange using the vCalendar content format. Even if a calendar application supports another content format, it must still support vCalendar. However, if a device does not have a calendar application, it need not support vCalendar.
- **Messaging applications**—These support data exchange using the vMessage content format. Even if a messaging application supports

another content format, it must still support vMessage. However, if a device does not have a messaging application, it need not support vMessage.

- **Notes applications**—These support data exchange using the vNote content format. Even if a notes application supports another content format, it must still support the vNote. However, if a device does not have a notes application, it need not support vNote.

A Push Client should not try to send objects of a format the Push Server does not support. The content formats supported by a Push Server are identified in the Service Discovery Database.

## OBEX Operations

There are various OBEX operations that both the Push Client and Push Server support in the Object Push Profile. The CONNECT command is used to establish the initial connection between Push Client and Push Server, while the DISCONNECT command takes down that connection when the Object Push application is no longer needed. The PUT command is used to push objects from the client to the server, while the GET command is used to transfer objects from the server to the client. The ABORT command is used to terminate the request. Table 7.30 summarizes the OBEX operations available in the Object Push Profile.

**TABLE 7.30**

OBEX Operations Available in the Object Push Profile Source: Bluetooth Specification 1.0

| Operation | Push Client | Push Server |
|---|---|---|
| Connect | M | M |
| Disconnect | M | M |
| Put | M | M |
| Get | O | M |
| Abort | M | M |

## Service Discovery Procedures

To support the Object Push service, the following service records must be put into the Service Discovery Database. Table 7.31 summarizes the information in the database, indicating whether the presence of the field is mandatory (M) or optional (O).

**TABLE 7.31**

Entries in the Service Discovery Database per the Object Push Profile Source: Bluetooth Specification 1.0

| Item | Value | Status |
|---|---|---|
| Service Class ID List | | M |
|    Service Class #0 | OBEX Object Push | M |
| Protocol Descriptor List | | M |
|    Protocol #0 | L2CAP | M |
|    Protocol #1 | RFCOMM | M |
|       Parameter for Protocol #0 | Server channel number (varies) | M |
|    Protocol #2 | OBEX | M |
| Service Name | Displayable text name | O |
| Bluetooth Profile Descriptor List | | O |
|    Profile #0 | OBEX Object Push | |
|       Version #0 | Profile version (varies) | |
| Supported Formats List | vCard 2.1 | |
| | vCard 3.0 | |
| | vCalendar 1.0 | |
| | vCalendar 2.0 | |
| | vNote | |
| | vMessage | |
| | Other object types | M |

# Synchronization Profile

The Synchronization Profile defines the requirements for the protocols and procedures used by applications providing the Synchronization usage model. The most common devices implementing this usage model are notebook PCs, PDAs, and mobile phones. The model provides device-to-device synchronization of personal information management (PIM) information, which typically consists of phonebook, calendar, message, and note information to be transferred and processed by devices utilizing a common protocol and format. The usage model also includes the use of a mobile phone or PDA by a computer, to start synchronization automatically when either of the devices comes within range of the computer.

As shown in Figure 7.21, the Synchronization Profile is dependent on both the Generic Object Exchange Profile and the Serial Port Profile, but uses the Generic Object Exchange Profile (GOEP) as a base profile to define the interoperability requirements for the protocols needed by the applications. Figure 7.22 shows the protocols and entities used in the Synchronization Profile.

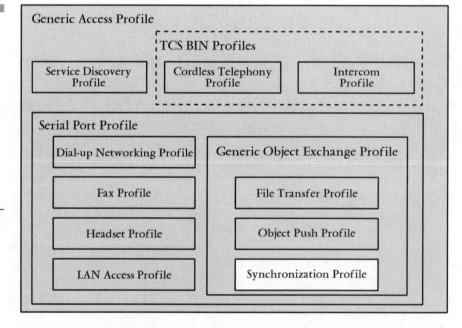

**Figure 7.21**
The Synchronization Profile is dependent on both the Serial Port Profile and the Generic Access Profile, but uses the Generic Object Exchange Profile (GOEP) as a base profile.
Source: Bluetooth Specification 1.0.

With regard to Figure 7.22, baseband corresponds to OSI Layer 1, while LMP and L2CAP correspond to OSI Layer 2. RFCOMM is the Bluetooth specification's adaptation of GSM TS 07.10 and SDP is the Bluetooth Service Discovery Protocol. OBEX is the Bluetooth specification's adaptation of the Infrared Object Exchange (IrOBEX) protocol standardized by the Infrared Data Association (IrDA).

The Synchronization Profile specifies two device roles: the IrMC client and IrMC server. The IrMC client device contains a sync engine

# Bluetooth Profiles for Usage Models

and also pulls and pushes the PIM data from and to the IrMC server. Usually, the IrMC client device is a desktop or notebook PC. However, because the IrMC client must also provide functionality to receive the initialization command for synchronization, it also temporarily acts as a server. The IrMC server device provides an object exchange server. Typically, this device is a mobile phone or PDA. If the IrMC server also provides the functionality to initiate the synchronization process, it also temporarily acts as a client.

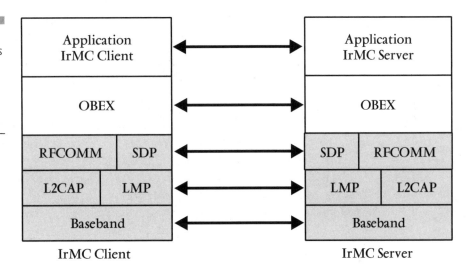

**Figure 7.22**
Protocols and entities used in the Synchronization Profile.
Source: Bluetooth Specification 1.0.

In the Synchronization Profile, because both the IrMC client and IrMC server can act as the other temporarily, both can initiate link and channel establishment; that is, create a physical link between themselves. This profile does not mandate that the IrMC server or client enter into any discoverable or connectable modes automatically, even if they can do so. This means that end-user intervention may be needed on both the devices when, for example, the synchronization is initiated on the IrMC client device.

## Basic Operation

The two modes associated with the Synchronization Profile are Initialization Sync and General Sync. In the Initialization Sync mode, the IrMC server is in the Limited Discoverable or the General Discoverable mode, and Connectable and Pairable modes. The IrMC client does not enter the Initialization Sync mode in this profile. The Bluetooth specification recommends that the Limited Inquiry procedure be used by the IrMC client when discovering the IrMC server.

Both the IrMC client and server can enter the General Sync mode. In this mode, the device is in the Connectable mode. For the IrMC server, this mode is used when the IrMC client connects the server and starts the synchronization after pairing. For the IrMC client, the mode is used when the synchronization is initiated by the IrMC server. The devices are not required to enter these modes automatically without user intervention, even if they can do so.

When an IrMC client wants to synchronize with an IrMC server for the first time, the IrMC server device must be in the General Sync mode. Otherwise, the user must activate this mode on the device. Once in General Sync mode, the user activates an application for synchronization, whereupon a list of devices within range of the IrMC client is displayed. The user selects a device to be connected and synchronized. If the user is alerted that the target device does not support the Synchronization feature, another possible device must be selected. Next, the Bluetooth PIN code is requested from the user and entered on both devices. If OBEX authentication is used, the user enters the password for the OBEX authentication on both devices. Upon completion of these steps, the first synchronization is processed and the user is usually notified of the results.

At subsequent times, when bonding is done, the following steps are taken. The IrMC server device is assumed to be in the General Sync mode. If the device is not in this mode, the user must activate this mode on the device. The user of the IrMC client selects the Synchronization feature on the device, or another event triggers the synchronization to start in the IrMC client. The synchronization is processed and the user is notified of the results.

When an IrMC server wants to initiate synchronization, and when bonding and the possible OBEX initialization are done, the following steps are taken. The IrMC client should be in the General Sync mode without user intervention, otherwise this operation is not applicable. The user selects the Sync Command feature in the IrMC server, and the synchronization is initiated with the IrMC client. On the IrMC server device, the user has earlier configured the IrMC client to which the sync command is sent. The synchronization is processed and the user is notified of the results.

When automatic synchronization of an IrMC server and client is desired, and when the bonding and (possible) OBEX initialization are done, the following steps are taken. The IrMC server comes into range of the IrMC client. The client detects it and starts the synchronization without any notification to the user. The IrMC server must be constantly in the General Sync mode so that the IrMC client can detect the server's presence in its radio frequency vicinity. The synchronization is processed and the user is notified of the results.

## Features

The units active in the Synchronization use case are required to support three features: Synchronization, Sync Command, and Automatic Synchronization.

Bluetooth Synchronization must support at least one of the following application classes:

- Synchronization of phonebooks
- Synchronization of calendars
- Synchronization of messages
- Synchronization of notes

To achieve application-level interoperability, specific content formats are defined (i.e., vCard, vCalendar, vMessage, and vNote), which are dependent on application classes. The supported application classes must be identified in terms of the data stores in the Service Discovery Database of the IrMC server.

The Sync Command feature allows the IrMC client device to work temporarily as a server and receive a Sync Command from the IrMC server, which in this case acts temporarily as a client. This Sync Command orders the IrMC client to start synchronization with the IrMC server. After sending the sync command and getting the response for it, the IrMC server terminates the OBEX session and the RFCOMM data-link connection. This feature must be supported by the IrMC client, but it can as an option be supported by the IrMC server.

With a feature known as Automatic Synchronization, the IrMC client can start the synchronization when the IrMC server comes within range of the IrMC client. On the baseband level, this means the IrMC client pages the IrMC server at intervals and, when it detects that the IrMC server is within range, the IrMC client can begin synchronization. The support of this feature is optional for the IrMC client, but mandatory for the IrMC server. This means that the IrMC server must have the capability to put itself into the General Sync mode so that it does not leave this mode automatically.

These application-layer features are summarized in Table 7.32, with indication of whether support is mandatory (M) or optional (O) at the IrMC client and IrMC server.

**TABLE 7.32**

Application-layer Features of the Synchronization Profile
Source: Bluetooth Specification 1.0

| Features | Support in Push Client | Support in Push Server |
|---|---|---|
| Synchronization of one or more of the following cases: phonebooks, calendars, messages, notes | M | M |
| Sync Command | M | O |
| Automatic Synchronization | O | M |

## OBEX Operations

There are various OBEX operations that both the IrMC Client and IrMC Server support in the Synchronization Profile in terms of their ability to send and respond. The CONNECT command is used to establish the initial connection between IrMC client and IrMC server, while the DISCONNECT command takes down that connection when it is no longer needed. The PUT command is used to push objects from the client to the server, while the GET command is used to transfer objects from the server to the client. The ABORT command is used to terminate the request. Table 7.33 summarizes the OBEX operations required in the Synchronization Profile.

## Service Discovery Procedures

To support the Synchronization Profile, service records must be put into the Service Discovery Database. However, since there are two separate services related to this profile, the actual synchronization server (i.e., IrMC server) and the sync command server (i.e., IrMC client), both must have service records in the Service Discovery Database. Table 7.34 summarizes the information in the database for the IrMC server, indicating whether the presence of the field is mandatory (M) or optional (O).

# Bluetooth Profiles for Usage Models

**TABLE 7.33**

OBEX Operations Required in the Synchronization Profile
Source: Bluetooth Specification 1.0

| Operation | Ability to Send | | Ability to Respond | |
|---|---|---|---|---|
| | IrMC Client | IrMC Server* | IrMC Client* | IrMC Server |
| Connect | M | O | M | M |
| Disconnect | M | O | M | M |
| Put | M | O | M | M |
| Get | M | X | X | M |
| Abort | M | O | M | M |
| Set Path | X | X | X | X |

*Support of the Sync Command feature in the IrMC Server is optional.*

**TABLE 7.34**

Entries in the Service Discovery Database for the IrMC Server Implementing the Synchronization Service
Source: Bluetooth Specification 1.0

| Item | Value | Status |
|---|---|---|
| Service Class ID List | | M |
|   Service Class #0 | IrMC Sync | M |
| Protocol Descriptor List | | M |
|   Protocol #0 | L2CAP | M |
|   Protocol #1 | RFCOMM | M |
|     Parameter for Protocol #0 | Server channel number (varies) | M |
|   Protocol #2 | OBEX | M |
| Service Name | Displayable text name | O |
| Bluetooth Profile Descriptor List | | O |
|   Profile #0 | IrMC Sync | |
|   Version #0 | Profile version (varies) | |
| Supported Data Stores List | Phonebook<br>Calendar<br>Notes<br>Messages | M |

The Sync Command Service is used for initiating the synchronization from the IrMC server device. The following service records must be put into the Service Discovery Database by the application which provides this service. Table 7.35 summarizes the information in the database for the IrMC Client.

**TABLE 7.35**

Entries in the Service Discovery Database for the IrMC Client Implementing the Sync Command Service
Source: Bluetooth Specification 1.0

| Item | Value | Status |
|---|---|---|
| Service Class ID List | | M |
|    Service Class #0 | IrMC Sync Command | M |
| Protocol Descriptor List | | M |
|    Protocol #0 | L2CAP | M |
|    Protocol #1 | RFCOMM | M |
|       Parameter for Protocol #0 | Server channel number (varies) | M |
|    Protocol #2 | OBEX | M |
| Service Name | Displayable text name | O |
| Bluetooth Profile Descriptor List | | O |
|    Profile #0 | IrMC Sync | |
|       Version #0 | Profile version (varies) | |

## Summary

The profiles discussed in this chapter define the protocols and features that support a particular usage model. As such, they draw from the processes defined in the four general profiles discussed in the previous chapter: the Generic Access Profile (GAP), the Serial Port Profile (SPP), the Service Discovery Application Profile (SDAP), and the Generic Object Exchange Profile (GOEP). The purpose of these profiles and usage models is to define the specific messages and procedures used to implement various features, some of which are mandatory, optional, and conditional. If a feature is implemented, it must be implemented in a manner specified in the appropriate profile to ensure that it works the same way for each device, regardless of manufacturer. If devices from different manufacturers conform to the same profile, they will be able to communicate with each other when used for that particular usage case. Given that over 1,800 companies now belong to the Bluetooth SIG, and that many are building products compliant with the Bluetooth specification, a high degree of interoperability among vendor offerings seems assured.

CHAPTER 8

# Bluetooth Security

Bluetooth signals can be easily intercepted, as can any other type of wireless signal. Therefore, the Bluetooth specification calls for the use of built-in security to discourage eavesdropping and attempts to falsify the origin of messages, which is called "spoofing." Specifically, link-level security features are available that implement authentication and encryption.

Authentication prevents spoofing and unwanted access to critical data and functions, while encryption protects link privacy.[1] In addition to these link-level functions, the frequency-hopping scheme used with the spread-spectrum signal, as well as the limited transmission range of devices using Bluetooth wireless technology, also make eavesdropping difficult.

However, in providing link-level authentication and encryption, enforcing security at only this level inhibits user-friendly access to more public-oriented usage models, such as discovering services and exchanging virtual business cards. Because there will be different demands on data security, applications and devices must have more flexibility in the use of link-level security. To meet these differing demands, the Bluetooth specification defines three security modes that cover the functionality and application of devices.

## Security Modes

Mode 1 refers to the absence of security and is used when the devices have no critical applications. In this mode, the devices bypass the link-level security functions, making them suitable for accessing databases containing non-sensitive information. The automatic exchange of business cards and calendars (i.e., vCard and vCalendar) are typical examples of non-secure data transfers.

Mode 2 provides service-level security, allowing for more versatile access procedures, especially for running parallel applications which may each have a different security requirement. Mode 3 provides link-level security, whereby the Link Manager (LM) enforces security at a common level for all applications at the time of connection setup. Although less flexible, this mode enforces a common security level, and it is easier to implement than Mode 2.

---

[1] While the wireless signal can be intercepted with special scanners, encryption makes the data unintelligible.

## Link-level Security

All link-level security functions are based on the concept of link keys, which are 128-bit random numbers stored individually for each pair of devices. Authentication requires no user input. It involves a device-to-device challenge and response scheme that employs a 128-bit common secret link key, a 128-bit challenge, and a 32-bit response. Consequently, this scheme is used to authenticate devices, not users.

Each time the same two devices communicate via Bluetooth transceivers, the link key is used for authentication and encryption, without regard for the specific piconet topology. The most secure type of link key is a *combination key*, derived from the input of both devices. For devices with low data-storage capabilities, there is also the option of choosing a *unit key*, which may be used for several remote devices. For broadcasting, a temporary key is required, which is not used for authentication, but prevents eavesdropping from outside the piconet. Only members of the piconet share this temporary key.

The first time two devices attempt to communicate, an initialization procedure called *pairing* is used to create a common link key in a safe manner. The standard way of doing this assumes that the user has access to both devices at the same time. For first-time connection, pairing requires the user to enter a Bluetooth security code of up to 16 bytes (or 128 bits) into the paired devices. However, when this is done manually, the length of the code will usually be much shorter.

Although the Bluetooth security code is often referred to as a PIN (Personal Identification Number), it is not a code the user must memorize to keep secret, since it is only used once. If for some reason a link key is deleted and the initial pairing must be repeated, any Bluetooth security code can be entered by the user again. In the case of low-security requirements, it is possible to have a fixed code in devices having no user interface to allow pairing. Per the Bluetooth specification, the pairing procedure is as follows:

- A common random number initialization key is generated from the user-entered Bluetooth security code in paired devices, which is used once and then discarded.

- Through authentication, the Bluetooth security code is checked to see if it is identical in the paired devices.

- A common 128-bit random-number link key is generated, which is stored temporarily in the paired devices. As long as this current link key is stored in both devices, no repetition of pairing is necessary, in which case only the normal procedure for authentication is implemented.

- Encryption for the baseband link requires no user input. After successful authentication and retrieval of the current link key, a new encryption key is generated from the link key for each communication session. The encryption key length ranges between 8 and 128 bits, depending on the level of security desired. The maximum encryption length is limited by the capacity of the hardware.

## A Matter of Trust

With security Mode 2 it is possible to define security levels for devices and services. Devices have two levels of "trust," which determines the level of access to services. A trusted device is one that has a fixed relationship (paired) and enjoys unrestricted access to all services.

An untrusted device is one that has no permanent fixed relationship, and is therefore not trusted. There may be cases where the device has a fixed relationship, but is still not trusted. In these instances, access to services is restricted. A possible variation in this scheme of things is to set the trust level of a device so that it can access a specific service or a group of services.

The requirement for authorization, authentication, and encryption are set independently, according to the type of service access the device must have:

- For services that require authorization and authentication, automatic access is granted only to trusted devices, while all other devices must undergo manual authorization.
- Services that require authentication only.
- Services open to all devices.

An approach to security might include a security manager, which is queried during connection establishment. The security manager grants access based on the trust level of the device and the security level of the service. Both of these levels are taken from internal databases.

A default security level is available to serve the needs of "legacy" applications. These are applications that are unable to make calls to the security manager on their own. Instead, an "adapter" application that can use the Bluetooth specification is required to make security-related calls to the Bluetooth security manager on behalf of the legacy application. In such cases, the default policy will be used unless other settings are found in a security database related to the service. The settings in the database will take precedence over the default policy.

Bluetooth security is not intended to replace existing network security features. For extremely sensitive requirements such as e-commerce or specialized requirements such as personal- instead of device-oriented authorization, additional application-level security mechanisms can be implemented. In the Bluetooth profiles, this approach is used for synchronization, for example, where OBEX authentication is implemented.

## Flexible Access

The Bluetooth security architecture allows for selective access to services, so that access can be granted to some services and denied to other services. The architecture supports security policies for devices with some services communicating with changing remote devices for applications such as file transfer and business card exchange. Access granted to a service on these devices does not open up access to other services on the device and does not grant future access automatically or in an uncontrolled way to services on the device.

A characteristic of the Bluetooth security architecture is that user intervention for access to services is avoided as much as possible. User intervention is needed only to allow devices limited access to services or for setting up a trusted relationship with devices, allowing unlimited access to services.

## Implementation

The security architecture accounts for Bluetooth multiplexing protocols at and above L2CAP, particularly RFCOMM. All other protocols are not specific to the Bluetooth architecture, and some may have their own security features.

The security architecture allows different protocols to enforce the security policies. For example, L2CAP enforces the Bluetooth security policy for cordless telephony, RFCOMM enforces the Bluetooth security policy for dialup networking, and OBEX uses its own security policy for file transfer and synchronization.

The architecture can work entirely through security Mode 2 of the Generic Access Profile, especially since there are no changes to baseband and LMP functions for authentication and encryption. Authentication and encryption are set for the baseband level. Lower layers are not cognizant of service/application layer security, and the enforcement policy for authentication, authorization, or encryption might be different for the client and server roles.

## Architecture Overview

The security architecture provides a flexible security framework that dictates when to involve a user (e.g., to provide a PIN) and what actions the underlying Bluetooth protocol layers need to follow to support the desired security checks. The Bluetooth security architecture is built on top of the link-level security features of the Bluetooth system. The general architecture is shown in Figure 8.1.

The key component of the Bluetooth security architecture is the security manager, which is responsible for carrying out the following tasks:

- Storing security-related information on services.
- Storing security-related information on devices.
- Responding to access requests by protocol implementation or application.
- Enforcing authentication and/or encryption before connecting to the application.
- Initiating or processing input from an External Security Control Entity (ESCE), the device user, to set up trusted relationships at the device level.
- Initiating pairing and query PIN entry by the user. PIN entry might also be performed by an application.

Typically, an ESCE represents an entity with the authority and knowledge to make decisions on how to proceed in a manner consistent with the Bluetooth security architecture. The entity could be a device user, or a utility application executed on behalf of the user and based on preprogrammed security policies. In the latter case, the utili-

# Bluetooth Security

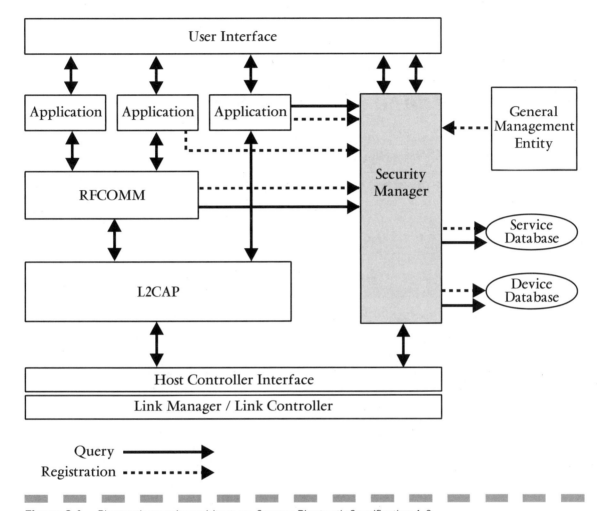

**Figure 8.1** Bluetooth security architecture. Source: Bluetooth Specification 1.0.

ty could reside within or outside a particular device using Bluetooth wireless technology.

A centralized security manager allows for easy implementation of flexible access policies because the interfaces to protocols and other entities are kept simple and are limited to query/response and registration procedures. The policies for access control are encapsulated in the security manager, so that implementation of more complex policies does not affect implementation of other policies. Implementations may decide where the registration task is performed—the application itself or a general management entity responsible for setting the path in the protocol stack and/or registering the service at the time of service discovery.

## Security Level of Services

As described in the Bluetooth specification, the security level of a service is defined by three attributes:

- **Authorization required**—Access is granted automatically only to trusted devices (i.e., those devices marked as such in the device database), or untrusted devices after an authorization procedure. Authorization always requires authentication to verify that the remote device is the correct one.
- **Authentication required**—Before connecting to an application, the remote device must be authenticated.
- **Encryption required**—The link must be changed to encrypted mode before access to the service is permitted.

This attribute information is stored in the service database of the security manager. If no registration has taken place, a default security level is used. For an incoming connection, the default is authorization and authentication required. For an outgoing connection, the default is authentication required.

## Connection Setup

To meet different service availability requirements without user intervention, authentication must be performed after determining the security level of the requested service. Thus, authentication cannot be performed when the Asynchronous Connectionless (ACL) link is established. The authentication is performed when a connection request to a service is submitted. Figure 8.2 illustrates the information flow for access to a trusted service.

The following sequence of events takes place to access a trusted device:

1. Connection request goes through L2CAP.
2. L2CAP requests access from the security manager.

# Bluetooth Security

3. Security manager performs a lookup in the service database.
4. Security manager performs a lookup in the device database.
5. If necessary, security manager enforces authentication and encryption.
6. Security manager grants access.
7. L2CAP continues to set up the connection.

Authentication can be performed in either direction, allowing the client to authenticate the server, or the server to authenticate the client.

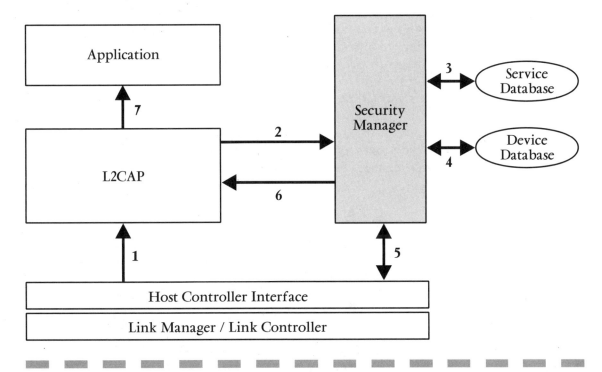

**Figure 8.2** Information flow for access to a trusted service. Source: Bluetooth Specification 1.0.

## Authentication on Baseband Link Setup

Although not targeted to security Mode 3, link-level enforced security, the Bluetooth security architecture can support this mode as well. The security manager can command the Link Manager to enforce authentication before accepting a baseband connection using the Host Controller Interface (HCI). However, before transitioning from Mode 2 to Mode 3, steps must be taken to ensure that untrusted devices do not obtain unwanted access. To keep this from happening, the security manager removes any link keys for untrusted devices stored in the radio module. For this, the security manager can use the HCI.

## Protocol Stack Handling

For incoming connections, the access control procedure is summarized in Figure 8.3. Access control is required at L2CAP and, in some cases, also at the multiplexing protocols above (e.g., RFCOMM). When receiving a connection request, the protocol entity queries the security manager, providing any multiplexing information it received with the connection request. The security manager determines whether access is granted or refused, and then replies to the protocol entity. If access is granted, the connection setup procedure is continued. If access is denied, the connection is terminated. If no access control is performed on a protocol layer, no interaction takes place with the security manager or other entities.

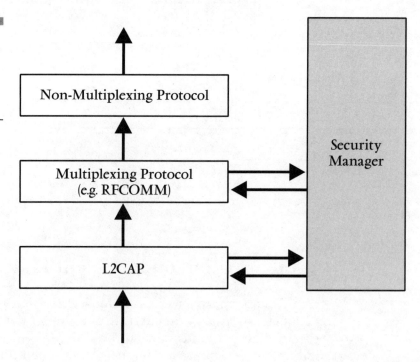

**Figure 8.3**
Protocol behavior for incoming connections.
Source: Bluetooth Specification 1.0.

# Bluetooth Security

The security manager stores information on existing authentications. This avoids multiple authentication procedures at the LMP level (i.e., over the air) within the same session. Thus, RFCOMM will do a policy check to the security manager. This requires an additional function call, but not necessarily an additional authentication. For outgoing connections, a security check might also be required to achieve mutual authentication, in which case, a similar procedure is carried out. The most efficient way to submit requests to the security manager is for the applications to submit them directly. If this is not possible, as is the case with legacy applications, queries to the security manager may be submitted by any multiplexing protocol (Figure 8.4).

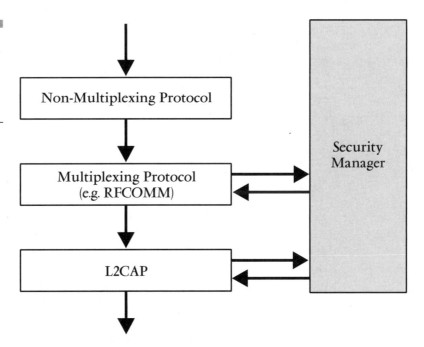

**Figure 8.4**
Protocol behavior for outgoing connections.
Source: Bluetooth Specification 1.0.

## Registration Procedures

As noted, the security manager maintains security information for services in security databases. Applications must register with the security manager before becoming accessible (Figure 8.5).

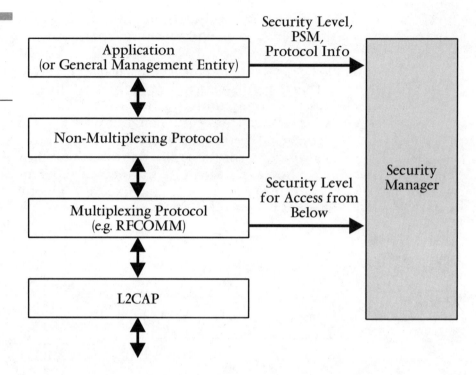

**Figure 8.5**
Registration processes.
Source: Bluetooth Specification 1.0.

Applications or their security surrogates must provide their security level and multiplexing information to the security manager. This is similar to the information that is registered for service discovery. The security manager needs this information to make a grant/deny decision on an access request submitted by a protocol entity, since this entity may not always be aware of the final application.

Multiplexing protocol implementations performing queries at the security manager register the policy for accessing them from below. Both registrations can also be done by the entity that is responsible for setting the path in the Bluetooth protocol stack. Which entity actually initiates the registration procedure depends on the implementation.

If no registration has taken place, the security manager will assume the default security level to make a decision on access. In this case, L2CAP does not require registration. For the first multiplexing protocol in the Bluetooth stack, there will be a query for every connection request.

### External Key Management

The Bluetooth security architecture does not exclude the use of external key management procedures. Such key management applications might be able to distribute PINs or the link keys directly, for example.

### Access Control Procedures

Figures 8.6 to 8.8 provide flow-chart examples of how access control can be handled in the security manager. According to the Bluetooth specification, variations on these decision trees are possible.

## Connectionless L2CAP

Since it is not practical to perform a security check on each connectionless data packet, a general policy of handling connectionless packets must be made at the L2CAP level. L2CAP can also be used to block connectionless traffic. Blocking can be applied to a single protocol on top of L2CAP, a list of protocols, or all protocols. The same options are possible for enabling connectionless traffic.

The security manager checks if there is any service in the database that does not permit connectionless data packets. Accordingly, the security manager will then enable or disable the use of connectionless data packets. If connectionless packets are allowed, there will be no security check. It is then up to the protocol above L2CAP to make sure that no security problem occurs. It will always be known whether the data came in via connectionless or connection-oriented mode, but for connectionless packets the originator may not be known or verifiable.

## Security Manager

The security manager that implements the security architecture has to maintain several databases. The service database maintains security-related entries for each service, which are either mandatory (M) or optional (O), as listed in Table 8.1.

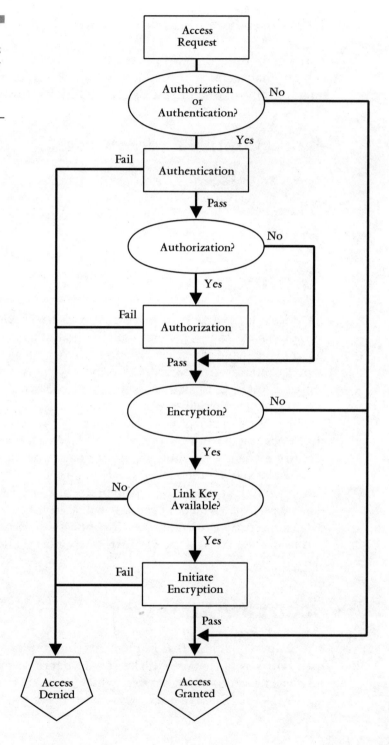

**Figure 8.6**
Flow chart for access check by the security manager.
Source: Bluetooth Specification 1.0.

# Bluetooth Security

**Figure 8.7**
Flow chart for the authentication procedure.
Source: Bluetooth Specification 1.0.

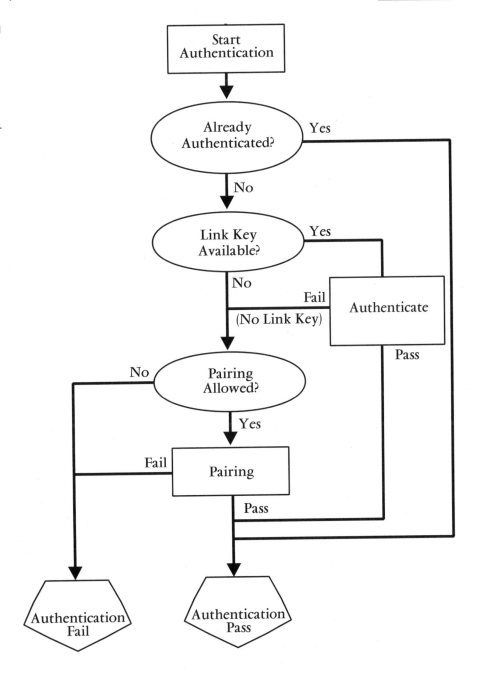

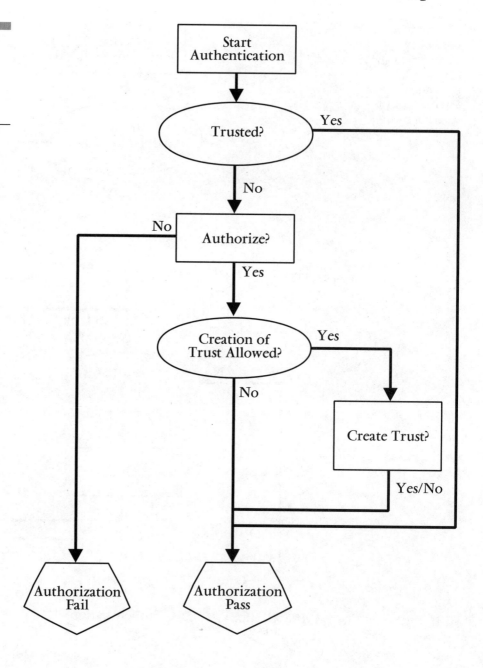

**Figure 8.8**
Flow chart for the authorization procedure.
Source: Bluetooth Specification 1.0.

# Bluetooth Security

**TABLE 8.1**

Security-related Entries in the Service Database Source: Bluetooth Specification 1.0

| Entry | Status |
|---|---|
| Authorization required | M |
| Authentication required | M |
| Encryption required | M |
| PSM Value | M |
| Broadcasting allowed | O |

The trust information is stored in the non-volatile memory of the Bluetooth devices. If entries are deleted for any reason, the respective device is treated as unknown and set to the default access level. The mandatory (M) and optional (O) entries in the device database are listed in Table 8.2.

**TABLE 8.2**

Security-related Entries in the Device Database Source: Bluetooth Specification 1.0

| Entry | Status | Contents |
|---|---|---|
| BD_ADDR | M | 48-bit MAC address |
| Trust level | M | Trusted/untrusted |
| Link key | M | Bit field (up to 128 bits) |
| Device name | O | String (can be used to avoid name request) |

Some data should be stored to reduce overhead on the air interface. Accordingly, for each baseband link, authentication status and encryption status are stored.

## Interface to L2CAP

L2CAP asks the security manager for access rights to a service on incoming and outgoing connection requests. Since L2CAP is mandatory in Bluetooth protocol stacks, there is no registration procedure.

## Interface to Other Multiplexing Protocols

Other multiplexing protocols, including RFCOMM, which need to make decisions on access to services, query the security manager much as L2CAP does. There is an additional registration procedure, which sets the access policy for connection to the multiplexing protocol itself.

## Interface to ESCE

The Bluetooth security architecture contains user interaction for authorization purposes. This includes access permission to services and setting up a trusted relationship to a remote device. The security manager calls the ESCE (e.g., the user interface); incoming parameters are the information submitted with the request, outgoing parameters hold the response.

If the security manager requests a PIN, a call to the ESCE can be used. The PIN entry can also be requested directly from the Link Manager. The security manager then requests authentication and, if no valid link is available, the Link Manager performs the necessary actions.

## Registration Procedures

As noted, applications must register with the security manager before becoming accessible. Among the parameters used for registering the application and multiplexing protocol are:

- Plain-language name of the application, which is intended for user queries
- Security level
- Protocol/Service Multiplexer (PSM), a value used at the L2CAP level
- Protocol identification
- Channel identification

# Bluetooth Security

Registration can be done by the entity responsible for setting the path in the Bluetooth protocol stack. Which entity performs the actual registration is implementation dependent. Without registration, default settings apply.

## Interface to HCI/Link Manager

Per the Bluetooth specification, the Host Controller Interface (HCI) provides a command interface to the baseband controller and Link Manager, as well as access to hardware status and control registers. This interface provides a uniform method of accessing the Bluetooth baseband capabilities. Among other things, the security manager can command the Link Manager to enforce authentication before accepting a baseband connection using the HCI.

### AUTHENTICATION REQUEST
The command `HCI_Authentication_Requested` is used for requesting an authentication of a remote device. For the answer, the Authentication Complete event is returned.

### ENCRYPTION CONTROL
For encryption control, the command `HCI_Set_Connection_Encryption` is used. For the answer, the Encryption Change event is returned, which enables and disables the link-level encryption.

### NAME REQUEST TO REMOTE DEVICE
For a name request to a remote device, the `HCI_Remote_Name_Request` command is used to obtain the user-friendly name of another Bluetooth device. The user-friendly name enables the user to distinguish one Bluetooth device from another. The `BD_ADDR` command parameter is used to identify the device for which the user-friendly name is to be obtained. In answer to the Remote Name Request command, `Remote_Name_Request_Complete` is returned.

### SET ENCRYPTION POLICY AT LINK LEVEL
The general encryption policy at the link level can be set by the `HCI_Write_Encryption_Mode` command, which is answered by the Command Complete event. The Encryption Mode parameter controls

whether the Bluetooth radio will require encryption at the link level for each connection with other Bluetooth radios.

**SET AUTHENTICATION POLICY AT LINK LEVEL**
The general authentication policy at the link level can be set by the `HCI_Write_Authentication_Enable` command. The `Authentication_Enable` parameter controls whether the Bluetooth radio will require authentication at the link level for each connection with other Bluetooth radios. At connection setup, only the device(s) with the `Authentication_Enable` parameter enabled will try to authenticate the other device. The `Read_Authentication_Enable` command will read the value for the `Authentication_Enable` parameter. When the `Read_Authentication_Enable` command is completed, a Command Complete event is generated.

## Summary

The Bluetooth security architecture is not without its share of limitations. There is no provision for allowing legacy applications to make calls directly to the security manager. In such cases, an adapter application is required to act on behalf of the legacy application. Then, only a device can be authenticated, not its user. Unless application-level security is used, there is always the possibility of unauthorized access. Third, there is no mechanism defined for presetting authorization per service without modifying the security manager and the registration processes. Fourth, the security approach only allows access control at connection setup. Finally, the security architecture does not specifically address end-to-end security, but assumes the existence of a higher-level solution that may draw upon the Bluetooth security architecture for controlling access to devices and services at both ends of a Bluetooth link.

CHAPTER 9

# Bluetooth in the Global Scheme of 3G Wireless

The next generation of cellular radio systems for mobile telephony is referred to as Third Generation (3G). Due to come on stream from 2001 onwards, 3G will be the first cellular radio technology designed from the outset to support wideband data communications just as well as it supports voice communications. One of the most important attributes of IMT-2000 is the capability for global roaming with a single low-cost terminal, which will be the basis for a wireless information society where access to information and services such as electronic commerce are available anytime and anywhere to anybody. The technical framework for 3G has been defined by the International Telecommunication Union (ITU) with its International Mobile Telecommunications 2000 (IMT-2000) program. The Bluetooth specification will support 3G systems in the delivery of a wide range of services by extending their reach to localized devices, such a handheld computers and PDAs, wherever they are and wherever they are going (Figure 9.1).

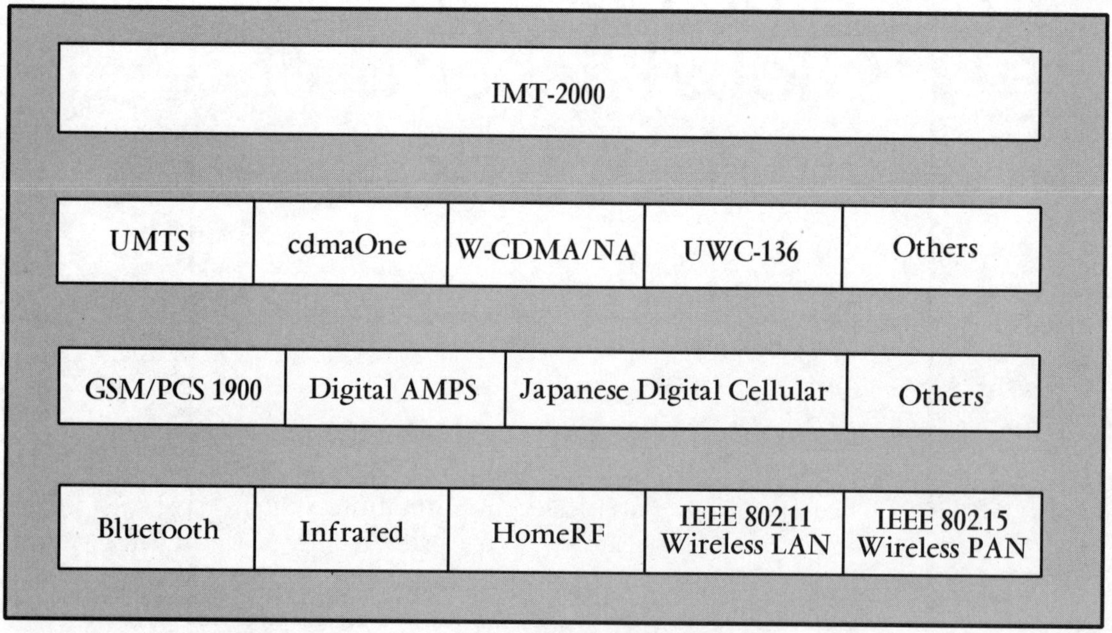

| IMT-2000 | | | | |
|---|---|---|---|---|
| UMTS | cdmaOne | W-CDMA/NA | UWC-136 | Others |
| GSM/PCS 1900 | Digital AMPS | Japanese Digital Cellular | | Others |
| Bluetooth | Infrared | HomeRF | IEEE 802.11 Wireless LAN | IEEE 802.15 Wireless PAN |

**Figure 9.1** The role of the Bluetooth specification in the global 3G wireless infrastructure.

## The IMT-2000 Vision

IMT-2000 is described by the ITU as a global family of third-generation mobile communications systems that are capable of bringing high-quality mobile multimedia telecommunications to a worldwide mass market, based on a set of standardized interfaces. The IMT-2000 framework encompasses a small number of frequency bands, available on a globally harmonized basis, that make use of existing mobile and mobile-satellite frequency allocations. This provides a high degree of flexibility to allow operators in each country to migrate toward IMT-2000 according to market and other national considerations.

IMT-2000 is the largest telecommunications project ever attempted, involving regulators, operators, manufacturers, media, and IT players from all regions of the world as they attempt to position themselves to serve the needs of an estimated two billion mobile users worldwide by 2010.

Originally conceived in the early 1990s when mobile telecommunications provided only voice and circuit-switched low-speed data, the IMT-2000 concept has adapted to the changing telecommunication environment as its development progressed. In particular, the advent of Internet, intranet, e-mail, e-commerce, and video services has significantly raised user expectations concerning the responsiveness of the network and the terminals, and hence the bandwidth of the mobile channel. It is expected that declining equipment costs and dropping airtime price will make 3G services the prime driver in subscriber growth.

## Spanning the Generations

Over the years, mobile telecommunications systems have been implemented with great success all over the world. Many are still first-generation systems—analog cellular systems such as the Advanced Mobile Phone System (AMPS), Nordic Mobile Telephone (NMT), and the Total Access Communication System (TACS). Most systems are now in the second generation, which is digital in nature. Examples of digital cellular systems include Global System for Mobile (GSM) communications, Digital AMPS (DAMPS), and Japanese Digital Cellular (JDC). Although both first- and second-generation systems were designed primarily for speech, they offer low-bit-rate data services as well. However, there is little or no compatibility between the different systems, even within the same generation.

The spectrum limitations and various technical deficiencies of second-generation systems have led to the development of 3G systems. The proliferation of standards and the potential fragmentation problems they could cause in the future led to research on the development and standardization of a global 3G platform. The ITU and regional standards bodies came up with a "family of systems" concept that would be capable of unifying the various technologies at a higher level to provide users with global roaming and voice-data convergence, leading to enhanced services and support for innovative multimedia and e-commerce applications.

The result of this activity is IMT-2000, a modular concept that takes full account of the trends toward convergence of fixed and mobile networks and voice and data services. The third-generation platform represents an evolution and extension of current GSM systems and services available today, optimized for high-speed packet data-rate applications, including high-speed wireless Internet services, videoconferencing, and a host of other data-related applications.

Vendor compliance with IMT-2000 enables a number of sophisticated applications to be developed. For example, a mobile phone with color display screen and integrated 3G communications module becomes a general-purpose communications and computing device for broadband Internet access, voice, video telephony, and conferencing (Figure 9.2). These applications can be used by mobile professionals on the road, in the office, or at home. The number of IP networks and applications is growing fast. Most obvious is the Internet, but private IP networks (i.e., intranets and extranets) show similar or even higher rates of growth and usage. With an estimated billion Internet users worldwide expected in 2010, there exists tremendous pent-up demand for 3G capabilities.

3G networks will become the most flexible means of broadband access because they allow for mobile, office, and residential use in a wide range of public and non-public networks. Such networks can support both IP and non-IP traffic in a variety of transmission modes including packet (i.e., IP), circuit switched (i.e., PSTN) and virtual circuit (i.e., ATM).

3G networks will be able to benefit from the standards work done by the Internet Engineering Task Force (IETF), which has extended the basic set of IP standards to include QoS capabilities that are essen-

# Bluetooth in the Global Scheme of 3G Wireless

tial elements for mobile operation.[1] Developments in new domain-name structures also are taking place. These new structures will increase the usability and flexibility of the system, providing unique addressing for each user, independent of terminal, application, or location.

**Figure 9.2**
This prototype of a 3G mobile phone from Nokia supports digital mobile multimedia communications, including video telephony. Using the camera eye in the top right corner of the phone, along with the thumbnail screen below it, the local user can line up his or her image so it can appear properly centered on the remote user's phone.

The promises of 3G networks could easily have gone unrealized. The objective of a 3G platform is to harmonize these and other technologies to provide users with seamless global connectivity and enhanced services and features at a reasonable cost. This is complicated by the fact that these and other technologies are not fully compatible with each other, and harmonization necessitates compromise. Initially, progress toward 3G had been hampered by a number of obstacles. Regulators wanted to protect their own countries' markets,

---

[1] The IETF's work on QoS-enabled IP networks has led to two distinct approaches: the Integrated Services (int-serv) architecture and its signaling protocol RSVP, and the Differentiated Services (diff-serv) architecture.

operators wanted to protect their capital investments in one technology or another, while some vendors sought to protect their patents.

The last of these hurdles was cleared in March 1999 with the agreement of Ericcson and Qualcomm to end their patent and licensing disputes. Now all the industry players and standards organizations can safely focus on the development and implementation of 3G networks. Although 3G networks are in various stages of planning, development, and implementation all over the world—especially the U.S., Europe, and Japan—IMT-2000 will be phased in globally over the course of a decade. It is expected that IMT-2000-compliant 3G networks will start becoming available commercially in 2001 and that enhancements will continue to embrace new technologies and concepts to address future applications and market demands.

## Current 2G Networks

First-generation analog cellular technology operates in the 800-MHz frequency band. Cellular systems conforming to the AMPS standard divide the 12.5 MHz of available bandwidth into 30-KHz channels, with one user assigned to each channel. Newer technologies such as Time Division Multiple Access (TDMA) and Code Division Multiple Access (CDMA) can support more users in the same region of the spectrum as well as provide a number of advanced features. In fact, TDMA and CDMA, along with the GSM, are the primary technologies contending for acceptance among cellular and Personal Communications Service (PCS) operators worldwide.

### Time Division Multiple Access

TDMA increases the call-handling capacity of AMPS by dividing the available 30-KHz channels into three time slots, with each user having access to one time slot at regular intervals in round-robin fashion. This time-slot scheme enables several calls to share a single channel without interfering with one another. In using the time slots for short bursts, as many as three calls can be handled on the same 30-KHz channel that supported only one call on an AMPS system.

The current version of TDMA, IS-136, is a revision of the original version of TDMA, IS-54, which was based on the technology available in the 1970s, which had limited system performance. The revised IS-136 standard was published in 1994, and took into account such later developments as digital control channels.

TDMA IS-136 is offered in North America at both the 800-MHz and 1900-MHz bands. IS-136 TDMA normally coexists with analog channels on the same network. One advantage of this dual-mode technology is that users can benefit from the broad coverage of established analog networks while IS-136 TDMA coverage grows incrementally, and at the same time takes advantage of the more advanced technology of IS-136 TDMA where it is available.

IS-136 TDMA, also known as Digital AMPS (DAMPS), specifies the addition of a DQPSK digital control channel (DCCH) to the existing FSK-based control channel used in AMPS and dual-mode (IS-54B) cellular service. The FSK control channel is now referred to as the Analog Control Channel (ACC) and the FM voice channel is referred to as the Analog Voice Channel (AVC). Signaling has been added to the AMPS channels to allow a dual-mode mobile terminal to switch between digital and analog channels to find the one that offers the best service.

At the physical layer, the Digital Traffic Channel (DTC) is very similar to the DCCH in that it is a slotted 48 Kbps DQPSK channel. The DTC was introduced with IS-54B, but has been greatly improved in IS-136 with a new voice coder and enhanced signaling capabilities. As before, the subdivided 30-KHz TDMA traffic channel allows three simultaneous conversations. IS-136 has also added new control messages and expanded previous messages on the DTC, providing new services and supporting transparent extension of cellular services into the PCS band.

In addition, IS-136 allows the mobile terminal to go into low-power mode (i.e., "sleep" mode), which extends standby battery life. Compared to AMPS, talk time is extended as well because the mobile terminal is only transmitting 33% of the time during a conversation.

From the operator's perspective, the TDMA digital channels provide a three-fold capacity improvement over AMPS. Among the factors that contribute to this increase in capacity are:

- Three TDMA Digital Traffic Channels use the same spectrum as one AMPS voice channel.
- TDMA supports a wider range of power levels.
- Service selection incorporates information about service capability in addition to signal strength.
- Digital communication allows more efficient reuse of cellular spectrum.

Service is also improved as the result of the mobile terminal's capability to constantly monitor signal strengths and provide these measurements to the base station when requested.

## Code Division Multiple Access

Instead of dividing the available radio spectrum into separate user channels by time slots, CDMA uses spread-spectrum technology to separate users by assigning them digital codes within a much broader spectrum. This results in higher channel capacity and provides each channel protection against interference from other signals. Like TDMA IS-136, CDMA operates in the 1900-MHz band as well as the 800-MHz band.

With spread-spectrum technology, the information contained in a particular signal is spread over a much greater bandwidth than the original signal. CDMA assigns a unique code to each user and spreads the transmissions of all users in parallel across a wide band of frequencies. Individual conversations are recovered by the respective mobile terminals based on their assigned code. All other conversations look like random noise and are ignored. In the IS-95 implementation of CDMA, the bandwidth of a single channel is 1.25 MHz; it is claimed that this supports 10 to 20 times as many users as the equivalent spectrum dedicated to analog cellular.

From its introduction into commercial communications systems in the mid-1980s, the great attraction of CDMA technology has been the promise of extraordinary capacity increases over the capabilities of narrowband cellular technologies. Early models suggested that the capacity improvement could be as much as 40 times that of the existing narrowband cellular standards, such as AMPS in North America, NMT in Scandinavia, and TACS in the United Kingdom. However, such performance claims were based on the theory of operation under ideal conditions, rather than on measurements from real-world installations. In reality, cell coverage areas are highly irregular, offered load varies greatly by time of day, and system engineering is often subject to uncontrollable influences, such as terrain and zoning laws, which may result in the final installation exhibiting suboptimal performance.

Even though the idealized performance claims of CDMA have not stood up in the real world—operators rarely achieve more than a twelve-fold performance gain over first-generation analog systems—

the technology offers other advantages. For example, the CDMA standard's handoff procedure improves reliability by minimizing the chance of dropped calls.

In all cellular systems, communication between base stations and mobile stations is established by a negotiation upon call origination. In first-generation systems, once communication is established between the base and mobile stations, movement of the mobile station is detected and the service is handed over from one base station to another for the duration of the call. In the CDMA standards, the handoff concept is extended to a multi-way simultaneous "soft" handoff, during which the base station exhibiting the strongest signal will take the call.

## CDMA versus TDMA

Throughout their histories, there have been conflicting performance claims for TDMA and CDMA. Advocates of CDMA believe it is a dramatically superior digital technology that will carry the cellular industry forward into the next century. The advantages of CDMA are summarized as follows:

- **Call clarity**—CDMA's digital encoding improves call quality by eliminating static and reducing background noise.
- **Network capacity**—CDMA offers an initial call capacity advantage of up to 10 times that of analog technology, reducing the need for additional cell sites, thereby minimizing equipment and maintenance costs.
- **Service provisioning**—CDMA provides greater opportunities to offer economical, personal communications services along with advanced digital features such as voice mail alert and digital messaging, as well as an array of wireless data applications.
- **Privacy**—CDMA's unique coding for each phone conversation makes unauthorized interception of a call extremely difficult, providing users with greater privacy than first-generation systems.
- **Reliability**—With its soft handoff capability, CDMA promises fewer dropped calls than first-generation systems. CDMA handsets are "smart" phones.
- **Environmental**—Because existing cell sites can be upgraded to serve 10 times the customers they do today, the need for new cell site towers will significantly diminish.

Not surprisingly, advocates of TDMA claim that CDMA confers no advantage over TDMA-based technology; specifically: the call quality offered by TDMA is indistinguishable from that of CDMA and the greater channel capacity claimed for CDMA is misleading. The principal arguments for TDMA are summarized as follows:

- **Proven technology**—TDMA is a field-proven technology that meets industry growth objectives.
- **Economy**—While analog networks grow by adding cells and equipment, TDMA enables the same equipment to be shared by multiple users. This results in the kind of efficiencies needed for mass-market penetration.
- **Evolutionary approach**—TDMA offers an evolutionary growth path, mitigating the capacity claims for CDMA.

Initially, TDMA and CDMA systems provided only basic voice services. The current direction for both types of systems embraces data services and inclusion into the global IMT-2000 family of systems.

## GSM

Global System for Mobile (GSM) telecommunications—formerly known as Groupe Spéciale Mobile, for the group that developed it—was designed from the beginning as an international digital cellular service. GSM's air interface is based on TDMA technology. It was intended from the start that GSM subscribers should be able to roam across national borders and find that their mobile services and features crossed with them.

The European version of GSM operates at 900 MHz as well as 1800 MHz. In North America, GSM is used for PCS 1900 service, which is currently available mostly in the Northeast and in California and Nevada. Since PCS 1900 uses the 1900-MHz frequency, the phones are not interoperable with GSM phones that operate on 900-MHz or 1800-MHz networks. However, this problem can be overcome with multiband GSM phones that operate in all the supported frequencies, eliminating the need to rent phones while traveling in different countries.

### GSM DEVELOPMENT

In the early 1980s, the market for analog cellular telephone systems was experiencing rapid growth in Europe. Each country developed its own cellular system independently of the others'. The uncoordinated development of national mobile communication systems meant that it was not possible for subscribers to use the same portable terminal when they traveled throughout Europe. Not only was the mobile equipment limited to operation within national boundaries, there was a very limited market for each type of equipment, so cost savings from economies of scale could not be realized. Without a sufficiently large home market with common standards, it would not be possible for any vendor to be competitive in world markets. Furthermore, government officials realized that incompatible communication systems could hinder progress toward achieving their vision of an economically unified Europe.

With these considerations in mind, the then 26-nation Conference of European Posts and Telegraphs (CEPT) formed a study group called the Groupe Spéciale Mobile in 1982 to study and develop a future pan-European mobile communication system. By 1986 it was clear that some of the current analog cellular networks would run out of capacity by the early 1990s. CEPT recommended that two blocks of frequencies in the 900-MHz band be reserved for the new system. The GSM standard specifies the frequency bands of 890 to 915 MHz for the uplink band, and 935 to 960 MHz for the downlink band, with each band divided up into 200 KHz channels.

The mobile communications system envisioned by CEPT had to meet the following performance criteria:

- Offer high quality speech
- Support international roaming
- Support hand-held terminals
- Support a range of new services and facilities
- Provide spectral efficiency
- Offer compatibility with ISDN
- Offer low terminal and service cost

In 1989, oversight responsibility for developing GSM specifications was transferred from CEPT to the European Telecommunication Standards Institute (ETSI). ETSI was set up in 1988 to set telecommunications standards for Europe and, in cooperation with other standards organizations, the related fields of broadcasting and office information technology.

ETSI published GSM Phase I specifications in 1990. Commercial service was started in mid-1991. By 1993 there were 36 GSM networks in 22 countries, with 25 additional countries having already selected or begun consideration of GSM. Since then, GSM has been adopted in South Africa, Australia, and many Middle and Far Eastern countries. In North America, GSM is being used to implement PCS. By January 2000, 359 GSM networks in 132 countries were serving 303.5 million subscribers. worldwide

## NETWORK ARCHITECTURE

A GSM network consists of the following architectural elements (Figure 9.3):

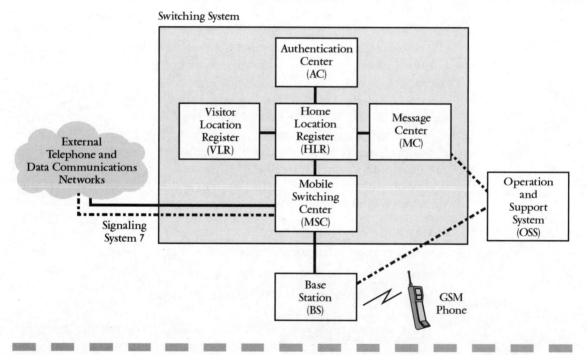

**Figure 9.3** Architecture of a GSM switching system.

- **Mobile Station**—This is the hand-held device carried by the subscriber, which is used for voice and/or data calls. It includes the removable smart card, or Subscriber Identity Module (SIM), which contains subscriber and authentication information.
- **Base Station Subsystem**—This subsystem controls the radio link with the Mobile Station and monitors call status for handoff purposes.

- **Network Subsystem**—The main part of this subsystem is the Mobile Services Switching Center (MSC), which sets up calls between the mobile unit and other fixed or mobile network users, and provides management services such as authentication.
- **Air Interface**—The air interface is a radio link over which the Mobile Station and the Base Station Subsystem communicate. The interface is a radio link over which the Base Station Subsystem communicates with the MSC.

Each GSM network also has an Operations and Maintenance center which oversees the proper operation and setup of the network.

### CHANNEL DERIVATION AND TYPES

Since radio spectrum is a limited resource shared by all users, a method must be devised to divide up the bandwidth among as many users as possible. The method used by GSM is a combination of Time and Frequency Division Multiple Access (TDMA/FDMA).

The FDMA part involves the division by frequency of the total 25-MHz bandwidth into 124 carrier frequencies of 200 KHz bandwidth. One or more carrier frequencies are then assigned to each base station. Each of these carrier frequencies is then divided in time, using a TDMA scheme, into eight time slots. One time slot is used for transmission by the mobile device and one for reception. They are separated in time so that the mobile unit does not receive and transmit at the same time.

Within the framework of TDMA, two types of channels are provided. Traffic channels carry voice and data between users, while control channels carry information that is used by the network for supervision and management. Among the control channels are the following:

- **Fast Associated Control Channel (FACCH)**—This channel is created by robbing slots from a traffic channel to transmit power control and handoff-signaling messages.
- **Broadcast Control Channel (BCCH)**—This continually broadcasts, on the downlink, information including base station identity, frequency allocations, and frequency-hopping sequences.
- **Standalone Dedicated Control Channel (SDCCH)**—This is used for registration, authentication, call setup, and location updating.
- **Common Control Channel (CCCH)**—This comprises three control channels used during call origination and call paging.

- **Random Access Channel (RACH)**—This is used to request access to the network.
- **Paging Channel (PCH)**—This is used to alert the mobile station of an incoming call.

### FOLLOW-UP PHASES

Phase II of GSM has been implemented by most operators. This phase offers additional services such as a call charge tally, calling line identity, call waiting, call hold, conference calling and closed user groups. Phase II+ provides support for multiple service profiles, allowing a user with a single handset to adopt different roles, such as private person or business executive. Through the introduction of private numbering plans, users can internetwork with other staff within their organization as if they were all on the same PBX. Business users can also access Centrex services, where available, through the GSM network, giving them PBX-type facilities while on the move. Phase II+ also adds internetworking specifications so that users can communicate via DECT, DCS1800, and PCS networks.

### HIGH-SPEED CIRCUIT-SWITCHED DATA

HSCSD, introduced in 1998, boosts data transmission capacity to 57.6 Kbps, with the potential for higher speeds. To accomplish this, the GSM standard was modified so that 14.4-Kbps channel coding replaced the original 9.6-Kbps coding used to support mobile telephony. The four channels of 14.4 Kbps each are combined into a single channel of 57.6 Kbps—almost the speed of a fixed ISDN channel of 64 Kbps. The extra bandwidth allows GSM phones and mobile terminals to handle multimedia applications.

With this much bandwidth, users can access the Web and download pages with graphics content in seconds. They can also take advantage of the higher speeds for accessing in-house LANs via corporate intranets. Because HSCSD has an integral bandwidth on demand capability, it will not matter what speed users need for any particular activity—the service will provide whatever speed they require, up to the maximum rate of 57.6 Kbps. With end-to-end compression between a client and a server, even higher speeds can be achieved.

### GENERAL PACKET RADIO SERVICES

GPRS for GSM became available in 1999. Like HSCSD, GPRS provides higher-speed data services for mobile users. However, as a packet-switching technology, GPRS is better suited to the "bursty" nature of most data applications than HSCSD, making it ideal for e-mail and database access services, for example, where users do not want to pay high call charges for short transmissions. GPRS will also permit the user to receive voice calls simultaneously with sending or receiving data calls. Messages are delivered direct to the user's phone, without the need for a full end-to-end connection. When they switch on their phones, users receive notification of messages waiting. The user can then choose to have the messages downloaded immediately or saved for later.

GPRS also offers faster call setup than HSCSD and more efficient connectivity with networks that use the IP protocol, including corporate intranets and LANs, as well as the Internet. Through various combinations of TDMA time slots, GPRS can handle all types of transmission from slow-speed short messages to the higher speeds needed for browsing Web pages. GPRS provides peak packet data rates of over 100 Kbps. The maximum rate is 171.2 Kbps over eight 21.4 Kbps channels.

With circuit-switched transmission, a channel is allocated to a single user for the duration of a call. With a packet-switched network like GPRS, the available radio spectrum is shared by all users in a cell. The spectrum is used only when the user has something to send. When there are no data to be transmitted, the spectrum is free to be used for another call. Thus, if the data are bursty in nature—as are LAN data—the network resources can be balanced more efficiently, because the operator can use the gaps in transmission to handle other calls.

As with HSCSD, GPRS works within the existing GSM infrastructure, allowing GPRS to be introduced quickly. GPRS provides two key benefits for providers and users of data services. First, it reduces the costs of providing connectivity, since GPRS uses radio and network resources more efficiently. With GPRS the applications occupy the network only when data are actually being transferred. Compared with today's circuit-switched methodology, the costs of providing connectivity are reduced. Second, GPRS provides transparent IP support. By tunneling the IP protocol transparently from the mobile terminal to the Internet or intranet and giving the terminal the same status as IP hosts on a LAN, GPRS enables mobile access to corporate intranets—without the traditional remote modem. As demand develops, higher data rates of up to 2 Mbps in packet mode will be supported through the use of Asynchronous Transfer Mode (ATM) technology.

## ENHANCED DATA FOR GLOBAL EVOLUTION

The last incremental development before the arrival of 3G will be the introduction of Enhanced Data for Global Evolution (EDGE) technology, which uses an alternative modulation scheme to provide data rates of 300 Kbps and beyond using the standard GSM 200-KHz carrier. EDGE will support both circuit- and packet-switched data and offers a solution for GSM operators who do not win third-generation licenses but wish to offer competitive wideband services.

The enhanced modulation scheme will automatically adapt to the current radio environment to offer the highest data rates for users in good propagation conditions close to the base station sites, while ensuring wider area coverage at lower data speeds. EDGE can also be implemented over TDMA IS-136 networks.

## GSM AND IP

GSM and IP are easily combined for value-added applications. Nokia's GSM Intranet Office, for example, combines GSM and IP telephony in the same network. This solution can substantially cut call costs for larger corporations by using the existing LAN (Figure 9.4).

Voice calls can be made via the company's intranet to various mobile phones, fixed phones, or PCs throughout an office by using any standard GSM mobile phone. In addition, the same mobile phones can be used outside the office, where calls are routed as normal using the operator's GSM network.

By connecting a dedicated base station—specifically, the Nokia InSite Base Station—to the corporate LAN, all intra-office phone calls can be routed locally through the corporate intranet. The InSite Base Station is a single-transceiver base station optimized for indoor picocellular applications. By implementing the mobile phone as the preferred phone for employees, whether in the office or traveling, a corporation can bring down its telecom costs and increase its productivity and efficiency.

GSM Intranet Office also includes a management system to monitor the performance of the IP and GSM networks. It stores information about all network element configurations and software versions. Data from all the network elements can be forwarded to other management systems, such as the Nokia NMS 2000, for further processing and analysis.

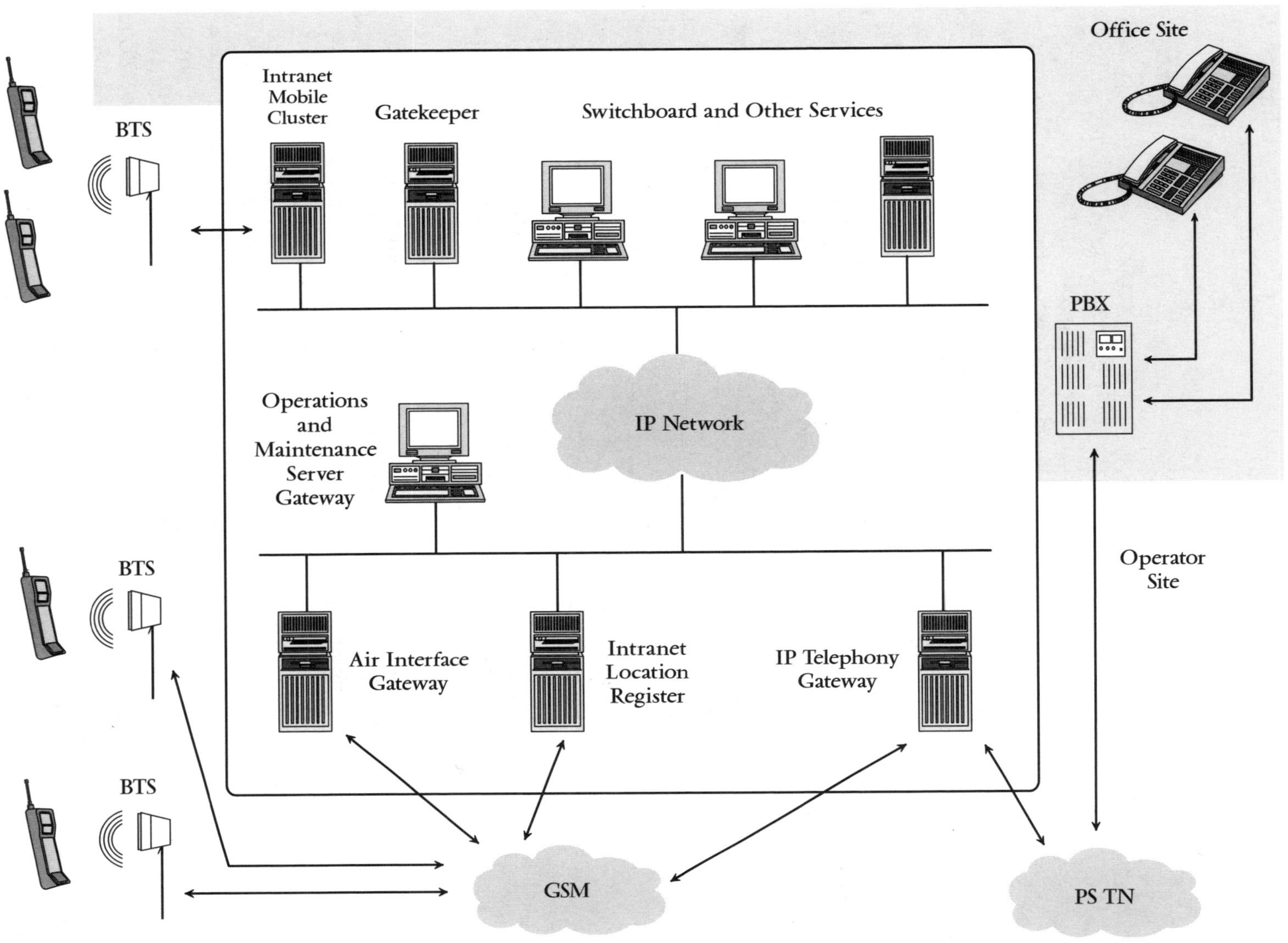

**Figure 9.4** Nokia's GSM Intranet Office architecture.

# Global 3G Initiative

As noted, IMT-2000 will result in wireless access to the global telecommunication infrastructure through both satellite and terrestrial systems, serving fixed and mobile users in public and private networks. This initiative is being developed as a "family of systems" framework, defined as a federation of systems providing advanced service capabilities in a global roaming offering. The initiative aims to facilitate the evolution from today's national and regional 2G systems that are incompatible with one another towards 3G systems that will provide users with genuine global service capabilities and interoperability. The role of the ITU is to provide direction to and coordinate the many related technological developments in this area to assist the convergence of competing national and regional wireless access technologies. Toward these ends, the ITU considered over a dozen national and regional proposals in an effort to select the key characteristics of the IMT-2000 set of radio interfaces.

## Standards Development

In 1992, the World Radio Conference (WRC) identified the frequency bands 1885-2025 MHz and 2110-2200 MHz for future IMT-2000 systems. Of these, the bands 1980-2010 MHz and 2170-2200 MHz were intended for the satellite part of these future systems. With the agreement on frequency bands, along with other important standards in place, work at the ITU focused next on selecting the all-important air interface technology for the system, known as the radio transmission technology (RTT). The need for additional spectrum to cope with the increasing amount of broadband and increasingly interactive traffic of third-generation systems will be addressed at the World Radio Conference in 2000.

For the radio interface technology, the ITU considered 15 submissions from organizations and regional bodies around the world. These proposals were examined by special independent evaluation groups, which submitted their final evaluation reports to the ITU in September 1998. The final selection of key characteristics for the IMT-2000 radio interfaces occurred in March 1999. This led to the development of more detailed ITU specifications for IMT-2000.

The decision of the ITU was to provide essentially a single flexible standard with a choice of multiple-access methods, which include

CDMA, TDMA, and combined TDMA/CDMA—all potentially in combination with Space Division Multiple Access—to meet the many different mobile operational environments around the world. Although second-generation mobile systems involve both TDMA and CDMA technologies, very little use is currently being made of SDMA. However, the ITU expects the advent of adaptive antenna technology linked to systems designed to optimize performance in the space dimension to significantly enhance the performance of future systems.

The IMT-2000 key characteristics are organized, for both the terrestrial and satellite components, into the RF part (front end), where impacts are primarily on the hardware part of the mobile terminal, and the baseband part, which is largely defined in software. In addition to RF and baseband, the satellite key characteristics also cover the architecture and system aspects. According to the ITU, the use of common components for the RF part of the terminals, together with flexible capabilities which are primarily software defined in baseband processing, should provide the mobile terminal functionality to cover the various radio interfaces needed in the twenty-first century as well as provide economies of scale in their production.

The ITU noted that there are already many multi-mode/multi-band mobile units appearing on the market to meet the evolution needs of today's systems and soon there should be negligible impact in areas such as power consumption, size, or cost due to the flexibility defined within the IMT-2000 standard.

The key characteristics of the radio transmission technology by themselves do not constitute an implementable specification but establish the major features and design parameters that will make it possible to develop the detailed specifications.

However, due to the constraints on satellite system design and deployment and because it was felt that little was to be gained at this time from harmonizing any of the satellite proposals, since they were already global, several satellite radio interfaces were included in the March 1999 agreement. Commonalities among elements of the satellite and terrestrial radio interfaces were sought and terrestrial/satellite commonality can be expected to increase further when the second phase of the IMT-2000 satellite component is ready for introduction.

The flexible approach to IMT-2000 implementation provides choice among multiple access methods within a single standard that will address the needs of the worldwide wireless community. Specifically, this approach allows operators to select those radio interfaces that will best address their specific regulatory, financial, and customer needs, while minimizing the impact of this flexibility on end-users.

## Goals of IMT-2000

Under the IMT-2000 model, mobile telephony will no longer be based on a range of market-specific products, but will be founded on common standardized flexible platforms, which will meet the basic needs of major public, private, fixed, and mobile markets around the world. This approach should result in a longer product life cycle for core network and transmission components, and offer increased flexibility and cost effectiveness for network operators, service providers, and manufacturers.

In developing the family of systems that would be capable of meeting the future communications demands of mobile users, the architects of IMT-2000 identified several key issues to be addressed to ensure the success of the third generation of mobile systems.

### HIGH SPEED
Any new system must be able to support high-speed broadband services, such as fast Internet access or multimedia-type applications. Demand for such services is already growing fast and—by some industry projections—the market for broadband services could be worth up to $10 billion by 2010. Users will expect to be able to access their favorite services just as easily from their mobile equipment as they can from their wireline equipment.

### FLEXIBILITY
The next generation of integrated systems must be as flexible as possible, supporting new kinds of services such as universal personal numbering and satellite telephony, while providing for seamless roaming to and from IMT-2000-compatible terrestrial wireless networks. These and other features will greatly extend the reach of mobile systems, benefiting consumers and operators alike.

### AFFORDABILITY
The system must be as affordable as today's mobile communications services, if not more so. The ITU recognizes that the economies of scale achievable with a single global standard would have the benefit of driving down the price to users. While important for all con-

sumers, affordability is vital to extending the penetration of telephony in developing countries. For third-generation equipment to be taken up quickly by consumers, it must deliver at least the same or better service than current systems, and it must be cheap. Even though economies of scale will inevitably bring prices down once sufficient volumes are achieved, if the systems are more expensive and do not initially offer much greater functionality, consumers will not buy.

**COMPATIBILITY**
Any new-generation system has to offer an effective evolutionary path for existing networks. While the advent of digital systems in the late 1980s and early 1990s often prompted the shutting down of first-generation analog networks, the enormous investments which have been made in developing the world's second-generation cellular networks over the last decade makes a similar scenario for adoption of third-generation systems completely untenable.

**DIFFERENTIATION**
In coordinating the design of the IMT-2000 framework, the ITU is also aware of the need to preserve a competitive domain for manufacturers in order to foster incentive and stimulate innovation—mindful that industrial organizations need to have the freedom to compete when it comes to technology. Accordingly, the aim of IMT-2000 standards is not to stifle the evolution of better technologies or innovative approaches, but to accommodate them.

## Universal Mobile Telecommunications System

One of the major new 3G mobile systems being developed within the IMT-2000 framework is the Universal Mobile Telecommunications System (UMTS), which has been standardized by the European Telecommunications Standards Institute (ETSI). UMTS makes use of UTRA (UMTS Terrestrial Radio Access) as the basis for a global terrestrial radio access network. Europe and Japan are implementing UTRA in the paired bands 1920-1980 MHz and 2110-2170 MHz. Europe has also decided to implement UTRA in the unpaired bands 1900-1920 MHz and 2010-2025 MHz.

UMTS combines key elements of TDMA—about 80 percent of today's digital mobile market is TDMA-based—and CDMA technologies with an integrated satellite component to deliver wideband multimedia capabilities over mobile communications networks. The transmission rate capability of UTRA will provide at least 144 Kbps for full-mobility applications in all environments, 384 Kbps for limited-mobility applications in the macro and micro cellular environments, and 2.048 Mbps for low-mobility applications particularly in the micro- and pico-cellular environments. The 2.048-Mbps rate may also be available for short-range or packet applications in the macro cellular environment.

Because the UMTS incorporates the best elements of TDMA and CDMA, this 3G system provides a glimpse of how future wireless networks will be deployed and what possible services might be offered within the IMT-2000 family of systems.

**UMTS OBJECTIVES**

UMTS makes possible a wide variety of mobile services ranging from messaging to speech, data, and video communications, Internet and intranet access, and high bit-rate communication up to 2 Mbps. As such, UMTS is expected to take mobile communications well beyond the current range of wireline and wireless telephony, providing a platform that will be ready for implementation and operation in 2002.

UMTS is intended to provide globally available, personalized, and high-quality mobile communication services. Its design objectives include:

- Integration of residential, office, and cellular services into a single system, requiring one user terminal.
- Speech and service quality at least comparable to current fixed networks.
- Service capability up to multimedia.
- Separation of service provisioning and network operation.
- Number portability independent of network or service provider.
- The capacity and capability to serve over 50 percent of the population.
- Seamless and global radio coverage and radio bearer capabilities up to 144 Kbps and further to 2 Mbps.
- Radio resource flexibility to allow for competition within a frequency band.

## DESCRIPTION

UMTS separates the roles of service provider, network operator, subscriber, and user. This separation of roles makes possible innovative new services, without requiring additional network investment from service providers. Each UMTS user has a unique network-independent identification number and several users and terminals can be associated with the same subscription. This enables one subscription and bill per household to include all members of the family as users with their own terminals. This would give children access to various communication services under their parents' account. This application would also be attractive for businesses, which require cost-efficient system operation—from subscriber/user management down to radio system—as well as adequate subscriber control over the user services.

UMTS supports the creation of a flexible service rather than standardizing the implementations of services in detail. The provision of services is left to service providers and network operators to decide, according to the market demand. The subscriber—or the user when authorized by the subscriber—selects services into individual user service profiles, either with the subscription or interactively with the terminal.

UMTS supports its services with networking, broadcasting, directory, localization, and other system facilities, giving UMTS a clear competitive edge over mobile speech and restricted data services of earlier-generation networks. Being adept at providing new services, UMTS is also competitive in the cost of speech services and as a platform for new applications.

UMTS offers a high-quality radio connection capable of supporting several alternative speech codecs at 2 Kbps to 64 Kbps, as well as image, video, and data codecs. Also supported are advanced data protocols covering a large portion of Integrated Services Digital Network (ISDN) offerings. The concept includes variable and high bit rates up to 2 Mbps.

## FUNCTIONAL MODEL

The UMTS functional model relies on distributed databases and processing, leaving room for service innovations without the need to alter implemented UMTS networks or existing UMTS terminals. This service-oriented model provides three main functions: management and operation of services, mobility and connection control, and network management.

- **Management and Operation of Services**—A service data function (SDF) handles storage and access to service related data. A service control function (SCF) contains overall service and mobility control logic and service related data processing. A service switching function (SSF) invokes service logic—to request routing information, for example. A call control function (CCF) analyzes and processes service requests in addition to establishing, maintaining, and releasing calls.

- **Mobility and Connection Control**—Drawing on the contents of distributed databases, UMTS provides for the real-time matching of user service profiles to the available network services, radio capabilities, and terminal functions. This function will handle mobile subscriber registration, authentication, location updating, handoffs, and call routing to a roaming subscriber.

- **Network Management**—Under UMTS, the administration and processing of subscriber data, maintenance of the network, and charging, billing, and traffic statistics will remain within the traditional Telecommunications Management Network (TMN).

TMN consists of a series of interrelated national and international standards and agreements, which provide for the surveillance and control of telecommunications service provider networks on a worldwide scale. The result is the ability to achieve higher service quality, reduced costs, and faster product integration. TMN is also applicable in wireless communications, cable television networks, private overlay networks, and other large-scale, high-bandwidth communications networks. With regard to UTMS (and other 3G wireless networks) TMN will be enhanced to accommodate new requirements. In areas such as service profile management, routing, radio resource management between UMTS services, networks, and terminal capabilities, new TMN elements will be developed.

## BEARER SERVICES

Under UMTS, four kinds of bearer services will be provided to support virtually any current and future application:

- **Class A**—This bearer service offers constant bit-rate (CBR) connections for isochronous (real-time) speech transmission. It provides a steady supply of bandwidth to ensure the highest quality speech.

- **Class B**—This bearer service offers variable bit-rate connections, which are suited for bursty traffic, such as transaction processing applications.

- **Class C**—This bearer service is a connection-oriented packet protocol, which can be used to support time-sensitive legacy data applications such as those based on IBM's SNA (Systems Network Architecture).
- **Class D**—This is a connectionless packet bearer service suitable for accessing data on the public Internet or private intranets.

## TECHNOLOGY APPROACHES

In developing the UMTS standard, there had been ongoing disagreement within the UMTS Forum about whether to use TD-CDMA (Time Division) or W-CDMA (Wideband) for the radio interface portion of the network.

TD-CDMA uses CDMA signal-spreading techniques to enhance the capacity offered by conventional TDMA technology. Digitized voice and data would be transmitted on a 1.6 MHz-wide channel using time-segmented TDMA technology. Each time slot of the TDMA channel would be individually coded using CDMA technology, thus supporting multiple users per time slot. The proposal establishes an economical and smooth network migration for existing GSM customers to the next-generation cellular standard. At the same time, the TD-CDMA solution allows CDMA technology to be integrated into the TDMA-based GSM structure worldwide, enabling GSM operators to compete for wideband multimedia services, while protecting their current and future investments.

W-CDMA, on the other hand, not only has the advantage of providing high capacity, but is the most widely deployed cellular technology. Proponents of W-CDMA had insisted that this be the air interface standard for UMTS. In January 1998, members of the UMTS Forum, which coordinates standards development, agreed to combine key elements of both TD-CDMA and W-CDMA cellular technologies into a unified solution, called UTRA. In the paired band (FDD—Frequency Division Duplex) of UMTS, the system adopts the radio access technique advocated by the W-CDMA group. In the unpaired band (TDD—Time Division Duplex) the UMTS system adopts the radio access technique advocated by the TD-CDMA group. UTRA offers a competitive continuation for GSM evolution to UMTS and will position UMTS as a leading member of the IMT-2000 family of systems.

## APPLICATIONS

UMTS will comprise a new air interface and new radio components. The aim is to combine these in a modular way with new network components and components from pre-UMTS fixed and mobile networks. This approach will allow new entrants to establish UMTS networks and afford existing operators a smooth migration by reusing parts of their existing infrastructure to the extent possible.

For the user, UMTS will provide adaptive multimode/multi-band terminals or terminals with a flexible air interface to enable global roaming across locations and with second generation systems. Software download to terminals may offer additional flexibility.

By harnessing the best in cellular, terrestrial, and satellite wideband technology, UMTS will guarantee access, from simple voice telephony to high-speed, high-quality multimedia services. It will deliver information directly to users and provide them with access to new and innovative services and applications. It will offer mobile personalized communications to the mass market regardless of location, network, or terminal used.

## U.S. Participation in 3G

United States proposals submitted to the ITU for consideration as the RTT in the IMT-2000 framework included wideband versions of CDMA of which there are three competing standards in North America: wideband cdmaOne, WIMS W-CDMA, and WCDMA/NA. All three have been developed from second-generation digital wireless technologies, and are evolving to third-generation technologies that will best fit their networks. However, early on, WIMS W-CDMA and WCDMA/NA were merged into a single proposed standard and, along with wideband cdmaOne, submitted to the ITU for inclusion into its IMT-2000 family of systems concept for globally interconnected and interoperable 3G networks. A proposal for a TDMA solution for the RTT was also submitted by the Universal Wireless Communications Consortium (UWCC) in the United States.

### CDMA Proposals

As noted, initially there were three competing WCDMA standards in North America: wideband cdmaOne, WIMS (Wireless Multimedia and Messaging Services) W-CDMA, and W-CDMA/NA (W-CDMA North America). Most wireless operators have chosen one of these in building out and enhancing their networks. Competition between these three viable standards has brought innovation in technologies, features, and services, as well as lowered prices.

Wideband cdmaOne technology was submitted to the ITU by the CDMA Development Group (CDG) as cdmaOne-2000. The WIMS W-CDMA technology was submitted to the ITU by AT&T Wireless, Hughes Network Systems, and InterDigital Communications Corporation, among others. The North American GSM Alliance, a group of 12 U.S. and one Canadian digital wireless PCS carriers, submitted the WCDMA/NA technology to the ITU.

There had been talk of combining all these technologies into a single, unified ITU submission. However, only the W-CDMA/NA and WIMS W-CDMA proposals were merged into what was referred to as the enhanced W-CDMA/NA proposal. This aligned proposal offers enhanced data capabilities, such as enabling packet data to be delivered to up to ten times as many users.

Supporters of the enhanced W-CDMA/NA, however, declined to unify their proposal with wideband cdmaOne, claiming that the necessary changes would have caused a significant degradation in system capacity and performance, affected additional capabilities, and probably raised the price of services to customers.

## WIDEBAND cdmaONE

Under the proposed standard of the CDMA Development Group (CDG), wideband cdmaOne will use a CDMA air interface based on the existing TIA/EIA-95-B standard to provide wireline quality voice service and high-speed data services, ranging from 144 Kbps for mobile users to 2 Mbps for stationary users. It will fully support both packet- and circuit-switched communications such as Internet browsing and landline telephone services, respectively.

Support for wideband cdmaOne is not limited to North America. It is found among major wireless carriers in Japan. Korean carriers and manufacturers have also contributed to the development of wideband cdmaOne.

Advanced services require more capacity, robustness, and flexibility than narrowband technologies can provide. Accordingly, CDG and its members have completed work on a specification for a 64-Kbps data rate service. The 64-Kbps data capability will provide high speed Internet access in a mobile environment, a capability that cannot be matched by other narrowband digital technologies, including second-generation CDMA.

The CDG believes that mobile data rates of up to 144 Kbps and fixed peak rates beyond 1.5 Mbps are possible without degrading the system's voice-transmission capabilities or requiring additional spectrum. In other words, cdmaOne is expected to double capacity and provide a 1.5-Mbps data rate capability—all within the existing 1.25 MHz channel structure. At the same time, cdmaOne supports existing second-generation CDMA-based services, including speech coders, packet data services, circuit data services, fax services, Short Messaging Service (SMS), and over-the-air service activation and provisioning.

cdmaOne uses a physical-layer channel structure that shares much of the fundamental/supplemental channel structure from TIA/EIA-95-B. This design provides for simultaneous voice/data structure and procedures that are upwardly compatible with TIA/EIA-95-B.

cdmaOne extends support for multiple simultaneous services far beyond the services in TIA/EIA-95-B by providing much higher data rates and a sophisticated multimedia QoS control capability to support multiple voice/packet data/circuit data connections with differing performance requirements.

The cdmaOne system Medium Access Control (MAC) layer provides extensive enhancements to negotiate multimedia connections, operates multiple concurrent services, and manages QoS tradeoffs between multiple active services in an efficient, structured, and extensible manner. Delivery of these multiple concurrent data streams over the radio interface is accomplished by the cdmaOne Layer 1 (Physical Layer).

Layer 1 supports multiple supplemental channels that can be operated with varying QoS characteristics tailored to the individual service's requirements. For example, one channel can carry circuit data with low bit-error rate (BER) and low latency transmission requirements, while another channel carries packet data that can tolerate a much higher BER and relatively unconstrained latency.

The cdmaOne Physical Layer also supports a dedicated control channel (DCCH) that can be utilized in a number of flexible configurations to provide for independence for competing services (e.g., voice and data), while maintaining a high level of performance. High-speed data service negotiation procedures are extended far beyond TIA/EIA-95-B to include ATM/B-ISDN QoS parameters, including:

- Data rate requirements (CBR, ABR, VBR, etc.).
- Data rate symmetry/asymmetry requirements.
- Tolerable delay/latency characteristics.

The QoS negotiation procedures provide a service functionally equivalent to B-ISDN Q.2931 procedures. This affords ease of implementing transparent multimedia call service via a gateway to ATM/B-ISDN networks (e.g. landline ATM networks).

Additionally, cdmaOne packet data services (i.e., IP) supports QoS negotiation upper-layer protocols such as the Resource ReSerVation Protocol (RSVP) that performs end-to-end service negotiation procedures to provide multimedia call support.

The cdmaOne system also provides extensive capabilities to support highly efficient and cost-effective wireless local loop (WLL) implementations. Optimal radio interface capacity and high single-user throughput are provided. Delay and cell/sector capacity can be traded off to optimize for the desired environment.

Optimized packet data service modes of operation provide Internet service that is highly competitive with wireline and other wireless environments. Voice quality is equal to or better than toll quality and high system capacity provides for a highly competitive replacement for landline voice systems. Improved capacity and single-user throughput using the same cell footprint as an existing TIA/EIA-95-B system permits the integration of WLL services with general cellular high-mobility traffic using the same infrastructure.

**ENHANCED W-CDMA/NA**

As noted, Enhanced W-CDMA/NA represents a merger of WIMS (Wireless Multimedia and Messaging Services), W-CDMA, and W-CDMA/NA (W-CDMA North America). The merger was relatively painless because WIMS W-CDMA had already incorporated key WCDMA/NA elements into its technology, including the chip rate, frame length, an adaptive multi-rate vocoder, and support for asynchronous base stations. This merger offers a technology that allows the acquisition of packet data within 10 milliseconds, which is much faster than other 3G technologies submitted to the ITU for consideration.

The enhanced W-CDMA/NA proposal incorporated two key WIMS elements into its technology: use of multiple parallel orthogonal codes for higher data rates, and a pilot/header structure that enables very rapid packet acquisition and release. This improves data performance and throughput to address the growing marketplace and the demanding requirements of multimedia and Internet based services.

Citing the significant commonality of key technical parameters of chip rate, frame length, asynchronous base station operation, and vocoders, among others, the GSM Alliance endorsed the merging of the two technologies. The GSM Alliance supports multiple 3G standards in the U.S. and other countries.

The third generation of GSM will be an evolution and extension of current GSM systems and available services, optimized for high-speed packet data-rate applications, including high-speed wireless Internet services, video on demand, and other data-related applications. Specifically, third-generation GSM adds the use of CDMA multiplexing at the radio interface level. This is one of the reasons why the GSM Alliance supported the merger of WIMS W-CDMA and W-CDMA/NA.

## TDMA Proposal

As noted, the Universal Wireless Communications Consortium (UWCC) submitted a Time Division Multiple Access (TDMA IS-136) solution for the RTT portion of IMT-2000. Its proponents say that UWC-136 meets the high-speed data application requirements of IMT-2000, enabling telecommunications carriers and vendors to provide these capabilities with minimal costs and maximum benefits. The UWC-136 proposal was developed by the Global TDMA Forum, a technical forum of the UWCC.

UWC-136 is a 100 percent pure TDMA digital solution that provides an evolutionary path to the next generation from IS-136 to IS-136+ to IS-136HS (the high-speed component of UWC-136). IS-136+ will provide extremely high-fidelity voice services and higher-rate packet-data services, up to 64-Kbps using the existing 30-KHz bandwidth. IS-136HS provides user-data rates up to 384 Kbps for wide area coverage in all environments and more than 2 Mbps for in-building coverage.

UWC-136 is a market-driven solution for TDMA IS-136 that enables carriers to implement high-speed data, multimedia, and other applications incrementally to meet market demand. Carriers can retain infrastructure investments and implement data applications while providing quality service delivery to customers—with little impact on the networks and existing spectrum.

In addition to the flexibility of just-in-time build-out, the advantages of UWC-136 include in-building coverage, increased capacity, wireline voice quality, and tightly integrated voice and data services. IS-136+, combined with adaptive channel allocation, advanced modulation, and vocoder enhancements, increases capacity to 10x AMPs while maintaining wireline voice quality.

The hierarchical cell structures of UWC-136 will enable carriers to provide in-building coverage and seamless voice and data services delivery for end-users. UWC-136 supports a common physical level of compatibility across the major TDMA-based technologies for delivery of wideband services. UWC-136 also creates a common technology base with GSM for the provision of wideband services and forms a foundation for the development of a world phone that provides wireless multimedia applications on both D-AMPS IS-136 and GSM networks in all bands.

## Role of Bluetooth

The next generation of cellular telephony, exemplified by the evolution of existing digital systems (GPRS, EDGE, and HSCSD) or the development of entirely new systems (UMTS/IMT 2000), will offer users greater flexibility and more capabilities than ever before. The Bluetooth specification will open up new applications to extend the role of the mobile phone far beyond today's conventional phone service. In fact, the commercial viability of the new developments may very well depend on the success of Bluetooth wireless technology in supporting wireless delivery mechanisms such as cellular telephony. While national networks are suited to delivering communication on the move or providing wireless access to any location, purely local interconnection is better handled by a multi-faceted system like that provided by the Bluetooth specification.

To deliver telephony services from one undefined location to another, and to distribute the services and functions at those locations, requires a hybrid solution, at the core of which is a cellular handset with a built in Bluetooth transceiver. Cellular telephones today are primarily intended for speech and are not particularly good at delivering data. Enhancements to existing second-generation systems, so called 2.5G systems, will allow data to be carried much more easily and at higher rates, typically up to 64 Kbps, though higher rates are possible. The data may be packet switched rather than circuit switched, when required. The third generation of cellular telephony has been designed to carry packet data, and speech is treated as just another data application. Data rates of up to several hundred Kbps will be readily available to the cellular telephone, making it a genuine multimedia terminal. The Bluetooth specification will support enhanced 2G as well as 3G systems in the delivery of a wide range of services from WAN to LANs and PDAs. The Bluetooth specification can provide all the local interconnection, plus a gateway to the national networks when required.

Cordless phone systems are used to access national networks and 3G systems can offer a combined national/cordless service. However, additional handsets are usually needed to provide intercom functionality, and while a national network could directly support this, it would be inefficient and uneconomic to provide this capability for two people in close proximity.

Bluetooth wireless technology is able simultaneously to interconnect up to eight devices in a piconet over a short range. Several piconets can operate in close proximity and Bluetooth devices can rapidly move from one piconet to another. In fact, Bluetooth devices need only remain members of a piconet for the period of time required to complete a communication transaction. Devices can join and leave a local piconet quickly and frequently, effectively overcoming the eight-device limit. Two interconnected piconets forming a scatternet are able to operate within the same area because they use different hopping sequences. This makes it possible to have several small groups of Bluetooth devices communicating with each other in the same area without interference.

Bluetooth devices can transmit and receive data at up to about 1 Mbps, though in reality to allow multiple applications to simultaneously communicate, data rates will be somewhat lower than this. Bluetooth devices not currently part of a piconet are constantly listening for other Bluetooth devices; when they are close enough to become part of a piconet, they identify themselves so that other devices can communicate with them if required.

3G terminals will provide access to many different forms of information and communication such as Web browsing, e-mail transmission and reception, video, and voice, making them true multimedia terminals. Voice will remain a major form of communication and this is recognized in the Bluetooth specifications, which provide specific support for high-quality 64-Kbps speech channels. With the ability to support packet data as well as speech—at the same time if required—the Bluetooth specification can provide full local support for these multimedia applications.

Bluetooth devices can support multiple data connections and up to three voice connections simultaneously, providing the functionality for a three-handset cordless multimedia/intercom system. The limit of three interconnected terminals applies specifically to speech; the limit for the number of terminals exchanging data would be eight per piconet.

A base station using Bluetooth wireless technology would provide the interconnection for the local similarly equipped terminals and also for telephone line connection. When an external connection is required, only two local handsets can participate in voice connections. The base station forms a gateway between the local environment and the various national networks and services. Any two handsets can connect to each other without the base station, making the arrangement very flexible. In some cases, the home PC would be connected to the base station via the Bluetooth specification, so that data could be retrieved from it by calling the base station from any remote location.

This scenario underscores the complementary functionality of Bluetooth and 3G cellular systems. The 3G system is used to provide a connection to a specific location, while Bluetooth wireless technology is used for final delivery to local devices. This considerably reduces the amount of unnecessary traffic on the 3G network, creating a cost-effective solution for the convergence of fixed and mobile services, while keeping RF interference to a minimum.

## Summary

IMT-2000 addresses the key needs of the increasingly global economy—cross-national interoperability, global roaming, high-speed transmission for multimedia applications and Internet access, and customizable personal services. The markets for all these exist now and will grow by leaps and bounds through the next millennium. IMT-2000 puts into place standards that permit orderly migration from current 2G networks to 3G networks, while providing a growth path to accommodate more-advanced mobile services.

Through its small size, considerable functionality and flexibility, and very low cost, Bluetooth wireless technology will find its way into many handheld devices and appliances, offering control and information easily and simply. The new generation of cellular telephony systems, while offering national coverage and mobility, could never provide cost-effective interconnection for so many locally-conected devices. Coupled with the Bluetooth specification, localized devices can be interconnected wherever they are and wherever they are going. Bluetooth wireless technology will extend the reach of cellular systems well beyond today's boundaries.

# APPENDIX A

## Contributors to the Bluetooth Specification

**3Com Corp.**
Jon Burgess
Todor Cooklev
David Kammer
Paul Moran
Ken Morley
Ned Plasson
Richard Shaw

**ComBit, Inc.**
Lawrence Jones

**Convergence**
John Avery
Jason Kronz

**Ericsson Mobile Communications AB**
Christian Andersson
Olof Dellien
Johannes Elg
Ayse Findikli
Jaap Haartsen
Magnus Hansson
Robert Hed
Sven Jerlhagen
Christian Johansson
Gert-jan van Lieshout
Patrik Lundin
Sven Mattisson
Mårten Mattsson
Fisseha Mekuria
Tobias Melin
Ingemar Nilsson
Lars Nord

Lars Novak
Patrik Olsson
Mats Omrin
Joakim Persson
Stefan Runesson
Gerrit Slot
Erik Slotboom
Dan Sönnerstam
Johan Sörensen
Goran Svennarp
Anders Svensson
Fredrik Töörn

**Extended Systems**
Dave Suvak

**IBM Corp.**
Tohru Aihara
Troy Beukema
Chatschik Bisdikian
Brian Gaucher
Parviz Kermani
Edgar Kerstan
Nathan Lee
Brent Miller
Akihiko Mizutani
Dick Osterman
Apratim Purakayastha
Gary Robinson
Aron Walker

**Intel Corp.**
Les Cline
Bailey Cross
Kris Fleming
Uma Gadamsetty
Brad Hosler
John Howard
Robert Hunter
Jon Inouye
Srikanth Kambhatla

# Contributors to the Bluetooth Specification

Kosta Koeman
Steve C. Lo
John McGrath
Shridar Rajagopal
Ramu Ramakeshavan
Jeffrey Schiffer
Vijay Suthar
John Webb
Chunrong Zhu

**Motorola**
Jay Eaglstun
Dale Farnsworth
Patrick Kane
Greg Muchnik
Brian Redding
Jean-Michel Rosso

**Nokia Mobile Phones**
Daniel Bencak
Paul Burgess
Thomas Busse
Michael T. Camp
Uwe Gondrum
Jan Grönholm
Olaf Joeressen
Arno Kefenbaum
Rauno Makinen
Riku Mettälä
Petri Morko
Thomas Müller
Dong Nguyen
Petri Nykänen
Peter Ollikainen
Thomas Sander
James Scales
Roland Schmale
Markus Schetelig
Kevin Wagner
Christian Zechlin

**Puma Technology**
Steve Rybicki
John Stossel

**Toshiba Corp.**
Warren Allen
Allen Huotari
Kazuaki Iwamura
Katsuhiro Kinoshita
Masahiro Tada
Jun'ichi Yoshizawa

**Xtraworx**
Bob Pascoe

The Bluetooth Specification was compiled and edited by Dan Sonnerstam, Pyramid Communication AB.

To stay updated on the status of the Bluetooth specification, including vendor product announcements, check the following Web page:

**www.bluetooth.com**

# APPENDIX B

## Terms and Definitions

**Application Layer**  The group of protocols at the user level. The application layer in the Bluetooth protocol layers will contain those protocols involved with the user interface (UI).

**Asymmetrical**  A type of Asynchronous Connectionless (ACL) link that operates at two different speeds in the upstream and downstream directions. For asymmetrical connections, the Bluetooth specification specifies a maximum data rate of up to 723.2 Kbps in the downstream direction, while permitting 57.6 Kbps in the upstream direction. *See also* Symmetrical.

**Asynchronous**  A form of data communication that brackets each byte with a start bit and a stop bit as the means of synchronizing the transmission between the send and receive devices. With this method of data communication, the transmitter and receiver do not explicitly coordinate each transmission with a clocking mechanism. Instead, the use of start and stop bits determines the boundary of a character. The Bluetooth specification supports one asynchronous data channel that provides an aggregate bandwidth of 778 Kbps. *See also* Synchronous.

**Authentication**  The process of verifying the identity of a device at the other end of the link. In Bluetooth wireless technology systems, this is achieved by the authentication procedure based on the stored link key or by pairing (entering a PIN).

**Authorization**  The act of granting a specific Bluetooth device access to a specific service. It may be based upon user confirmation or the existence of a trusted relationship.

**Baseband**  Describes the specifications of the digital signal processing part of the Bluetooth hardware; specifically, the link controller, which carries out the baseband protocols and other low-level link routines. Baseband refers to the physical layer of the Bluetooth protocol, which among other things manages the physical channels and links. The Baseband specification defines two link types: Synchronous Connection-Oriented (SCO) links and Asynchronous Connectionless (ACL) links. SCO links support real-time voice traffic using reserved bandwidth, while ACL links support data traffic on a best-effort basis. *See also* Asynchronous and Synchronous.

**Beacon Channel** To support parked slaves, the master establishes a beacon channel when one or more slaves are parked. The beacon channel consists of one beacon slot or a train of equidistant beacon slots, and is transmitted periodically with a constant time interval. When parked, the slave will receive the beacon parameters through a Link Management Protocol (LMP) command. *See also* Park Mode.

**Bit** Short for binary digit; a 1 or 0, which when strung together as 8 bits, equals a byte. *See also* Byte.

**Bluetooth** The specification for the wireless communication link, operating in the unlicensed ISM band at 2.4 GHz using a frequency-hopping transceiver. It allows real-time voice and data communications between Bluetooth hosts. The link protocol is based on time slots.

**Bluetooth Development Kit** A set of development tools to enable early adopters of the technology to accelerate the production of prototype applications quickly and easily. A development kit provides a flexible design environment within which engineers can seamlessly integrate the Bluetooth open wireless standard into a range of digital devices for volume production. A variety of interfaces allows for quick development of final applications.

**Bluetooth Device Class** A parameter that indicates the type of device and which types of services are supported. The class is received during the discovery procedure. The parameter contains the major and minor device class fields. The term "Bluetooth device class" is used on the user interface level.

**Bluetooth Device Name** The name of the device (248 bytes maximum).

**Bluetooth Host** A computing device, peripheral, cellular telephone, access point to PSTN network, etc. A Bluetooth host attached to a Bluetooth unit may communicate with other Bluetooth hosts attached to their Bluetooth units as well. The communication channel through the Bluetooth units provides almost wire-like transparency.

**Bluetooth Qualification Body (BQB)** A specific entity authorized by the BQRB to be responsible for checking declarations and documents against requirements, reviewing product test reports, and listing conforming products in the official database of Bluetooth qualified products.

# Terms and Definitions

**Bluetooth Qualification Review Board (BQRB)** The entity responsible for managing, reviewing, and improving the Bluetooth qualification program through which vendor products are tested for conformance. The original Bluetooth SIG companies appoint the BQRB members. *See also* Conformance.

**Bluetooth Radio** A transceiver that transmits and receives modulated electrical signals wirelessly between peer Bluetooth devices.

**Bluetooth Service Type** One or more services a device can provide to other devices. The service information is defined in the service class field of the Bluetooth device class parameter.

**Bluetooth Session** The activity and participation of a device on a piconet.

**Bluetooth Unit** Voice/data circuit equipment for a short-range wireless communication link. It allows voice and data communications between Bluetooth hosts.

**Bonding** A procedure that creates a relationship between two Bluetooth devices based on a common link key. The link key is created and exchanged during the bonding procedure and is stored by both Bluetooth devices for use in future authentication.

**Byte** A sequence of eight bits, providing a binary representation of a character, number, symbol, or keyboard function. *See also* Bit.

**Calling Line Identification Presentation (CLIP)** Refers to the ability to view the number of the calling party before deciding whether to accept the call. This feature is supported in the Cordless Telephony Profile.

**Carrier Sense Multiple Access/Collision Avoidance** A network access method in which each device signals its intent to transmit data before it actually does so. This keeps other devices from sending information, thus preventing collisions between the signals of two or more devices. CSMA/CA is used in Home Radio Frequency (HomeRF) networks, which operate in the same unlicensed 2.4-GHz ISM band as the Bluetooth specification.

**Channel Establishment** Procedure for establishing a Bluetooth channel (a logical link) between two Bluetooth devices using the Bluetooth File Transfer Profile Specification. Channel establishment starts after link establishment is completed and the initiator sends a channel establishment request.

**Class of Device**  A parameter used during the device discovery procedure. This parameter is conveyed from the remote device, indicating the type of device it is and types of services it supports.

**Client**  The Bluetooth device that can push or/and pull data object(s) to and from the server. *See also* Server.

**Clock Offset**  The difference between the slave's clock and the master's clock, the value of which is included in the payload of the Frequency Hop Synchronization (FHS) packet. The clock offset is updated each time a packet is received from the master. The master can request this clock offset anytime during the connection. By saving this clock offset, the master knows on what RF channel the slave wakes up to PAGE SCAN after it has left the piconet. This can be used to speed up the paging time when the same device is paged again.

**Conformance**  When conformance to a Bluetooth profile is claimed by a vendor all mandatory capabilities for that profile must be supported in the specified manner (process-mandatory). This also applies for all optional and conditional capabilities for which support is indicated. All mandatory, optional, and conditional capabilities for which support is indicated are subject to verification as part of the Bluetooth certification program. *See also* Bluetooth Qualification Review Board.

**Connecting**  A phase in the communication between devices when a connection between them is being established. The connecting phase follows after the link establishment phase is completed.

**Connection Establishment**  A procedure for establishing a connection between the applications on two Bluetooth devices. Connection establishment starts after channel establishment is completed. At that point, the initiating device sends a connection establishment request. The specific request used depends upon the application.

**Connectivity Modes**  With respect to paging, a Bluetooth device must be in either the connectable mode or non-connectable mode. When a Bluetooth device is in connectable mode it puts itself into the "page scan" state. Upon receipt of a page, it can respond. When a Bluetooth device is in non-connectable mode it does not put itself into the page-scan state. Since it does not receive pages, it never responds to paging. *See also* Paging.

**Conscious**  Refers to a process that requires the explicit intervention of the device user for its implementation. *See also* Unconscious.

**Continuously Variable Slope Delta (CVSD)**  Modulation technique for converting an analog signal (voice) into a serial bit stream.

This method of encoding voice was chosen for Bluetooth wireless technology for its robustness in handling dropped and damaged voice samples. *See also* Pulse Code Modulation.

**Cordless Telephony Profile**   A "3-in-1 phone" solution for providing an extra mode of operation to cellular phones, using Bluetooth wireless technology as a short-range bearer for accessing fixed-network telephony services via a base station. The 3-in-1 phone solution can also be applied generally for wireless telephony in a residential or small-office environment, such as cordless-only telephony or cordless telephony services in a multimedia PC. *See also* Intercom Profile.

**Coverage Area**   The area where two Bluetooth units can exchange messages with acceptable quality and performance.

**Cyclic Redundancy Check (CRC)**   A computational means to ensure the accuracy of a data packet. The mathematical function is computed before the packet is transmitted at the originating device. Its numerical value is computed based on the content of the packet. This value is compared with a recomputed value of the function at the destination device.

**Delay Variation**   The difference, in microseconds, between the maximum and minimum possible delay that a packet will experience as it goes out over a channel. This value is used by applications to determine the amount of buffer space needed at the receiving side in order to restore the original data transmission pattern.

**Device Discovery**   The mechanism to request and receive the Bluetooth address, clock, class of device, used page scan mode, and names of devices.

**Device Security Level**   Access to a device can be denied based on the required device security level. There are two levels of device security: trusted device and untrusted device *See also* Service Security Level.

**Dialup Networking Profile**   Defines the protocols and procedures used by Bluetooth devices such as modems and cellular phones for implementing the usage model called "Internet Bridge." Among the possible scenarios for this model are the use of a cellular phone as a wireless modem for connecting a computer to a dialup Internet access server, or the use of a cellular phone or modem by a computer to receive data.

**Digital Enhanced Cordless Telephone (DECT)**   A standard that defines the radio connection between two points and can be used for remote access to public and private networks. Extensions of DECT are used in Home Radio Frequency (HomeRF) networks, which operate in the same unlicensed 2.4-GHz ISM band as Bluetooth wireless technology components.

**Discoverable Device**  A Bluetooth device in range that can respond to an inquiry is said to be in a discoverable mode. There are two types of discoverable modes: limited and general. In the first case, a device may be available for discovery for a limited period of time, during temporary conditions, or for a specific event. In the second case, a device may be available for discovery on a continuous basis.

**Discovery**  A term used to describe the protocols and mechanisms by which a network connected device or software service becomes aware of the network to which it is connected and discovers which network services are available. For example, a palmtop needs to discover the home network and find a service that will provide palmtop-to-PC synchronization capabilities. *See also* Salutation.

**Dual Tone Multi-Frequency (DTMF)**  Refers to the ability, when placing external calls, to send a DTMF signal over the external network to call another party. For example, pressing the digit 7 on the device's keypad sends an 852-hertz tone and a 1209-hertz tone simultaneously. The central office switch recognizes these different frequencies and associates them with the numbers dialed. DTMF is supported in the Cordless Telephony Profile.

**Electronic Commerce**  The ability to engage in financial transactions over a network, such as the public Internet, a private intranet, or shared extranet. Bluetooth facilitates mobile electronic commerce by offering a secure link over which goods and services can be ordered and paid for via a handheld device that contains an electronic wallet.

**Events**  Incoming messages to the L2CA layer along with any timeouts. Events are categorized as Indications and Confirms from lower layers, Requests and Responses from higher layers, data from peers, signal Requests and Responses from peers, and events caused by timer expirations.

**Fax Profile**  Defines the protocols and procedures used by Bluetooth devices implementing the fax part of the usage model called "Data Access Points, Wide Area Networks." A Bluetooth cellular phone or modem may be used by a computer as a wireless fax-modem to send or receive fax messages.

**Flow Control**  A procedure for controlling the transfer of data between two connected devices. Wire-based serial port connections implement flow control between devices either through software control using characters such as Transmitter On/Transmitter Off (XON/XOFF) or by using signals such as Request to Send/Clear to Send (RTS/CTS) or Data Terminal Ready/Data Set Ready (DTR/DSR). These methods may be used by both sides of a wired link, or may be used only in one direction. In the Bluetooth specification, the RFCOMM protocol used for emulated serial connections over a wireless link uses its own flow control commands—Flow Control On/Flow Control Off (FCON/FCOFF)—that operate on the aggregate data flow between two devices.

**Frequency Hopping**  The Bluetooth specification uses frequency-hopping spread-spectrum technology, which entails the transmitter's jumping from one frequency to the next at a specific hopping rate in accordance with a pseudo-random code sequence. Bluetooth wireless technology uses 79 hops per second displaced by 1 MHz, starting at 2.402 GHz and stopping at 2.480 GHz. Frequency hopping makes the transmission more secure and resistant to noise and fade. *See also* Spread Spectrum.

**Gateway**  A base station using Bluetooth wireless technology, which is connected to an external network.

**Generic Access Profile (GAP)**  One of four general profiles issued by the Bluetooth SIG, the GAP describes the use of the lower layers of the Bluetooth protocol stack—Link Control (LC) and Link Manager Protocol (LMP). Also defined in this profile are security-related procedures, in which case higher layers—L2CAP, RFCOMM, and OBEX—come into play.

**Generic Object Exchange Profile (GOEP)**  Defines the protocols and procedures used by the applications providing the usage models which need object exchange capabilities. Examples of such usage models include Synchronization, File Transfer, and Object Push. The most common devices using these usage models include notebook PCs, PDAs, smart phones, and mobile phones.

**Global System for Mobile (GSM)**  A cellular digital service standard developed in 1982 that supports both voice and data. RFCOMM is the Bluetooth adaptation of GSM TS 07.10, which provides a transport protocol for serial port emulation.

**Group Abstraction**  A function performed at the Logical Link Control and Adaptation Protocol (L2CAP) layer. The baseband protocol supports the concept of a piconet, a group of devices synchronously hopping together using the same clock. Group abstraction permits implementations to map protocol groups efficiently onto piconets. Without a group abstraction, higher-level protocols would need to be exposed to the baseband protocol and Link Manager (LM) functionality to manage groups efficiently. *See also* Logical Link Control and Adaptation Protocol.

**Headset**  A combination microphone and earpiece used to conduct conversations. Headsets can be connected directly to a cellular device or remotely using Bluetooth communications technology.

**Headset Profile**  Defines the protocols and procedures used by devices implementing the usage model called "Ultimate Headset." The most common examples of such devices are headsets, personal computers, and cellular phones. The headset can be wirelessly connected for the purpose of acting as a remote device's audio input and output interface. The headset increases the user's freedom of movement while maintaining call privacy. In a common scenario, a headset is used with a cellular handset, cordless handset, or personal computer for audio input and output.

**Hidden Computing**  The ability to access and control the functionality of a computer from a peer mobile device. Examples of hidden computing tasks are remote control, data retrieval from mobile devices, and waking up devices from a suspended state.

**Hold Mode**  A mode of operation typically entered into by a device when there is no need to send data for a relatively long time. The transceiver can then be turned off in order to save power. The hold mode can also be used if a device wants to discover or be discovered by other Bluetooth devices, or wants to join other piconets. With the hold mode, capacity can be freed up to do other things like scanning, paging, inquiring, or attending another piconet.

**Home Audio-Video (HAVi)**  A specification for home networks comprised of consumer electronics devices such as CD players, televisions, VCRs, digital cameras, and set-top boxes. The network configuration is automatically updated as devices are plugged in or removed. The IEEE 1394 protocol, also known as FireWire, is used to connect devices on the wired HAVi network at up to 400 Mbps.

**Home RF**  A standard for wireless connectivity in the home, which uses frequency-hopping spread-spectrum technology in the same 2.4 GHz ISM band as Bluetooth wireless technology. Home RF uses 50 hops per second, whereas Bluetooth wireless technology uses 79 hops per second. There is the possibility of interference between devices using the two technologies.

# Terms and Definitions

**Host Terminal Interface** The interface between a Bluetooth host and Bluetooth unit.

**HyperText Transfer Protocol (HTTP)** An application-level protocol for distributed, collaborative, hypermedia information systems. It is a generic, stateless protocol that can be used for many tasks beyond its use for hypertext, such as name servers and distributed object management systems, through extension of its request methods, error codes, and headers. A feature of HTTP is the typing and negotiation of data representation, allowing systems to be built independently of the data being transferred. HTTP has been in use by the World Wide Web global information initiative since 1990. The Object Exchange (OBEX) authentication scheme is based on HTTP, but does not support all of its features and options.

**Idle Mode** A device is in idle mode when it has no established links to other devices. In this mode, the device may discover other devices. In general, a device sends inquiry codes to other devices. Any device that allows inquiries will respond with information. If the devices decide to form a link, then bonding will occur.

**Industrial, Scientific, Manufacturing (ISM)** These bands of the electromagnetic spectrum include frequency ranges at 902-928 MHz and 2.4-2.484 GHz, which do not require an operator's license from the Federal Communications Commission (FCC) or the regulatory agencies of other countries. Bluetooth wireless technology operates in the upper band of 2.4 GHz.

**Infrared** A method of communication that uses 850-nanometer (nm) infrared light for data transfers between electronic devices. This type of signal must have a clear, straight path from one device to another. Typically, the distance between devices is no more than 3 feet. The data rate can be as high as 16 Mbps.

**Initiator** The Bluetooth device initiating an action to another Bluetooth device. The device receiving the action is called the acceptor. The initiator is typically part of an established link.

**Inquiry** A procedure whereby a unit transmits messages in order to discover the other units that may be active within the coverage area. The units that capture inquiry messages may send responses to the inquiring unit. These responses contain information about the unit itself and its Bluetooth host.

**Inquiry Procedure** The inquiry procedure enables a device to discover other devices that are in range and determine the addresses and clocks for the devices. After the inquiry procedure has been completed, a connection can be established using the paging procedure.

**Inter-Piconet Capability** The capability of a master device to keep the synchronization of a piconet while page scanning in free slots and allowing new members to join the piconet. While a new unit is in the process of joining the piconet, and until the master-slave switch is performed, operation may be temporarily degraded for the other members. A gateway that supports multiple terminals has the inter-piconet capability. The terminals also may have the inter-piconet capability. *See also* Piconet.

**Intercom Call** Refers to a call originating from a terminal toward another terminal. An intercom call between two terminals can be set up with gateway support if the two terminals are members of the same Wireless User Group, or WUG. *See also* Cordless Telephony Profile, Intercom Profile, Wireless User Group.

**Intercom Profile** Defines the protocols and procedures that would be used by Bluetooth devices for implementing the intercom part of the usage model called "3-in-1 phone," also known as the "walkie-talkie" application of the Bluetooth specification.

**Internet Protocol (IP)** The layer 3 protocol that delivers datagrams between different TCP/IP networks through routers that process packets from one autonomous system (AS) to another.

**Isochronous User Channel** The channel used for time-bounded information like compressed audio.

**Java** A network programming language originally developed by Sun Microsystems. It is intended for cross-platform networks, particularly those that call for the thin-client model of computing. Java is actually a scaled down version of the C++ programming language that omits many seldom-used features, while adding an object orientation. Java provides a cleaner, simpler language that can be processed faster and more efficiently than C or C++ on nearly any microprocessor, including memory-constrained embedded devices equipped with Bluetooth wireless technology.

**Jini** A connection technology developed by Sun Microsystems that provides simple mechanisms which enable devices to plug together to form an impromptu community—a community put together without any planning, installation, or human intervention. Each device provides services that other devices in the community may use. These devices provide their own interfaces, which ensures reliability and compatibility. Jini works at higher layers, while the Bluetooth specification works at much lower layers.

**LAN Access Profile** Defines how the Point-to-Point Protocol (PPP) is supported with regard to LAN access for a single Bluetooth device, LAN access for multiple Bluetooth devices, and PC-to-PC communication using PPP networking over serial cable emulation. *See also* Point-to-Point Protocol.

**Latency** The maximum acceptable delay between transmission of a bit by the sender and its initial transmission over the air, expressed in milliseconds or microseconds.

**Legacy Application** In link-level security scenarios, a legacy application is one that cannot make calls to the security manager on its own. Instead, an "adapter" application that conforms to the Bluetooth specification is required to make security-related calls to the Bluetooth security manager on behalf of that application.

**Link Establishment** A procedure is used to setup a physical link—specifically, an Asynchronous Connectionless (ACL) link—between two Bluetooth devices using procedures from the Bluetooth IrDA Interoperability Specification and Generic Object Exchange Profile.

**Link Manager** Software that carries out link setup, authentication, link configuration, and other management functions. It provides services like encryption control, power control, and quality of service (QoS) capabilities. It also manages devices in different modes of operation (park, hold, sniff, and active).

**Link Supervision** Each Bluetooth link has a timer used for link supervision. This timer is used to detect link loss caused by devices moving out of range, a device's power-down, or failure cases. The scheme for link supervision is described in Bluetooth's Baseband Specification.

**Local Device (LocDev)** The device that initiates the service discovery procedure under the Service Dicovery Profile (SDP). To accomplish this, the LocDev must contain at least the client portion of the Bluetooth SDP architecture, specifically the service discovery application (SrvDscApp), which enables the user to initiate discoveries and display the results of these discoveries. *See also* Remote Device.

**Logical Channel** A channel that rides over a physical link.

**Logical Link Control and Adaptation Protocol (L2CAP)** A Bluetooth protocol that resides at the data-link layer (Layer 2) of the OSI reference model and provides connectionless and connection-oriented data services to upper-layer protocols with protocol multiplexing capability, segmentation and reassembly (SAR) operation, and group abstractions. L2CAP permits higher-level protocols and applications to transmit and receive L2CAP data packets of up to 64 Kbps in length.

**MAC Address**  The medium access control (MAC) address used to distinguish between units participating in a piconet.

**Master**  The device in a piconet whose clock and hopping sequence are used to synchronize all other devices (i.e., slaves) in the piconet. The unit that carries out the paging procedure and establishes a connection is by default the master of the connection. Knowledge about the clock will accelerate the setup procedure. *See also* Slave.

**Mode**  A set of directives that define how a Bluetooth device responds to certain events. There are modes for discoverability, connectability, and pairing. *See also* Discovery, Connectivity Mode, Pairing.

**Name Discovery**  A procedure that provides the initiating device with the device name of other connectable devices—specifically, Bluetooth devices within range that will respond to paging. The request is targeted at known devices for which the device addresses are available.

**OBEX**  A session-layer protocol for object exchange originally developed by the Infrared Data Association (IrDA) as IrOBEX. Its purpose is to support the exchange of objects in a simple and spontaneous manner over an infrared or Bluetooth wireless link.

**On Hook**  Refers to the ability of a terminal to terminate a call (i.e., hang up), thereby releasing all radio resources that supported the call. This capability is supported in the Cordless Telephony Profile.

**Open Systems Interconnection (OSI)**  The seven-layer reference model first defined in 1978 by the International Organization for Standardization (ISO). The lower layers (1 to 3) represent local communications, while the upper layers (4 to 7) represent end-to-end communications. Each layer contributes protocol functions necessary to establish and maintain the error-free exchange of information between network users.

**Packet**  In the Bluetooth specification, the format of aggregated bits that comprise a discrete unit of information that can be transmitted in 1, 3, or 5 time slots. The format of packets varies according to the transmission technology employed. Packets can vary in length up to about 12,000 bits for Ethernet and IP and up to about 32,000 bits for frame relay. By comparison, the standard data packet conveyed over a piconet using Bluetooth wireless technology varies in length up to about 2,800 bits.

**Page Scan**  The process whereby a device listens for page messages containing its own Device Access Code (DAC).

**Paging**  A procedure that involves transmitting a series of messages with the objective of setting up a communication link to a unit active within the coverage area. If a remote unit responds to the page, a connection is established.

**Paired Device**  A Bluetooth device with which a link key has been exchanged, either before connection establishment was requested or during the connecting phase.

**Pairing**  An initialization procedure whereby two devices communicating for the first time create a common link key that will be used for subsequent authentication. For first-time connection, pairing requires the user to enter a Bluetooth security code or PIN.

**Park Mode**  When a slave does not need to participate on the piconet channel, but still wants to remain synchronized to that channel, it can enter the park mode, which is a low-power mode with very little activity in the slave. The parked slave wakes up at regular intervals to listen to the channel in order to resynchronize and to check for broadcast messages.

**Payload**  As applied to a data field, a payload is a discrete package of information, also called a packet, which contains a header, user data, and a trailer. In the Bluetooth specification, the format for a data field consists of a payload header that specifies the logical channel, controls the flow on the logical channel, and indicates the length of the payload body. The payload body consists of the user data. The "trailer" consists of a Cyclic Redundancy Check (CRC) code for ensuring the accuracy of the packet.

**Peak Bandwidth**  Expressed in bytes per second, this limits how fast packets may be sent back-to-back from applications. Some intermediate systems can take advantage of this information, so that more efficient resource allocation results.

**Personal Area Network**  A network concept in which all the devices in a person's life communicate and work together, sharing each other's information and services.

**Personal Digital Assistant**  Hand-held computers that are equipped with an operating system, applications software, and communications capabilities for short-text messaging, e-mail, fax, news updates, and voice mail. They are intended for mobile users who require instant access to information, regardless of their location at any given time. Such devices are prime candidates for embedded Bluetooth functionality.

**Personal Identification Number**  The Bluetooth PIN is used to authenticate two devices that have not previously exchanged a link key. By exchanging a PIN, the devices create a trusted relationship. The PIN is used in the pairing procedure to generate the initial link used for further identification.

**Pervasive Computing**  Refers to the emerging trend toward numerous, casually accessible, often invisible computing devices. The computing devices are frequently mobile or embedded in the environment, and connected to an increasingly ubiquitous network structure. The objective is to facilitate computing anywhere it is needed.

**Physical Channel**  The synchronized radio frequency (RF) hopping sequence in a piconet. This baseband-level association between two devices is established using paging.

**Physical Link**  A sequence of transmission slots on a physical channel alternating between master and slave transmission slots.

**Piconet**  The units sharing a common channel constitute a piconet. Up to eight interconnected devices can be supported on a single piconet, consisting of one master and seven slaves. This relationship remains in place for the duration of the piconet connection. *See also* Inter-Piconet Capability.

**Personal Identification Number**  Used to authenticate two devices that have not previously exchanged a link key. By exchanging a PIN, the devices create a trusted relationship. The PIN is in the pairing procedure to generate the initial link used for further identification.

**Point-to-Point Protocol (PPP)**  A widely deployed means of allowing access to networks, PPP provides authentication, encryption, data compression, and support for multiple protocols. PPP over RFCOMM provides LAN access for Bluetooth devices. Although PPP is capable of supporting various networking protocols (e.g. IP, IPX, etc.), the LAN Access Profile does not require the use of any particular protocol. *See also* LAN Access Profile.

**Post-dialing**  Refers to the ability of a terminal to send dialing information after the outgoing call request setup message is sent.

**Power Mode**  The gateway is normally the master of the piconet in the Cordless Telephony Profile. As such, it controls the power mode of the terminals (i.e., slaves). The gateway is as conservative as possible when deciding what power mode to put the terminals in. When a terminal is not engaged in signaling, the gateway puts it in a low-power mode. The recommended low-power mode is referred to as park mode. The low-power mode parameters chosen are such that the terminal can always return to the active state within 300 milliseconds (ms). If the gateway can save power during a call, it may use the sniff mode. A terminal may also request to be put in the sniff mode.

**Primitives**  Abstract descriptions of particular functions used by a service. Examples of primitives used in conjunction with the Service Discovery Profile (SDP) are openSearch(.) and closeSearch(.), which indicate "open search" and "close search," respectively. The services offered by L2CAP are also described in terms of service primitives.

**Private**  A mode of operation whereby a device can only be found via Bluetooth baseband pages; it only enters into page scans. *See also* Public.

**Profile**  Defines the protocols and features that support a particular usage model. If devices from different manufacturers conform to the same Bluetooth SIG profile specification, they will interoperate when used for that particular service and usage case.

**Protocol**  Defines the communication functions between two peer devices at a certain layer of a given protocol stack.

**Protocol Analyzer**  A troubleshooting tool that analyzes all levels of the Bluetooth protocol stack, including the baseband. The device can capture specific Bluetooth traffic from a piconet in real time, allowing the technician to view and record traffic packet by packet or focus on a particular transaction level by filtering out certain fields, packets, errors, and other items. Filtering provides the technician with only the relevant information needed to identify and highlight abnormal network traffic conditions quickly and efficiently. This recorded data are then displayed using color and graphics in a Windows environment. Advanced search and viewing capabilities allow the technician to locate specific data, errors, and other desired conditions quickly.

**Protocol Data Unit (PDU)**  A generic term for the header-data package used at any layer of a protocol stack.

**Protocol Multiplexing**  A function performed at the Logical Link Control and Adaptation Protocol (L2CAP) layer. L2CAP must support protocol multiplexing because the baseband protocol does not support a "type" field to identify the higher-layer protocol being multiplexed above it. L2CAP must therefore be able to distinguish between upper-layer protocols such as the Service Discovery Protocol (SDP), Radio Frequency Communication (RFCOMM), and the Telephony Control Specification (TCS).

**Public** A mode of operation whereby a device can be found via Bluetooth baseband inquiries; that is, it enters into inquiry scans. A public device also enters into page scans. *See also* Private.

**Pulse Code Modulation (PCM)** A common method of encoding an analog voice signal into a digital bit stream in which the voice signal is sampled 8,000 times a second. Each sampled point is assigned a value, which is put into binary form as 8-bit bytes. The resulting 1s and 0s constitute the digital bit stream, which is conveyed over a wired or wireless link. At the receiving device, the 1s and 0s are used to reconstruct the original voice signal so it can be understood as ordinary speech. The Bluetooth specification supports both A-law and µ-law PCM. The U.S. has standardized on the µ-law version of PCM, while Europe and many other countries in Asia and South America and Africa have standardized on the A-law version of PCM. *See also* Continuously Variable Slope Delta modulation.

**Push-Pull** Under the Generic Object Exchange Profile (GOEP), a client sends (pushes) data objects to the server, or retrieves (pulls) data objects from the object exchange server.

**Quality of Service (QoS)** With regard to L2CAP signaling, guaranteed performance for an application may be requested during connection setup. Such parameters as delay variation (microseconds), peak bandwidth (bytes/second), and latency (microseconds) may be specified in a QoS configuration request. If no QoS option is specified in the request, best-effort service is assumed.

**Register Recall** Refers to the ability of the terminal to request "register recall," and of the gateway to transmit the request to the local network. Register recall means to seize a register (with dial tone) to permit input of further digits or perform other actions. This is also known as "flash hook." This feature is supported in the Cordless Telephony Profile.

**Remote Device (RemDev)** Any device that participates in the service discovery process by responding to the service inquiries generated by a local device (LocDev) under the Service Dicovery Profile (SDP). To accomplish this, the RemDev must contain at least the server portion of the Bluetooth SDP architecture. A RemDev contains a service records database, which the server portion of SDP consults to create responses to service discovery requests. *See also* Local Device.

**RFCOMM** A simple transport protocol with additional provisions for emulating the nine circuits of RS-232 (EIA/TIA-232-E) serial ports over the L2CAP protocol. This serial cable emulation protocol is based on ETSI TS 07.10.

**Salutation** A procedure for looking up, discovering, and accessing services and information. This architecture defines abstractions for devices, applications, and services; a capabilities exchange protocol; a service request protocol; standardized protocols for common services; and application programming interfaces (APIs) for information access and session management. *See also* Salutation Architecture.

**Salutation Architecture** A service discovery standard developed by the Salutation Consortium to help solve the problems of service discovery and utilization among the growing amount of appliances and equipment becoming available from a broad range of vendors. The architecture is processor, operating system, and communication protocol independent. As such the Salutation Architecture provides a standard method for applications, services, and devices to advertise and describe their capabilities to other applications, services, and devices, as well as to discover the capabilities of any other entity. Once the capabilities of other entities are known, interoperable sessions can be requested and established to utilize those capabilities. The Salutation Architecture has implementations modeled in TCP/IP, InfraRed Data Association standards, the Bluetooth specification, and the Internet Engineering Task Force's (IETF) Service Location Protocol (SLP), a protocol for automatic resource discovery on IP networks. *See also* Service Location Protocol.

**Scatternet** Two or more piconets colocated in the same area, with or without inter-piconet communication.

**Secure Sockets Layer (SSL)** An Internet protocol that provides encryption and authentication to safeguard communications between a client and server. *See also* Wireless Transport Layer Security.

**Security Manager** The module in a Bluetooth device that controls the security aspects of communications to other Bluetooth devices.

**Security Modes** There are three security modes for Bluetooth devices. When in security mode 1, the device never initiates a security procedure. When in security mode 2, the device does not initiate any security procedure before a channel establishment request has been received or it has initiated a channel establishment procedure. When in security mode 3, the device initiates security procedures before it sends a message indicating that link setup is complete.

**Segmentation and Reassembly**  A function performed at the Logical Link Control and Adaptation Protocol (L2CAP) layer, which allows Bluetooth devices to support protocols using packets larger than those supported by the baseband. Large L2CAP packets are segmented into multiple smaller baseband packets prior to their transmission over the air. On the receive side, the baseband packets are reassembled into a single larger L2CAP packet following a simple integrity check. *See also* Logical Link Control and Adaptation Protocol.

**Serial Port Profile (SPP)**  Defines the requirements for setting up emulated serial cable connections between two peer Bluetooth devices using RFCOMM, a simple transport protocol that emulates RS-232 serial ports between two peer devices.

**Server**  The data storage device to and from which data objects can be pushed and pulled, respectively. *See also* Client.

**Service Discovery**  The ability to discover the capability of connecting devices or hosts.

**Service Discovery Application Profile (SDAP)**  Defines the procedures whereby an application in a Bluetooth device can discover services registered in other Bluetooth devices and retrieve information pertinent to the implementation of these services.

**Service Discovery Application (SrvDscApp)**  The application in a local device (LocDev) that actually performs service inquiries against already connected remote devices.

**Service Discovery Protocol (SDP)**  One of four general profiles issued by the Bluetooth SIG, the SDP describes the protocol used by applications to discover what services are available to them and to determine the characteristics of those services.

**Service Location Protocol**  A standard issued by the Internet Engineering Task Force (IETF) that provides a scalable framework for the discovery and selection of network services. Using this protocol, computers on a LAN, a corporate intranet, or public Internet need little or no static configuration of network services for network based applications. This is an important concern, especially as computers become more portable, and users less tolerant of or able to fulfill the demands of network system administration.

**Service Security Level**  Access to services can be denied based on the required service security level. There are three levels of service security: authorization and authentication, authentication only, and no security. Encryption can be another security requirement for service use, but this is typically applied at the baseband or physical level. *See also* Device Security Level.

**Silent Device**  A Bluetooth device appears as silent to a remote device if it does not respond to its inquiries. A device may be silent due to being non-discoverable, or because of baseband congestion while it is discoverable.

**Slave**  A unit within a piconet that is synchronized to the master via its clock and hopping sequence. Up to seven slaves may be related in this way to a master within the same piconet. *See also* Master.

**Smart Phones**  Digital cellular phones that offer enhanced communications capabilities such as Internet access for e-mail and Web browsing for such items as news, stock quotes, sports scores, weather reports, and travel information. Smart phones usually support the Wireless Application Protocol (WAP) and are intended for mobile users who require instant access to information, regardless of their location at any given time. Such devices are prime candidates for embedded Bluetooth functionality.

**Sniff mode**  A mode of operation in which the duty cycle of a slave's listen activity can be reduced to save power. If a slave participates on an ACL link, it has to listen in every ACL slot to the master traffic. With the sniff mode, the time slots where the master can start transmission to a specific slave are reduced; that is, the master can only start transmission in specified time slots. These "sniff" slots are spaced at regular intervals.

**Spread Spectrum**  A digital coding technique in which the signal is taken apart or "spread" so that it sounds more like noise to the casual listener. Using the same spreading code as the transmitter, the receiver correlates and collapses the spread signal back down to its original form. With the signal's power spread over a larger band of frequencies, the result is a more robust signal less susceptible to impairment from electromechanical noise and other sources of interference. It also makes voice and data communications more secure. *See also* Frequency Hopping.

**Symmetrical**  A type of Asynchronous Connectionless Link (ACL) that offers the same data rate in both the send and receive directions. For symmetrical connections, the Bluetooth specification specifies a maximum data rate of 433.9 Kbps in both directions, send and receive. *See also* Asymmetrical.

**Synchronous** Describes transmission that relies on a common clock for synchronizing the flow of data between devices. Synchronization occurs at the beginning of the session so that timing (not start and stop bits, as in asynchronous communication) defines the data boundaries. A unique bit pattern embedded in the digital signal assists in maintaining the timing between the sending and receiving devices. *See also* Asynchronous.

**Telephony Control protocol—Binary (TCS Binary)** Defines the call control signaling for the establishment of speech and data calls between Bluetooth devices. In addition, it defines mobility management procedures for handling groups of Bluetooth TCS devices. The Bluetooth SIG has defined the set of AT commands by which a mobile phone and modem can be controlled in the multiple usage models.

**Third Generation** The next generation of cellular radio for mobile telephony. Due to come on stream from 2001 onward, 3G will be the first cellular radio technology designed from the outset to support wideband data communications just as well as it supports voice communications. It will be the basis for a wireless information society where access to information and information services such as electronic commerce is available anytime, anyplace, and anywhere to anybody. 3G's technical framework is being defined by the International Telecommunication Union (ITU) with its International Mobile Telecommunications 2000 (IMT-2000) program. Bluetooth wireless technology will support 3G systems on a localized basis in the delivery of a wide range of services as well the current 2G systems.

**Time Division Multiple Access** A method of dividing the total available bandwidth so that all the users sharing the physical resource have their own assigned, repeating time slot within a group of time slots called a frame. A time slot assignment is often called a channel. TDMA is used in Home Radio Frequency (HomeRF) networks, which operate in the same unlicensed 2.4-GHz ISM band as Bluetooth wireless technology devices.

**Time Slot** In the Bluetooth specification, the physical channel is divided into 625 µs long time slots.

**Token Bucket Size** The size of the token bucket (i.e., buffer) in bytes. If the bucket is full, then applications must either wait or discard data. For best-effort service, the application gets a bucket as big as possible. For guaranteed service, the maximum buffer space will be available to the application at the time of the request. *See also* Token Rate.

## Terms and Definitions

**Token Rate** The rate at which traffic credits are granted in bytes per second. An application may send data at this rate continuously. Burst data may be sent up to the token bucket size. Until that data burst has been drained, an application must limit itself to the token rate. For best-effort service, the application gets as much bandwidth as possible. For guaranteed service, the application gets the maximum bandwidth available at the time of the request. *See also* Token Bucket Size.

**Transmission Control Protocol (TCP)** Provides a reliable connection between devices at the transport layer with IP (Internet Protocol) in the network layer. IP provides protocol multiplexing and connections based on IP addresses. *See also* Internet Protocol.

**Trusted Device** A device using Bluetooth wireless technology that has been previously authenticated and allowed to access another Bluetooth device based on its link-level key. *See also* Untrusted Device.

**Unconscious** Refers to a process that requires no explicit intervention of the device user for its implementation. *See also* Conscious.

**Unknown Device** A Bluetooth device that is currently not connected with a local device and the local device has not paired with it in the past. No information about the device is stored, either by address, link key, or other information. An unknown device is also called a new device.

**Untrusted Device** A Bluetooth device that is unknown to another Bluetooth device; it may require authorization based on some type of user interaction before access is granted.

**User Datagram Protocol** A simple protocol that passes individual messages to IP for transmission across the network on a best-effort basis.

**vCalendar** A format for calendaring and scheduling information. The vCalendar specification was created by the Versit consortium and is now managed by the Internet Mail Consortium (IMC).

**vCard** A format for personal information such as would appear on a business card. The vCard specification was created by the Versit consortium and is now managed by the Internet Mail Consortium (IMC).

**Walkie-Talkie** A mode of voice communication in which one subscriber talks and other subscribers listen on the same talk group.

**Wireless Application Protocol (WAP)** A specification for sending and reading Internet content and messages on small wireless devices, such as cellular phones equipped with text displays. The Bluetooth specification supports WAP.

**Wireless Connections Made Easy** This basic Bluetooth philosophy encapsulates the following important end-user experiences: reliable high-quality radio links, interoperability between products of any brand, and easily understood product capabilities. According to the Bluetooth SIG, a reliable radio link experience depends on all products' demonstrating compliance with the Bluetooth radio link performance specifications. Interoperability is achieved by protocol and profile implementation conformance. Ease of use depends on clear, consistent documentation of Bluetooth capabilities in product literature. All these elements are addressed in the requirements for Bluetooth compliance.

**Wireless LAN** Usually refers to wireless network based on the IEEE 802.11 standard. An 802.11 wireless LAN provides the means to interconnect wireless stations to each other and to the conventional wired network at up to 11 Mbps using direct sequence spread spectrum technology, or 1 Mbps or 2 Mbps using frequency-hopping spread-spectrum technology. *See also* Home RF.

**Wireless Session Protocol (WSP)** Establishes a relationship between a client application and the WAP server. This session is relatively long lived and able to survive service interruptions. The WSP uses the services of the Wireless Transport Protocol (WTP) for reliable transport to the destination proxy/gateway. *See also* Wireless Transport Protocol.

**Wireless Transport Layer Security (WTLS)** An optional component of the Wireless Application Protocol (WAP) stack that provides a secure data pipe for the session between a client and server. WTLS is derived from the Secure Sockets Layer (SSL) specification. As such, it performs the same authentication and encryption services as SSL. *See also* Secure Sockets Layer.

**Wireless Transport Protocol (WTP)** Provides services that fill many of the same requirements as TCP. On the Internet, the Transmission Control Protocol (TCP) provides a reliable, connection-oriented, character-stream protocol based on IP services. In contrast, WTP provides both reliable and unreliable one-way, and reliable two-way message transports. The transport is optimized for the short request, long response dialogues used by the Wireless Application Protocol (WAP). WTP also provides message concatenation to reduce the number of messages transferred over the wireless link. *See also* Wireless Application Protocol.

# Terms and Definitions

**Wireless User Group**  A group of terminals restricted to communicating only with each other under the control of a master. In not allowing calls from non-group members, the WUG also enhances security. This is a feature supported by the Cordless Telephony Profile.

# APPENDIX C

## Acronyms

| | |
|---|---|
| 2G | Second Generation |
| 3G | Third Generation |

### A

| | |
|---|---|
| ABR | Available Bit Rate |
| ACC | Analog Control Channel |
| ACK | Acknowledge |
| ACL | Asynchronous Connectionless |
| ACO | Authenticated Ciphering Offset |
| ADSL | Asymmetrical Digital Subscriber Line |
| AG | Audio Gateway |
| AM_ADDR | Active Member Address |
| AMPS | Advanced Mobile Phone Service |
| AP | Access Point |
| API | Application Programming Interface |
| AR_ADDR | Access Request Address |
| ARQ | Automatic Repeat Request |
| AS | Autonomous System |
| ASCII | American Standard Code for Information Interchange |
| ATM | Automated Teller Machine; Asynchronous Transfer Mode |
| AVC | Analog Voice Channel |

### B

| | |
|---|---|
| B-ISDN | Broadband Integrated Services Digital Network |
| BB | Baseband |
| BBS | Bulletin Board System |
| BCCH | Broadcast Control Channel |
| BCH | Bose, Chaudhuri & Hocquenghem (Type of code; named for the people who discovered these codes in 1959 [H] and 1960 [B&C]) |
| BD_ADDR | Bluetooth Device Address |
| BER | Bit Error Rate |

| | |
|---|---|
| BQB | Bluetooth Qualification Body |
| BQTF | Bluetooth Qualification Test Facility |
| BT | Bandwidth Time |

## C

| | |
|---|---|
| CAC | Channel Access Code |
| CBR | Constant Bit Rate |
| CC | Call Control |
| CCCH | Common Control Channel |
| CCF | Call Control Function |
| CDG | CDMA Development Group |
| CDMA | Code Division Multiple Access |
| CHAP | Challenge Handshake Authentication Protocol |
| CID | Channel Identifier |
| CKPD | Control Keypad (an AT-based command issued from a keyboard) |
| CL | Connectionless |
| CLIP | Calling Line Identification Presentation |
| CO | Connection Oriented |
| CoD | Class of Device |
| CODEC | COder DECoder |
| COF | Ciphering Offset |
| CR | Carriage Return |
| CRC | Cyclic Redundancy Check |
| CSMA/CA | Carrier Sense Multiple Access with Collision Avoidance |
| CSMA/CD | Carrier Sense Multiple Access with Collision Detection |
| CTP | Cordless Telephony Profile |
| CTR | Clear to Send |
| CVSD | Continuous Variable Slope Delta |

## D

| | |
|---|---|
| DAC | Device Access Code |
| DAMPS | Digital Advanced Mobile Phone System |
| DC | Direct Current |
| DCCH | Digital Control Channel |
| DCE | Data Communication Equipment |
| DCI | Default Check Initialization |
| DECT | Digital Enhanced Cordless Telephone |
| DH | Data-High Rate |

# Acronyms

| | |
|---|---|
| DIAC | Dedicated Inquiry Access Code |
| DM | Data—Medium Rate |
| DNS | Domain Name System |
| DQPSK | Differential Quadrature Phase-Shift Keying |
| DSL | Digital Subscriber Line |
| DSR | Data Set Ready |
| DT | Data Terminal |
| DTC | Digital Traffic Channel |
| DTE | Data Terminal Equipment |
| DTMF | Dual Tone Multiple Frequency |
| DTR | Data Terminal Ready |
| DUT | Device Under Test |
| DV | Data-Voice |
| DVD | Digital Video Disc |

## E

| | |
|---|---|
| EDGE | Enhanced Data for Global Evolution |
| EIA | Electronic Industries Alliance (formerly, Electronic Industries Association) |
| EMV | Europay-MasterCard-Visa |
| ETSI | European Telecommunications Standards Institute |

## F

| | |
|---|---|
| FCC | Federal Communications Commission |
| FACCH | Fast Associated Control Channel |
| FCOFF | Flow Control Off |
| FCON | Flow Control On |
| FDMA | Frequency Division Multiple Access |
| FEC | Forward Error Correction |
| FH | Frequency Hopping |
| FHS | Frequency Hop Synchronization |
| FIFO | First In First Out |
| FSK | Frequency Shift Keying |
| FTP | File Transfer Protocol |

## G

| | |
|---|---|
| GAP | Generic Access Profile |
| Gbps | Gigabits per second |
| GEOP | Generic Object Exchange Profile |
| GFSK | Gaussian Frequency Shift Keying |

| | |
|---|---|
| GHz | Gigahertz (millions of hertz) |
| GIAC | General Inquiry Access Code |
| GM | Group Management |
| GOEP | Generic Object Exchange Profile |
| GPRS | General Packet Radio System |
| GPS | Global Positioning System |
| GSM | Global System for Mobile (telecommunications) |
| GW | Gateway |

## H

| | |
|---|---|
| HA | Host Application |
| HAPi | Home Application Programming interface |
| HAVi | Home Audio-Video interoperability |
| HCI | Host Controller Interface |
| HDLC | High-level Data-link Control |
| HEC | Header Error Check |
| HID | Human Interface Device |
| HS | Headset |
| HSCSD | High Speed Circuit-switched Data |
| HTML | HyperText Markup Language |
| HTTP | HyperText Transfer Protocol |
| HV | High-quality Voice |
| Hz | Hertz (cycle per second) |

## I

| | |
|---|---|
| I/O | Input/Output |
| IAC | Inquiry Access Code |
| ICC | Integrated Circuit Card |
| IEC | International Electrotechnical Commission |
| IEEE | Institute of Electronic and Electrical Engineers |
| IETF | Internet Engineering Task Force |
| IMC | Internet Mail Consortium |
| IMT-2000 | International Mobile Telecommunications 2000 |
| IP | Internet Protocol |
| IPX | Internet Protocol eXchange |
| IrDA | Infrared Data Association |
| IrMC | Infrared Mobile Communications |
| IrOBEX | Infrared Object Exchange |
| ISDN | Integrated Services Digital Network |
| ISM | Industrial, Scientific, Medical |
| ISO | International Organization for Standardization |

# Acronyms

| | |
|---|---|
| ITU | International Telecommunication Union |
| ITU-T | International Telecommunication Union-Telecommunications |

## J

| | |
|---|---|
| JDC | Japanese Digital Cellular |
| JVM | Java Virtual Machine |

## K

| | |
|---|---|
| KB | Kilobytes |
| Kbps | Kilobits per second |
| KHz | Kilohertz (thousands of hertz) |

## L

| | |
|---|---|
| L_CH | Logical Channel |
| L2CA | Logical Link Control and Adaptation |
| L2CAP | Logical Link Control and Adaptation Protocol |
| LAN | Local Area Network |
| LAP | LAN Access Point |
| LA | Lower Address Part |
| LC | Link Controller |
| LCP | Link Control Protocol |
| LCSS | Link Controller Service Signaling |
| LF | Line Feed |
| LFSR | Linear Feedback Shift Register |
| LIAC | Limited Inquiry Access Code |
| LLC | Logical Link Control |
| LM | Link Manager |
| LMP | Link Management Protocol |
| LocDev | Local Device |
| LSB | Least Significant Bit |

## M

| | |
|---|---|
| M_ADDR | Medium Access Control Address |
| mA | Milliamp |
| MAC | Medium Access Control |
| MAPI | Messaging Application Procedure Interface |
| ME | Management Entity |
| MM | Mobility Management |
| MMI | Man Machine Interface |

| | |
|---|---|
| MPEG | Moving Picture Experts Group |
| ms | Millisecond |
| MS | Mobile Station |
| MS | Multiplexing Sublayer |
| MSB | Most Significant Bit |
| MSC | Message Sequence Chart |
| MSC | Modem Status Command |
| MTU | Maximum Transmission Unit |
| MUX | Multiplexing Sublayer |
| mW | Milliwatt |

## N

| | |
|---|---|
| NAK | Negative Acknowledge |
| NA | Non-significant Address Part |
| NBC | Number-of-repetitions for Broadcast packets |
| NCP | Network Control Protocol |
| nm | Nanometer |
| NMT | Nordic Mobile Telephone |

## O

| | |
|---|---|
| OBEX | OBject EXchange protocol |
| OCF | Opcode Command Field |
| OSI | Open Systems Interconnection |

## P

| | |
|---|---|
| PAN | Personal Area Network |
| PC | Personal Computer |
| PCH | Paging Channel |
| PCM | Pulse Code Modulation |
| PCMCIA | Personal Computer Memory Card International Association |
| PCS | Personal Communications Service |
| PDA | Personal Digital Assistant |
| PDU | Protocol Data Unit |
| PDN | Packet Data Network |
| PIM | Personal Information Manager |
| PIN | Personal Identification Number |
| PM_ADDR | Parked Member Address |
| PN | Pseudo-random Noise |
| PnP | Plug and Play |

# Acronyms

| | |
|---|---|
| POTS | Plain Old Telephone Service |
| PPM | Part Per Million |
| PPP | Point-to-Point Protocol |
| PRBS | Pseudo Random Bit Sequence |
| PRNG | Pseudo Random Noise Generation |
| PSM | Protocol/Service Multiplexer |
| PSTN | Public Switched Telephone Network |

## Q

| | |
|---|---|
| QCIF | Quarter Common Intermediate Format |
| QoS | Quality of Service |

## R

| | |
|---|---|
| RACH | Random Access Channel |
| RAND | Random number |
| RAS | Remote Access Server |
| RemDev | Remote Device |
| RF | Radio Frequency |
| RFC | Request For Comments |
| RFCOMM | Radio Frequency Communication |
| RSSI | Received Signal Strength Indicator |
| RSVP | Resource ReSerVation Protocol |
| RTCON | Real Time Connection |
| RTP | Real Time Protocol |
| RTS | Request to Send |
| RTT | Radio Transmission Technology |
| RX | Receiver |

## S

| | |
|---|---|
| SAP | Service Access Points |
| SAR | Segmentation and Reassembly |
| SCF | Service Control Function |
| SCO | Synchronous Connection Oriented |
| SD | Service Discovery |
| SDAP | Service Discovery Application Protocol |
| SDCCH | Standalone Dedicated Control Channel |
| SDDB | Service Discovery Database |
| SDF | Service Data Function |
| SDP | Service Discovery Protocol |
| SDSL | Symmetrical Digital Subscriber Line |

| | |
|---|---|
| SeP | Serial Port |
| SEQN | Sequential Numbering |
| SET | Secure Electronic Transactions |
| SIG | Special Interest Group |
| SIM | Subscriber Identity Module |
| SIR | Serial Infrared |
| SLP | Service Location Protocol |
| SMS | Short Messaging Service |
| SNA | Systems Network Architecture |
| SPP | Serial Port Profile |
| SRES | Signed Response |
| SrvDscApp | Service Discovery Application |
| SS | Supplementary Services |
| SSF | Service Switching Function |
| SSI | Signal Strength Indication |
| SSL | Secure Sockets Layer; Service Security Level |
| SUT | System Under Test |
| SWAP | Shared Wireless Access Protocol |

# T

| | |
|---|---|
| TACS | Total Access Communications System |
| TAE | Terminal Adapter Equipment |
| TBD | To Be Defined |
| TC | Test Control |
| TCI | Test Control Interface |
| TCP/IP | Transmission Control Protocol/Internet Protocol |
| TCP/UDP | Transmission Control Protocol/User Datagram Protocol |
| TCS | Telephony Control Specification |
| TCS BIN | Telephony Control Specification Binary |
| TDD | Time Division Duplex |
| TDM | Time Division Multiplexing |
| TDMA | Time Division Multiple Access |
| TIA | Telecommunications Industry Association |
| TL | Terminal |
| TLo | Terminal Originating a Call |
| TLT | Terminal Terminating a Call |
| TMN | Telecommunications Management Network |
| TS | Technical Specification |
| TTP | Tiny Transport Protocol |
| TX | Transmit |

# Acronyms

## U

| | |
|---|---|
| UA | User Asynchronous |
| UAP | Upper Address Part |
| UART | Universial Asynchronous Receiver-Transmitter |
| UC | User Control |
| UDP | User Datagram Protocol |
| UDP/IP | User Datagram Protocol/Internet Protocol |
| UI | User Interface; User Isochronous |
| UIAC | Unlimited Inquiry Access Code |
| UMTS | Universal Mobile Telecommunications System |
| US | User Synchronous |
| USB | Universal Serial Bus |
| UTRA | UMTS Terrestrial Radio Access |
| UUID | Universally Unique Identifier |
| UWCC | Universal Wireless Communications Consortium |

## V

| | |
|---|---|
| VBR | Variable Bit Rate |
| VCR | Video Cassette Recorder |
| VFIR | Very Fast Infrared |
| VGA | Video Graphics Array |

## W

| | |
|---|---|
| W-CDMA | Wideband Code Division Multiple Access |
| WAE | Wireless Application Environment |
| WAN | Wide Area Network |
| WAP | Wireless Application Protocol |
| WEP | Wired Equivalent Privacy |
| WID | Wireless Information Device |
| WIMS | Wireless Multimedia and Messaging Services |
| WLL | Wireless Local loop |
| WML | Wireless Markup Language |
| WRC | World Radio Conference |
| WSP | Wireless Session Protocol |
| WTLS | Wireless Transport Layer Security |
| WTP | Wireless Transport Protocol |
| WUG | Wireless User Group |

## X

| | |
|---|---|
| XML | eXtensible Markup Language |

| | |
|---|---|
| **XOFF** | Transmitter Off |
| **XON** | Transmitter On |

# INDEX

Note: Boldface numbers indicate illustrations, italic t indicates a table.

## A

Acceptor devices, Serial Port Profile (SPP), 191
access code, 77—78, **78**
access control, 301
Acer, 16
ACK/NAK signals, 80, 97, 99
acronyms, 371—380
actions, Logical Link Control and Adaptation Layer (L2CAP), 161—163, 164, 165—168($t$)
Active Member Address (AM_ADDR), 79, 123, 138—139
address groups, Logical Link Control and Adaptation Layer (L2CAP), 152
addressing, 76—77, 98, 109—110
    Active Member Address, 79, 123, 138—139
    address groups, (L2CAP), 152
adopted protocols, 102($t$), 103, 107—118
Advanced Mobile Phone Service (AMPS), 71, 311, 314, 315, 316
advantages of Bluetooth, 13—14
Analog Control Channel (ACC), 315
Analog Voice Channel (AVC), 315
Antheil, George, spread-spectrum inventor, 61—62
appliances, home
    Home Application Programming interface (HAPi), 41—42
    remote control, 43
Application layer, OSI model, 91($t$), 92—93, **93**
application-level procedures, Serial Port Profile (SPP), 191—193
applications for Bluetooth, 26—37, 47($t$)
    behavior enforcement, 34
    card scanning, 27
    collaborative workgroups, 27
    communicator platforms, 29—30, **30**
    e-commerce, 34—37
    electronic books, 30—31, **31**
    home entertainment, 32
    in-car systems, 29
    law enforcement, 34
    payment systems, 32—33, **33**
    presentations, 26
    printing, 28—29
    remote synchronization, 28
    scanners, 33
    smart cards, 34, **35**
    synchronizing data, 27—28
    travel, 29, 31—32
ASCII code, 54, 95
asymmetrical links, 56
asynchronous connectionless (ACL) links, 17, 56, 74, 75, 77, 80, 82, 83, 104, 105, 123, 137, 143, 150, 188, 296
asynchronous transfer mode (ATM), 77, 312
asynchronous transmissions, 50, 53, 54—56, 67—68, 74—75, 77, 80, 82, 83, 104, 105, 123, 137, 143, 150, 188, 296
    ASCII code, 54
    asymmetrical links, 56
    asynchronous connectionless link (ACL), 56
    clock rates, 56
    Digital Subscriber Line (DSL), 56(f), 56
    error checking, 55
    even/odd parity, 55
    parity bits, 54, **54**
    start/stop bits, 54—55, **54**
    symmetrical links, 56
    user asynchronous (UA), 82, 83
AT commands
    Dialup Networking Profile, 250
    Fax Profile, 258
AT-Commands protocol, 102($t$)
Audio Gateway (AG), Headset Profile, 242
audio port emulation, Headset Profile, 241
authentication, 25, 26, 127—128, 127($t$), 290, 291, 292, 294, 296—298, 307, 308
    Cordless Telephony Profile, 240
    Dialup Networking Profile, 250
    File Transfer Profile, 269, 270
    Generic Object Exchange Profile (GOEP), 212
authentication, 127—128, 127($t$), 127
authorization, 296
automatic repeat request (ARQ), 80, 155
autonomous systems (AS), 109

## B

bandwidth, 25
    peak bandwidth, 174
Baseband protocol, 102($t$), 104, 119, 150, 297—298
    Dialup Networking Profile, 250
    Fax Profile, 258
    Generic Object Exchange Profile (GOEP), 210, 211
    security, 294
behavior enforcement using Bluetooth, 34
Blue-Connect module, 16
Blue-Share synchronization scheme, 16
BlueLinx Inc., 34
Bluetooth basics, 49—88
Bluetooth module, **12**, 16
Bluetooth protocol (See also protocols), 98—99, 101—118, **101**, 102—103($t$)
Bluetooth Qualification Body (BQB), 45
Bluetooth Qualification Test Facility (BQTF), 45
Bluetooth Special Interest Group (SIG), 8, 14—15, 19, 44

Bluetooth, Harald, king of Denmark, 15—16
bonding
    Generic Access Profile (GAP), 187—188
    Serial Port Profile (SPP), 191
bridges, 76
Broadcast Control Channel (BCCH), Global System for Mobile (GSM), 321
broadcasting, 142
bucket size, token, 174
Business Card Exchange function, Object Push Profile, 275, 277
Business Card Pull function, Object Push Profile, 275, 277
business cards (*See also* vCard)

## C

C Pen, 33
C Tech, 33
cable replacement protocol, 102(*t*), 103, 105—107
Cahners In-Stat Group, 44, 46
calendar applications (*See* vCalendar)
Call Control (CC) element
    Cordless Telephony Profile, 226, 232—234
    Intercom Profile, 219—220, 219
call failure, Intercom Profile, 223—225, 225(*t*)
call procedures, Intercom Profile, 221—223
calling line identification presentation (CLIP), Cordless Telephony Profile, 229
Capshare 910 scanner, **111**
card scanning, 27
Carrier Sense Multiple Access/Collision Avoidance (CSMA/CA), 11
CD-ROMs, 20
CDMA Development Group (CDG), 335
cdmaOne-2000, 335—337, 335
cell phones, **111**
cellular communications, 71, 311—314, 339
Centronics connectors, 51
Challenge Handshake Authentication (CHAP), 264
change of link key, 129—131, 130(*t*)
channel access code (CAC), 78
channel quality-driven change of data rate, 141—142, 142(*t*)
channels and channel identifiers (CIDs) (*See also* logical channels)
    channel establishment, 189
    Global System for Mobile (GSM), 321—322, 321
    Logical Link Control and Adaptation Layer (L2CAP), 153—154, **154**, 155—156, 163—164
checksums, 96
chipping code, 8
circuit switching, 66—68, 66
class of device, Generic Access Profile (GAP), 184
client/server architectures, 83—86, **84**, 110
    Generic Object Exchange Profile (GOEP), 211—212
    IrMC client/servers, 280—283
    local device (LocDev), 197—198
    Object Push Profile, 276
    remote device (RemDev), 197—198

Service Discovery Application Profile (SDAP), 197—198
Synchronization Profile, 280—283
clock offset, 133, 133(*t*)
clock signals, 51
coaxial cable, 85
Code Division Multiple Access (CDMA), 61, 71—73, 72, 314, 316—318, 327, 333
code, character codes, 95
collaborative workgroups, 27
collisions, 90
combination key, 291
ComBit, Inc., 343
Command Reject packet, 171
Common Control Channel (CCCH), Global System for Mobile (GSM), 321
communicator platforms, 29—30, **30**
compression, data, 95
Conference of European Posts and Telegraph (CEPT), Global System for Mobile (GSM), 319
configuration, 173—175
    Cordless Telephony Profile, 234—235
    flush timeout, 174
    Logical Link Control and Adaptation Layer (L2CAP), 194, 204
    maximum transmission unit (MTU), 174
    quality of service (QoS), 174—175
    Request Path packet, 175
    Response Path packet, 175
Configuration Request packet, 171
Configuration Response packet, 172
Configuration Response primitive, 177
Configure primitive, 177
Confirm actions, Logical Link Control and Adaptation Layer (L2CAP), 156, 161—163
Connect primitive, 176
Connect Response primitive, 177
Connectable mode, Synchronization Profile, 281
connection establishment, 146, 146(*t*), 189
Connection Request packet, 171
Connection Response packet, 171
connection setup, security, 296—297, **297**, 296
connection-oriented links, 17, 57, 74, 168—169
connectionless links, 17, 56, 74, 75, 169—170, **169**, 179
connectivity modes, Generic Access Profile (GAP), 185
connectivity solutions, 40—43
content formats (*See also* vCalendar; vCard)
    Object Push Profile, 277—278
    Synchronization Profile, 283
content generators, WAE, 115
Continuous Variable Slope Delta (CVSD) modulation, 17, 50, 58—60, 218, 243
cordless telephones, 10
Cordless Telephony Profile, 225—241, **226**, **228**
    ACKNOWLEDGE, 232, 233
    ALERTING, 232
    authentication, 240
    Baseband protocol, 225
    Call Control (CC) element, 226, 232—234

# Index

calling line identification presentation (CLIP), 229
clearing calls, 233
configuration, 234—235
CONNECT, 232, 233
device roles, 226—227
dual tone multi-frequency (DTMF), 228, 230
features, 229—230, 231(t)
gateways, 227, 230—231
Generic Access Profile (GAP), 225, 239—240, 239—241(t)
Group Management, 234—235
H.323 gateways, 227(f)
idle mode, 240, **241**
inter-piconet capability, 236
Link Control (LC), 238, 238—239(t)
link keys, 234—236
Link Manager (LM), 226
Link Manager Protocol (LMP), 225, 237, 237(t)
Logical Link Control and Adaptation Layer (L2CAP), 225, 226, 229, 230
modes, 239, 239(t)
quality of service (QoS), 229
security, 240, 240(t)
Service Discovery Protocol (SDP), 225, 236, 236(t)
SETUP, 232
synchronous connection-oriented (SCO) links, 226, 229
Telephony Control Protocol Specification Binary (TCS BIN), 225, 226, 230
terminal-to-gateway connection protocols, 230
terminal-to-terminal connection, 231—232, 231
terminals, 227, 230—232
typical call scenario, 227—229, **228**
Wireless User Group (WUG), 229, 234—235
core protocols, 102(t), 103—105
cost of Bluetooth, 13, 16, 44—45
credit card security, 35
current link keys, 130
cyclic redundancy code (CRC), 80, 82, 155

## D

Data Access Points, Wide Area Networks (*See* Fax Profile)
data communications, 11, 30, 44, 50, 339
data conversion, 95
Data Link layer, OSI model, 92(t), 98—99, **99**, 150
data rate of Bluetooth devices, 340
data synchronization, 27—28
  remote synchronization, 28
Data Terminal Ready/Data Set Ready (DTR/DSR), 193
data terminals
  Dialup Networking Profile, 249
  Fax Profile, 257
  LAN Access Profile, 261, 262
datagrams, 96, 107, 108—109
Data—High Rate (DH) packets, 141
Data—Medium Rate (DM) packets, 141
delay variation, 175
delimiting, frames, 99
detach links, 137

device access code (DAC), 78
device addresses, 76
device discovery, Generic Access Profile (GAP), 187
device names
  device discovery, 187
  Generic Access Profile (GAP), 184
  name discovery, 187
  name request to remote device, 307
dialup Internet access, 107—108, 119
  Dialup Networking Profile, 248—256, **249**
Dialup Networking Profile, 110, 119, 248—256, **249**
  AT commands, 250
  audio feedback, 253
  authentication, 250
  Baseband protocol, 250
  data terminals, 249
  encryption, 250
  gateways, 248—249, 251, 252—253(t)
  Generic Access Profile (GAP), 248, 250, 254—255, 256(t)
  idle mode, 255
  Link Control (LC), 254
  Link Manager Protocol (LMP), 248, 250, 253
  Logical Link Control and Adaptation Layer (L2CAP), 248
  modems, 248
  modes, 255, **255**
  Radio Frequency Communication (RFCOMM) protocol, 248
  restrictions, 250
  security, 250—251, 255
  Serial Port Profile (SPP), 248, 251
  Service Discovery Protocol (SDP), 248
  service discovery, 254, 254(t)
  services, 251, 251(t)
  synchronous connection-oriented (SCO) links, 248, 253
Digital Advanced Mobile Phone Service (D-AMPS), 71(f), 311, 315
digital cameras, JetSend wireless link, 40—41
digital control channel (DCCH), 315
Digital Enhanced Cordless Telephone (DECT), 10
Digital Subscriber Line (DSL), 9, 56(f)
Digital Traffic Channel (DTC), 315
direct sequencing spread-spectrum, 8, 63, 64
directory services, 93
Disable Connectionless Traffic primitive, 179
Disconnect primitive, 177
Disconnection Request packet, 172
Disconnection Response packet, 172
discovery (*See* service discovery)
discovery modes, Generic Access Profile (GAP), 185
DQPSK digital control channel (DCCH), 315
dual tone multi-frequency (DTMF), Cordless Telephony Profile, 228, 230
duplexing, 68—73, 95, 152
  Code Division Multiple Access (CDMA), 61, 71—73, **72**, 314, 316—318, 327, 333
  Frequency Division Multiple Access (FDMA), 71—73, **72**, 321

# Index

full-duplex transmissions, 68, 95
half-duplex transmissions, 69
master/slave devices, 69, 70
piconet, 69, 70
RF hop frequency, 70, 71,
Time Division Multiple Access (TDMA), 10, 11, 61, 71—73, 72, 314—318, 327, 333, 338—339
DVD, 20

## E

e-commerce using Bluetooth, 25, 34—37
   credit card security, 35
   Europay-MasterCard-Visa (EMV) protocol, 36
   secure electronic transactions (SET), 36
   wireless access protocol (WAP), 36—37
   Wireless Transport Layer Security (WTLS) protocol, 113
   wireless wallets, 36
e-mail, 43, 93, 111
EBCDIC, 95
Echo Request packet, 172
Echo Response packet, 172
electronic books using Bluetooth, 30—31, 31
Enable Connectionless Traffic primitive, 179
encapsulation, 108
encoding, 100
   Wireless Applications Environment (WAE), 115
encryption, 95, 131—133, 131($t$), 290, 291, 292, 294, 296, 307—308
   Dialup Networking Profile, 250
   File Transfer Profile, 269
   Generic Object Exchange Profile (GOEP), 212
Enhanced Data for Global Evolution (EDGE) technology, 324
entertainment using Bluetooth, 32
Ericsson Mobile Communications AB, 14—16, 30, 314, 343—344
error handling, 55, 99, 147—148, 207
   quantizing noise, 60
   Serial Port Profile (SPP), 195, 196
   Service Discovery Application Profile (SDAP), 210
establishment procedures, Generic Access Profile (GAP), 188—189
Ethernet LANs, 6—8, 22, 76, 77
Europay-MasterCard-Visa (EMV) protocol, 36
European Telecommunications Standard Institute (ETSI), 105, 319, 329—334
even/odd parity, 55
Event Indication primitive, 176
events, Logical Link Control and Adaptation Layer (L2CAP), 157—161, 164, 165—168($t$)
Extended Systems Corporation, 344
extensible markup language (XML), 37
External Security Control Entity (ESCE), 294, 306

## F

Fast Associated Control Channel (FACCH), Global System for Mobile (GSM), 321
fax, 119

Fax Profile, 256—261, 261, 262
   AT commands, 258
   audio feedback, 260
   Baseband protocol, 258
   data terminals, 257
   gateways, 257, 259
   Generic Access Profile (GAP), 256, 258, 260—261
   idle mode, 261
   Link Control (LC), 260
   Link Manager Protocol (LMP), 257, 258
   Logical Link Control and Adaptation Layer (L2CAP), 257
   modes, 261
   Radio Frequency Communication (RFCOMM) protocol, 257, 258
   restrictions, 257—258
   security, 261
   serial port emulation, 257
   Serial Port Profile, 256, 258
   Service Discovery Protocol (SDP), 257, 259, 260, 260($t$)
   services, 259, 259($t$)
   synchronous connection-oriented (SCO) links, 257
fiberoptics, 85
File Transfer Profile, 189, 268—273, 269
   applications supported, 268
   authentication, 269, 270
   Create Folder, 270, 271
   Delete Object, 271
   encryption, 269
   features, 271—272($t$)
   functions, 270—271
   Generic Access Profile, 268
   Generic Object Exchange Profile (GOEP), 210, 214, 215, 268
   IrOBEX protocol, 268
   Link Management Protocol (LMP), 268
   Logical Link Control and Adaptation Layer (L2CAP), 268
   Navigate Folders, 271
   Object Exchange Protocol (OBEX), 268, 270, 272, 272($t$)
   Pull Object, 270, 271
   Push Object, 270, 271
   Radio Frequency Communication (RFCOMM) protocol, 268
   security, 269, 270
   Select Server, 271
   Serial Port Profile, 268
   Service Discovery Protocol (SDP), 268, 273, 273($t$)
File Transfer Protocol (FTP), 103
first-generation wireless, 311—314
flow control, 96
Flow Control On/Flow Control Off (FCON/FCOFF), 193
flush timeout, 174, 194, 204
forward error correction (FEC), 141
fragmentation, frame, 98
frame division duplexing (FDD), 68—69, 69
frames, 98, 99

# Index

frequency, 16—17
    direct sequencing spread-spectrum, 8, 63, 64
    frequency-hopping spread-spectrum, 8, 9(*t*), 21, 61, 63, 65—66
    industrial, scientific, and medical (ISM) band, 16
    interference, 21—22
    RF hop frequency, 70, **71**
    spread-spectrum, 60—66, **62**
    wireless LANs, 8, 9(*t*), 8
Frequency Division Multiple Access (FDMA), 71—73, **72**, 321
Frequency Hop Synchronization (FHS), 133, 139
frequency-hopping spread spectrum, 8, 9(*t*), 21, 61, 63, 65—66
full-duplex transmissions, 68, 95

## G

gateways, wireless 10, 77
    Cordless Telephony Profile, 227, 230—231
    Dialup Networking Profile, 248—249, 251, 252—253(*t*)
    Fax Profile, 257, 259
General Discoverable mode, Synchronization Profile, 281
General Inquiry Access Codes (GIAC), 185, 186, 195—196, 209
General Packet Radio Services (GPRS), 322—323
general response messages, 127, 127(*t*)
General Sync mode, Synchronization Profile, 281—282
Generic Access Profile (GAP), 182—189, **183**, 286
    asynchronous connectionless (ACL) links, 188
    bonding, 187—188
    channel establishment, 189
    class of device, 184
    connection establishment, 189
    connectivity modes, 185
    Cordless Telephony Profile, 225, 239—240, 239—241(*t*)
    device discovery, 187
    device names, 184
    Dialup Networking Profile, 248, 250, 254—255, 256(*t*)
    discovery modes, 185
    establishment procedures, 188—189
    Fax Profile, 256, 258, 260—261
    File Transfer Profile Specification, 189, 268
    General Inquiry Access Codes (GIAC), 185, 186
    Generic Object Exchange Profile (GOEP), 188, 211
    Headset Profile, 241, 243, 246, 247, 247(*t*), 248(*t*)
    idle mode procedures, 186—187
    Intercom Profile, 218, 220
    LAN Access Profile, 262, 265—266, 265(*t*)
    Limited Inquiry Access Codes (LIAC), 185, 186
    Link Control (LC), 182
    link establishment, 188—189
    Link Manager Protocol (LMP), 182, 188
    Logical Link Control and Adaptation Layer (L2CAP), 182
    name discovery, 187
    Object Exchange Protocol (OBEX), 182
    pairing modes, 185
    parameters, 184—186
    personal identification number (PIN), 184
    Radio Frequency Communication (RFCOMM) protocol, 182, 189
    security, 182, 186, 294
    service discovery, 183
    Telephony Control Specification Binary (TCS BIN) protocol, 189
Generic Object Exchange Profile (GOEP), 182, 188, 210—214, **212**, 286
    authentication, 212
    Baseband protocol, 210
    Baseband protocol, 211
    client/server, 211—212
    encryption, 212
    establishing OBEX session, 213
    File Transfer Profile, 210, 214, 215, 268
    Generic Access Profile (GAP), 211
    IrOBEX, 211, 213
    Link Manager Protocol (LMP), 210, 211
    Logical Link Control and Adaptation Layer (L2CAP), 211
    Object Exchange Protocol (OBEX), 210, 211, 212—214
    Object Push Profile, 210, 214, 215, 273, 274, 275
    pulling data object, 213
    pushing data object, 213
    Radio Frequency Communication (RFCOMM) protocol, 211
    security, 212
    Service Discovery Protocol (SDP), 211
    Synchronization Profile, 210, 214, 215, 280
Get Group Membership primitive, 179
Get Info primitive, 179
global 3G wireless, 42—43, 309—341, **310**
    Advanced Mobile Phone System (AMPS), 311, 314, 315, 316
    Analog Control Channel (ACC), 315
    Analog Voice Channel (AVC), 315
    Bluetooth's role, 339—341
    CDMA Development Group (CDG), 335
    cdmaOne-2000, 335—337, 335
    Code Division Multiple Access (CDMA), 314, 316—318, 327, 333
    Digital AMPS (DAMPS), 311, 315
    digital control channel (DCCH), 315
    Digital Traffic Channel (DTC), 315
    DQPSK digital control channel (DCCH), 315
    Enhanced Data for Global Evolution (EDGE) technology, 324
    first-generation wireless, 311—314
    Frequency Division Multiple Access (FDMA), 321
    General Packet Radio Services (GPRS), 322—323
    global initiative, 326—334
    Global System for Mobile (GSM), 311, 314, 318—325, **320, 325**
    High-speed Circuit-switched Data (HSCSD), 322
    International Mobile Telecommunications 2000 (IMT-2000), 310, 311, 312, 314, 326—329
    Japanese Digital Cellular (JDC), 311
    Nordic Mobile Telephone (NMT), 311, 316

North American GSM Alliance, 335
Personal Communications Service (PCS), 314
personal digital assistants (PDAs), 310
second-generation wireless, 314—325
Space Division Multiple Access (SDMA), 327
standards development, 326—328
Time Division Multiple Access (TDMA), 314—321, 327, 333, 338—339
Total Access Communication system (TACS), 311, 316
U.S. participation, 334—339
Universal Mobile Telecommunications System (UMTS), 329—334
Universal Wireless Communications Consortium (UWCC), 334, 338
W-CDMA/W-CDMA/NA, 334—335, 337—338
WIDEBAND, 335—337
Wireless Multimedia and Messaging Services (WIMS), 334
global positioning system (GPS), 30, 32
Global System for Mobile (GSM), 71, 105, 311, 314, 318—325, **320**, **325**
  air interface, 321
  architecture, 320—321
  channel derivation and types, 321—322
  Conference of European Posts and Telegraph (CEPT), 319
  development, 318—320
  Enhanced Data for Global Evolution (EDGE) technology, 324
  European Telecommunications Standards Institute (ETSI), 319
  Frequency Division Multiple Access (FDMA), 321
  General Packet Radio Services (GPRS), 322—323
  High-speed Circuit-switched Data (HSCSD), 322
  Internet Protocol (IP), 324
  phases, 319—320, 322
  stations, 320—321
  subsystems, 320—321
  Time Division Multiple Access (TDMA), 318, 321
Group Add Member primitive, 178
Group Close primitive, 178
Group Create primitive, 178
Group Management, Cordless Telephony Profile, 234—235
Group Remove Member primitive, 178

# H

H.323 gateways, 227(f)
half-duplex transmissions, 69, 95
header error check (HEC) field, 81
header, Bluetooth packet, 79—81, **79**
Headset (HS) device, Headset Profile, 242
Headset Profile, 18—19, 241—248, **242**
  Audio Gateway (AG), 242
  audio port emulation, 241
  basic operation, 243
  connection handling, 246
  Continuously Variable Slope Delta (CVSD) modulation, 243
  device roles, 242
  features, 244, 244(t)
  Generic Access Profile (GAP), 241, 243, 246, 247, 247(t), 248(t)
  Headset (HS) device, 242
  idle mode, 247
  incoming/outgoing audio connections, 244—245, **245**
  Link Control (LC), 246, **246**
  Link Manager Protocol (LMP), 241
  Logical Link Control and Adaptation Layer (L2CAP), 241
  modulation, 243
  personal identification numbers (PIN), 243
  Radio Frequency Communication (RFCOMM) protocol, 241, 243, 246
  remote volume control, 245
  restrictions, 243
  serial port emulation, 243
  Serial Port Profile (SPP), 241, 246
  Service Discovery Protocol (SDP), 241
  synchronous connection-oriented (SCO) links, 241, 242, 243
  transfer of audio connection, 245
Hewlett-Packard, 40, 110
hidden computing, 113
High-level Data Link Control (HDLC) protocol, 108
High-speed Circuit-switched Data (HSCSD), Global System for Mobile (GSM), 322
hold mode, 137—138, 137(t), 193
Home Application Programming interface (HAPi), 41—42
Home Audio-Video interoperability (HAVi) specification, 41—42
home entertainment using Bluetooth, 32
  Home Application Programming interface (HAPi), 41—42
  Home Audio-Video interoperability (HAVi) specification, 41—42
  Home Radio Frequency (HomeRF) networks and, 8—13
HomeRF Consortium, 8
hop frequency, 70, **71**
hopping, frequency (See frequency hopping spread spectrum)
Host Controller Interface (HCI), 103, 154
  security, 297—298, 307—308,
Hypertext Markup Language (HTML), 67, 112
HyperText Transfer Protocol (HTTP), 37, 103, 110, 112—113, 211

# I

IBM, 14, 344
IBM Mobile Connect, 46
idle mode, 186—187
  Cordless Telephony Profile, 240, **241**
  Dialup Networking Profile, 255
  Fax Profile, 261

# Index

Headset Profile, 247
in-car systems using Bluetooth, 29
Indication actions, Logical Link Control and Adaptation Layer (L2CAP), 156, 161—163
industrial, scientific, and medical (ISM) band, 16
Information Request packet, 173
Information Response packet, 173
Infrared Data Association (IrDA), 19, 44, 87, 110
Infrared Mobile Communications (IrMC) protocol, 19, 103(t)
infrared wireless technology, 2—4, 44—47, 13
    Infrared Data Association (IrDA), 19, 44, 87, 110
    Infrared for Mobile Communications (IrMC) protocol, 19, 103
    Internet dial-up connections using, 5—6
    IrDA Interoperability Specification, 215—216
    IrOBEX protocol, 211, 213, 268, 274, 280
    Object Exchange Protocol (OBEX), 110—111
    RTCON (real time connection) links, 19
    Salutation Architecture, 87
    Serial Infrared (SIR) data link, 44
    speed of, 4
    Synchronization Profile, 279—286, **280, 281**
    Very Fast Infrared (VIFR) protocol, 4
    voice, 19
    Wireless Application Protocol (WAP), 111—115, **112**
    wireless LANs and, 6—8, **7**
    wireline connection to, 4—5, **5**
Initialization Sync mode, Synchronization Profile, 281—282
Initiator devices, Serial Port Profile (SPP), 191
inquiry access code (IAC), 78
inquiry scan
    Serial Port Profile (SPP), 196
    Service Discovery Application Profile (SDAP), 209
Intel, 14, 344—345
interactive voice communications, 11
interception of signals, spread-spectrum, 63
Intercom Profile, 119, 218—225, **218, 219**
    Call Control (CC) element, 219—220
    call failure, 223—225, 225(t)
    call procedures, 221—223
    Generic Access Profile (GAP), 218, 220
    Link Manager Protocol (LMP), 222, 222—223(t)
    Logical Link Control and Adaptation Layer (L2CAP), 220—222
    messaging, 223
    modulation, 218
    quality of service (QoS), 218
    synchronous connection-oriented (SCO) links, 220—222
    Telephony Control Specification Binary (TCS BIN) protocol, 219, 223, 224(t)
interference, 21—22, 63, 66
International Electrotechnical Commission (IEC), 20
International Mobile Telecommunications 2000 (IMT-2000), 42, 71—72, 310, 311, 312, 314, 326—329
International Organization for Standardization (ISO), 20

International Standards Organization (ISO), 91
International Telecommunications Union (ITU), 42, 310, 311, 312, 314, 326—329
International Telecommunications Union—Telecommunications (ITU—T), 107
Internet/Internet access, 5—6, 30, 46, 67, 90, 107—108, 112, 119
    Dialup Networking Profile, 248—256, **249**
Internet Bridge (*See also* Dialup Networking Profile), 110, 119
Internet Engineering Task Force (IETF), 87, 107, 261, 312—313
Internet Protocol (IP), 37, 76, 77, 108—110, 312, 324
IP addressing, 76, 77, 109—110, 264
IrDA Interoperability Specification, 188, 215—216
IrOBEX protocol, 211, 213
    File Transfer Profile, 268
    Object Push Profile, 274
    Synchronization Profile, 280
ISDN, 57
isochronous transmissions, user isochronous (UI), 82, 83

## J

jamming, spread-spectrum, 63
Japanese Digital Cellular (JDC), 311
Java/Java Virtual Machine (JVM), 37—38, 40, 87
JavaScript, 115
JetSend wireless link, 40—41
Jini, 38—40, **39**
Johnson Controls, 29

## K

K Virtual Machine (KVM), 38
keys, link keys, 127—131, 291, 301

## L

Lamarr, Hedy, spread-spectrum inventor, 61—62, **61**
LAN Access Point (LAP), LAN Access Profile, 262—264
LAN Access Profile, 261—268, **262, 263**
    applications supported, 263
    Challenge Handshake Authentication (CHAP), 264
    data terminals, 261, 262
    Generic Access Profile (GAP), 262, 265—266, 265(t)
    IP addressing, 264
    LAN Access Point (LAP), 262, 264
    Link Control (LC), 267
    Link Management Protocol (LMP), 262
    Logical Link Control and Adaptation Layer (L2CAP), 262, 264
    Management Entity (ME), 262, 267—268
    networks supported, 262
    Point-to-Point Protocol (PPP), 261, 262, 264
    Radio Frequency Communication (RFCOMM) protocol, 261, 262, 264
    Remote Access Server (RAS), 262
    restrictions, 263
    security, 265
    Serial Port Profile, 262
    Service Discovery Protocol (SDP), 262, 266—267, 267(t)

# Index

LAN adapters, 7
latency, 175
law enforcement using Bluetooth, 34
Layer 2/Layer 3 addressing, 76—77
legacy applications, security, 292—293
levels of security, 296
Limited Discoverable mode, Synchronization Profile, 281
Limited Inquiry Access Codes (LIAC), 185, 186
line of sight communications, 14
Link Control (LC), 82, 83, 123
    Cordless Telephony Profile, 238, 238—239($t$)
    Dialup Networking Profile, 254
    Fax Profile, 260
    Generic Access Profile (GAP), 182
    Headset Profile, 246, **246**
    LAN Access Profile, 267
    Serial Port Profile (SPP), 195
    Service Discovery Application Profile (SDAP), 208—210, 208—209($t$)
Link Control Protocol (LCP), 108
Link Manager (LM), 82, 83, 122—123, **122**, 146
    Cordless Telephony Profile, 226
    security, 290, 297—298, 307—308
    Serial Port Profile (SPP), 195—196
    Service Discovery Application Profile (SDAP), 206—207, 207($t$)
Link Manager Protocol (LMP), 74, 101, 102, 102($t$), 104, 122—123, **122**, 134—135, 135($t$), 188, 210, 211
    File Transfer Profile, 268
    LAN Access Profile, 262
    Object Push Profile, 274
    security, 294, 299
    Synchronization Profile, 280
    Cordless Telephony Profile, 225, 237, 237($t$)
    Dialup Networking Profile, 248, 250, 253
    Fax Profile, 257, 258
    Generic Access Profile (GAP), 182
    Headset Profile, 241
    Intercom Profile, 222, 222—223($t$)
    Logical Link Control and Adaptation Layer (L2CAP), 150, 152
link-level security, 290—292
links/link management (*See also* Logical Link Control and Adaptation Layer), 17, 77, 40—43, 74—75, 99, 121—148
    Active Member Address (AM_ADDR), 123, 138—139
    asymmetrical links, 56
    asynchronous connectionless (ACL) links, 17, 56, 74, 75, 77, 80, 82, 83, 104, 105, 123, 137, 143, 150, 188, 296
    authentication, 127—128, 127($t$)
    bonding, 187—188
    change of link key, 129—131, 130($t$)
    channel establishment, 189
    channel quality-driven change of data rate, 141—142, 142($t$)
    class of device, 184
    clock offset, 133, 133($t$)

connection establishment, 146, 146($t$), 189
connection-oriented links, 57, 74, 168—169
connectionless links, 56, 74, 75, 169—170, **169**, 179
connectivity modes, 185
current link keys, 130
detach, 137
device discovery, 187
device names, 184
discovery modes, 185
encryption, 131—133, 131($t$)
error behavior, 147—148, 195, 196, 207
establishment, 188—189
General Inquiry Access Codes (GIAC), 185, 186, 195—196, 209
general response messages, 127, 127($t$)
Generic Access Profile (GAP), 182
Generic Object Exchange Profile (GOEP), 188
hold mode, 137—138, 137($t$), 193
Home Application Programming interface (HAPi), 41—42
Home Audio-Video interoperability (HAVi) specification, 41—42
idle mode procedures, 186—187
JetSend wireless link, 40—41
keys, link keys, 127—128, 129—131, 234—236, 291, 301
Limited Inquiry Access Codes (LIAC), 185, 186
Link Control (LC), 82, 123, 182, 195, 208—210, 208—209($t$),238, 238—239($t$), 246, **246**, 254, 260, 267
Link Control Protocol (LCP), 108
Link Management Protocol (LMP), 74, 82, 83, 101, 1012, 102($t$), 122—123, **122**, 134—135, 135($t$), 182, 188, 195—196, 206—207, 207($t$), 210, 211, 222, 222—223($t$), 225, 237, 237($t$), 241, 248, 250, 253, 257, 258, 262, 268, 274, 280, 294
Link Manager (LM), 82, 83, 146, 226, 290), 297—298, 307—308
logical channels, 82—83
Logical Link Control and Adaptation Layer (L2CAP) protocol, 101, 102, 102($t$), 103, 104, 105, 119, 149—180, **150**, 182, 190, 192, 193, 194, 211, 220—222, 225, 226, 229, 230, 241, 248, 257, 262, 264, 268, 274, 280, 293, 297, 298, 300, 301, 305
lost links, link loss handling, 193
mandatory PDUs, 126
multi-slot packet control, 144, 145($t$)
name requests, 137, 137($t$), 187
optional PDUs, 126
packet data units (PDUs), 123—126, 124—126($t$)
paging mode, 145—146, 145($t$)
pairing, 128—129, 128($t$), 185, 199—200
park mode, 139—140, 193
personal identification number (PIN), 184
point-to-point connections, 67
policies, 195, 207
power control, 140—141, 141($t$), 193
protocols, 90—91
quality of service (QoS), 142, 142($t$)
radio link for Bluetooth, 21

# Index

security, 290, 291—292
semi-permanent link keys, 130
Serial Port Profile (SPP), 191—192
service discovery, 122—123, 183
slot offset, 133, 134(t)
sniff mode, 138—139, 139(t), 193
supervision of links, 146
supported features, 136, 136(t)
switching master/slave role, 136
symmetrical links, 56
synchronous connection-oriented (SCO) links, 17, 57, 74—75, 77, 80, 82, 83, 104, 105, 123, 143—144, 143(t), 150, 220—222, 226, 229, 241, 242, 243, 248, 253, 257
temporary link keys, 130—131
test modes, 147
timing accuracy, 134, 134(t)
Wireless User Group (WUG), 234—235
wireline-to-wireless connections, 106, **106**, 190
local area networks (LAN) (*See also* client/server architectures; wireless LANs), 6, 51, 83—86, **84**
local device (LocDev), 85—86, 197—198
logical channels, 82—83
Logical Link Control and Adaptation Layer (L2CAP) protocol, 101, 102, 102(t), 103, 104, 105, 119, 149—180, **150**
  actions, 161—163, 164, 165—168(t)
  address groups, 152
  automatic repeat request (ARQ), 155
  Baseband protocol, 150
  channel operational states, 163—164
  channels and channel identifiers (CIDs), 153—156, **154**, 163—164
  Command Reject packet, 171
  configuration parameter options, 173—175, 194, 204
  Configuration Request/Response packet, 171
  Configuration Response primitive, 177
  Configure primitive, 177
  Confirm actions, 156, 161—163
  Connect primitive, 176
  Connect Response primitive, 177
  Connection Request/Response packet, 171
  connection-oriented channels, 168—169
  connectionless channels, 169—170, **169**, 179
  Cordless Telephony Profile, 225, 226, 229, 230
  cyclic redundancy check (CRC), 155
  Data Link layer, OSI model, 150
  Dialup Networking Profile, 248
  Disable Connectionless Traffic primitive, 179
  Disconnect primitive, 177
  Disconnection Request/Response packet, 172
  duplexing, 152
  Echo Request/Response packet, 172
  Enable Connectionless Traffic primitive, 179
  Event Indication primitive, 176
  events, 164, 165—168(t), 157—161
  Fax Profile, 257
  File Transfer Profile, 268
  flush timeout, 174, 194, 204
  functions, 151—152
Generic Access Profile (GAP), 182
Generic Object Exchange Profile (GOEP), 211
Get Group Membership primitive, 179
Get Info primitive, 179
Group Add Member primitive, 178
Group Close primitive, 178
Group Create primitive, 178
Group Remove Member primitive, 178
Headset Profile, 241
Host Controller Interface (HCI), 154
Indication actions, 156, 161—163
Information Request/Response packet, 173
Intercom Profile, 220, 221—222
L2CAP-to-L2CAP data actions, 162
L2CAP-to-L2CAP data events, 158—159
L2CAP-to-L2CAP signaling actions, 161
L2CAP-to-L2CAP signaling events, 158—159
L2CAP-to-lower layer actions, 161
L2CAP-to-upper layer actions, 162—163
LAN Access Profile, 262, 264
Link Manager Protocol (LMP), 150, 152
lower-layer to L2CAP events, 158
mapping events to actions, 164, 165—168(t)
maximum transmission unit (MTU), 152, 155, 174, 194, 204
message sequencing, 157, **157**
multiplexing, 150, 151—152
Object Push Profile, 274
packets, 168—171, **168**, 173
Pending, 156
Ping primitive, 179
protocol data units (PDUs), 155, 205
protocol interfaces, 151, **151**
quality of service (QoS), 150, 152, 174—175, 194, 204
Radio Frequency Communication (RFCOMM) protocol, 151, 152
Read primitive, 178
Request events, 156, 157—161
Request Path packet, 175
Response events, 156, 157—161
Response Path packet, 175
security, 293, 297, 298, 300, 301, 305
segmentation and reassembly (SAR) packets, 150, 152, 154—155
Serial Port Profile (SPP), 190, 192, 193, 194
Service Discovery Application Profile (SDAP), 196, 203—206
Service Discovery Protocol (SDP), 151, 152
service primitives, 176—179
signaling, 170—173, **170**, 194, 203
state machine, 155—157, **156**
Synchronization Profile, 280
Telephony Control Specification Binary (TCS BIN) protocol, 151, 152
timer events, 160—161
upper-layer to L2CAP events, 159—160
Write primitive, 177
Lucent, 15

## M

MAC addresses, 76
Management Entity (ME), LAN Access Profile, 262, 267—268
management, network, 93
mandatory PDUs, 126
mapping addresses, 98
market for Bluetooth, 45—46
master/slave devices (*See also* links/link management), 23—25, **24**, **69**, **70**
maximum transmission unit (MTU), 98, 152, 155, 174, 194, 204
Media Access Control (MAC) addresses, 76
Media Access Control (MAC) protocol, 85
media access management, 99
media, physical, 100
medium access control (MAC), 18
messaging services, 93, 96
    Intercom Profile, 223
    Object Push Profile, 277
    Service Discovery Application Profile (SDAP), 202
    Synchronization Profile, 283
Microsoft, 15, 33, 37
Modem Status Control (MSC), 193
modems, 106, 107, 119, 248
modes of security, 290
modulation, 17, 50, 57—60, **58**, **59**, 218
Motorola, 15, 345
Moving Pictures Experts Group (MPEG), 20
MPEG-1/2/3/7/21 video encoding, 20(f)
MPEG-4 video encoding, 19, 20
multi-slot packet control, 144, 145(t)
multimedia, 20(f)
multipath conditions, spread-spectrum, 63
multiplexing, 97
    Logical Link Control and Adaptation Layer (L2CAP), 150—152
    security, 299, 300

## N

name discovery, Generic Access Profile (GAP), 187
name requests, 137, 137(t)
network addresses, 76
Network Control Protocols (NCPs), 108
Network layer, OSI model, 91(t), 97—98, **98**
networking using Bluetooth, 17—18, 46
NeWeb, 16
Nokia, 14, 25, 110, 313, 345
Nordic Mobile Telephone (NMT), 311, 316
North American Digital Cellular, 71
North American GSM Alliance, 335
Notes, Object Push Profile, 278

## O

Object Exchange mode, Object Push Profile, 276
Object Exchange Protocol (OBEX), 102, 103(t), 110—111
    File Transfer Profile, 268, 270, 272, 272(t)
    Generic Access Profile (GAP), 182
    Generic Object Exchange Profile (GOEP), 210, 211, 212—214
    IrOBEX, 211, 213
    Object Push Profile, 274, 275, 278
    security, 294
    Synchronization Profile, 280, 282—284, 285(t)
Object Push function, Object Push Profile, 275—277
Object Push Profile, 215, 273—279, **274**, **275**
    Business Card Exchange function, 275, 277
    Business Card Pull function, 275, 277
    client/servers, 276
    content formats, 277—278
    features, 277, 277(t)
    Generic Object Exchange Profile (GOEP), 210, 214, 215, 273, 274, 275
    IrOBEX protocol, 274
    Link Management Protocol (LMP), 274
    Logical Link Control and Adaptation Layer (L2CAP), 274
    messaging, 277
    Object Exchange mode, 276
    Object Exchange Protocol (OBEX), 274, 275, 278
    Object Push function, 275—277
    phone book applications, 277
    Radio Frequency Communication (RFCOMM) protocol, 274
    Service Discovery Database, 278—279, 279(t)
    Service Discovery Protocol (SDP), 274, 278—279, 279(t)
    vCalendar, 277
    vCard, 277
    vMessage, 277
    vNotes, 278
odd parity, 55
open architecture, 91, 102
Open Systems Interconnection (OSI) model, 91—101, 91(t)
    Application layer, 91, 92—93, **93**
    Data Link layer, 92, 98—99, **99**, 150
    Network layer, 91, 97—98, **98**
    Physical layer, 92, 99—100, **100**
    Presentation layer, 91, 94—95, **94**
    Session layer, 91, 95—96, **95**
    Transport layer, 91, 96—97, **97**
optional PDUs, 126
origin of Bluetooth, 14—16

## P

packet data units (PDUs), 123—126, 124—126(t)
packets and packet switching, 66—68, 75—82, **77**, 96, 97, 312
    access code, 77—78, **78**
    ACK/NAK signals, 80
    active member address, 79
    addressing, 76—77
    asynchronous connectionless (ACL) packets, 77, 80
    automatic repeat request field, 80
    channel access code (CAC), 78
    CRC Code generator, 82
    cyclic redundancy code (CRC), 80, 82

# Index

Data—High Rate (DH) packets, 141
Data—Medium Rate (DM) packets, 141
device access code (DAC), 78
flow field of header, 80
header error check (HEC) field, 81
header, 79—81, **79**
inquiry access code (IAC), 78
Logical Link Control and Adaptation Layer
   (L2CAP), 168—170, **168**, 170—173
master/slave devices, 79
maximum transmission units (MTU), 152
multi-slot packet control, 144, 145(t)
packet data units (PDUs), 123—126, 124—126(t)
payload, 81—82
segmentation and reassembly (SAR) packets,
   150—155
sequence number field, 80—81
synchronous connection-oriented (SCO) packets, 77,
   80
type field of header, 80
page scan, 185, 210
paging, 145—146, 145(t)
   Serial Port Profile (SPP), 196
   Service Discovery Application Profile (SDAP), 210
Paging Channel (PCH), Global System for Mobile
   (GSM), 322
Pairable mode, Synchronization Profile, 281
pairing, 128—129, 128(t), 291
   Generic Access Profile (GAP), 185
   Service Discovery Application Profile (SDAP), 199—200
Palm, 37, 87
parallel transmissions, 50—52, **53**
parity bits, 54, **54**
park mode, 79, 139—140, 193
payload key/payload key generator, 52, **53**
payload, Bluetooth packets, 81—82
payment systems using Bluetooth, 32—33, **33**
PC cards, 16
PCS 1900 networks, 71
peak bandwidth, 174
Pending, Logical Link Control and Adaptation Layer
   (L2CAP), 156
performance characteristics of Bluetooth products,
   23(t)
personal area networks (PANs), 14, 22, 24
Personal Communications Service (PCS), 314
Personal Digital Assistants (PDA), 50, 88, 111, 310
personal identification number (PIN), 291, 294, 306
   Generic Access Profile (GAP), 184
   Headset Profile, 243
   Service Discovery Application Profile (SDAP),
      196—197
personal information management (PIM) applications,
   Synchronization Profile, 279—280
phone book applications, Object Push Profile, 277
Physical layer, OSI model, 92(t), 99—100, **100**
piconet, 23—25, **24**, 69, 70, 69, 236
Ping primitive, 179
point-to-point connections, 67

Point-to-Point Protocol (PPP), 102(t), 107—108, 119
   LAN Access Profile, 261, 262, 264
polling, 142
power control, 140—141, 141(t), 193
Presentation layer, OSI model, 91(t), 94—95, **94**
presentations using Bluetooth, 26
primitives (See service primitives)
printers and printing, 28
   JetSend wireless link, 40—41
   parallel transmission, 51—52
privacy issues, 290
problems with Bluetooth, 44—45
profiles, general (See also usage models), 181—216
   Cordless Telephony Profile, 225—241, **226**, **228**
   Dialup Networking Profile, 248—256, **249**
   Fax Profile, 256—261, **261**, **262**
   File Transfer Profile, 210, 214, 215, 268—273, **269**
   Generic Access Profile (GAP), 182—189, **183**, 211, 218,
      220, 225, 239—241, 239—241(t), 243, 247, 247(t),
      248(t), 248, 250, 254—255, 256(t), 260—262,
      265—266, 265(t), 268, 286, 294
   Generic Object Exchange Profile (GOEP), 182,
      210—214, **212**, 268, 273, 274, 275, 280, 286
   Headset Profile, 241—248, **242**
   IrDA Interoperability Specification, 215—216
   LAN Access Profile, 261—268, **262**, **263**
   Object Exchange Protocol (OBEX), 182
   Object Push Profile, 210, 214, 215, 273—279, **274**, **275**
   Serial Port Profile (SPP), 182, 190—196, **190**, 241, 246,
      248, 256, 258, 262, 268, 280, 286
   Service Discovery Application Profile (SDAP), 182,
      196—210, **199**, 286
   Synchronization Profile, 210, 214, 215, 279—286, **280**,
      **281**
   usage models, 217—287
promiscuous mode operation, 76
Promoter Group, Bluetooth, 15
protocol data units (PDUs), 155, 205
protocols, 90—91
   adopted protocols, 102(t), 103, 107—118
   AT-Commands protocol, 102(t)
   Baseband protocol, 102(t), 104, 119, 210, 211, 225, 250,
      258, 294, 297—298
   Bluetooth protocol stack, 101—103, **101**, 102—103(t)
   cable replacement protocol, 102(t), 103, 105—107
   content formats, 116—118
   core protocols, 102(t), 103—105
   File Transfer Protocol (FTP), 103
   High-level Data Link Control (HDLC) protocol, 108
   Host Controller Interface (HCI), 103, 154
   Hypertext Transfer Protocol (HTTP), 103, 110,
      112—113, 211
   Infrared Mobile Communications (IrMC) protocol,
      103(t)
   Internet Protocol (IP), 108, 109—110, 312, 324
   IrOBEX protocol, 268, 274, 280
   Link Control Protocol (LCP), 108
   Link Management Protocol (LMP), 74, 82, 83, 101,
      1012, 102(t), 122—123, **122**, 134—135, 135(t), 182,

188, 195—196, 206—207, 207(*t*), 210, 211, 222, 222—223(*t*), 225, 237, 237(*t*), 241, 248, 250, 253, 257, 258, 262, 268, 274, 280, 294
Logical Link Control and Adaptation Layer (L2CAP), 101—105, 102(*t*), 119, 149—180, **150**, 190, 192, 193, 194, 211, 220—222, 225, 226, 229, 241, 248, 257, 262, 264, 268, 274, 280
Network Control Protocols (NCPs), 108
Object Exchange Protocol (OBEX), 102, 103(*t*), 110—111, 210—214, 268, 270, 272, 272(*t*), 274, 275, 278, 280, 282—284, 285(*t*), 294
open architecture, 102
Point-to-Point Protocol (PPP), 102(*t*), 107—108, 119, 261, 262, 264
Radio Frequency Communication (RFCOMM) protocol, 102(*t*), 103, 105—106, **106**, 110, 119, 120, 151, 152, 182, 189, 190, 192, 193, 211, 241, 243, 246, 248, 257, 258, 261, 262, 264, 268, 274, 280, 283, 293, 298, 299, 306
security, 298—299, **298, 299**
Service Discovery Protocol (SDP), 102(*t*), 105, 151, 152, 192, 195, 197, 202, 204—206, 211, 225, 236, 236(*t*), 241, 248, 257, 259, 260, 260(*t*), 262, 266—268, 267(*t*), 273—274, 273(*t*), 278—280, 279(*t*), 284, 285(*t*)
TCP/IP, 20, 76, 87, 90, 96, 100, 102(*t*), 108—110
telephony control protocols, 102(*t*), 103, 107
Telephony Control Specification Binary (TCS BIN), 101, 102(*t*), 107, 103, 151, 152, 189, 219 223, 224(*t*), 225, 226, 230
Transmission Control Protocol (TCP), 108, 109, 112—113
usage models, 101, 118—120
User Datagram Protocol (UDP), 102, 102(*t*), 108, 109
vCalendar protocol, 103(*t*), 115, 116, **117**,
vCard protocol, 103(*t*), 115, 116—118, **118**
WAP Applications Environment (WAE), 113—115, **114**
Wireless Application Environment (WAE) protocol, 103(*t*)
Wireless Application Protocol (WAP), 36—37, 102, 103(*t*), 110, 111—115, **112**, 120
Wireless Transport Layer Security (WTLS) protocol, 113
Public Switched Telephone Network (PSTN), 9, 12, 28, 57, 66, 119, 225, 264, 312
pulling data object, Generic Object Exchange Profile (GOEP), 213
pulse code modulation (PCM), 17, 57—58, **58**, 218
Puma Technology, 346
pushing data object, Generic Object Exchange Profile (GOEP), 213
Pyramid Communication AB, 346

## Q

Q-Zone volume control, 34
Qualcomm, 314
qualification programs, 45
quality of service (QoS), 105, 142, 142(*t*), 174—175, 312—313

Cordless Telephony Profile, 229
delay variation, 175
Intercom Profile, 218
latency, 175
Logical Link Control and Adaptation Layer (L2CAP), 150, 152, 194, 204
peak bandwidth, 174
token bucket size, 174
token rate, 174
quality of service (QoS)
quantizing noise, 60
quarter common intermediate format (QCIF), 20

## R

radio frequency (RF), Home Radio Frequency (HomeRF) networks and, 8—13
Radio Frequency Communication (RFCOMM) protocol, 102(*t*), 103, 105—106, **106**, 110, 119, 120, 151, 152
Dialup Networking Profile, 248
Fax Profile, 257, 258
File Transfer Profile, 268
Generic Access Profile (GAP), 182, 189
Generic Object Exchange Profile (GOEP), 211
Headset Profile, 241, 243, 246
LAN Access Profile, 261, 262, 264
Object Push Profile, 274
security, 293, 298, 299, 306
Serial Port Profile (SPP), 190, 192, 193
Synchronization Profile, 280, 283
radio link for Bluetooth, 21
Random Access Channel (RACH), Global System for Mobile (GSM), 322
range (distance) of Bluetooth devices, 13—14, 25, 44, 340
Read primitive, 178
Real-time Transfer Protocol (RTP), 20
receiver signal strength indicator (RSSI) values, 140—141
registration of devices/services, 295—296, 299—300, **300**, 306—307
Remote Access Server (RAS), LAN Access Profile, 262
remote controls, 43
remote device (RemDev), 85—86, 197—198
remote synchronization of data, 28
Request events, Logical Link Control and Adaptation Layer (L2CAP), 156, 157—161,
Request Path packet, 175
Request to Send/Clear to Send (RTS/CTS), 193
Response events, Logical Link Control and Adaptation Layer (L2CAP), 156, 157—161
Response Path packet, 175
RF hop frequency, 70, **71**
routers/routing, 76, 98
RS-232 serial port emulation, 105—106, 190
RTCON (real time connection) links, 19

## S

safety issues, 22
Salutation Architecture, 86—88
scanners, 33, 110—111

# Index

second-generation wireless, 314—325
secure electronic transactions (SET), 36
security, 22, 25—26, 50, 95, 289—308
    access control, 293, 301
    architecture overview, 294—296, **295**
    asynchronous connectionless (ACL) links, 296
    authentication, 25, 26, 127—128, 127(*t*), 290—298, 307, 308
    authorization, 296
    Baseband protocol, 294, 297—298
    Challenge Handshake Authentication (CHAP), 264
    combination key, 291
    connection setup, 296—297, **297**
    connectionless L2CAP, 301
    Cordless Telephony Profile, 240, 240(*t*)
    credit card security, 35
    Dialup Networking Profile, 250—251, 255
    encryption, 131—133, 131(*t*), 290, 291, 292, 294, 296, 307—308
    Europay-MasterCard-Visa (EMV) protocol, 36
    external key management, 301
    External Security Control Entity (ESCE), 294, 306
    Fax Profile, 261
    File Transfer Profile, 269, 270
    Generic Access Profile (GAP), 182, 186, 294
    Generic Object Exchange Profile (GOEP), 212
    Host Controller Interface (HCI), 297—298, 307—308
    implementation of security, 293—294
    interface to L2CAP, 305
    keys, link keys, 291, 301
    LAN Access Profile, 265
    legacy applications, 292—293
    levels of security, 296
    Link Management Protocol (LMP), 294, 299
    Link Manager (LM), 290, 297—298, 307—308
    link-level, 290—292
    Logical Link Control and Adaptation Layer (L2CAP), 293, 297, 298, 300, 301, 305
    modes, 290
    multiplexing, 299, 300
    name request to remote device, 307
    Object Exchange Protocol (OBEX), 294
    pairing, 291
    personal identification number (PIN), 291, 294, 306
    privacy issues, 290
    protocol stack handling, 298—299, **298**, **299**
    Radio Frequency Communication (RFCOMM) protocol, 293, 298, 299, 306
    registration of devices/services, 295—296, 299—300, **300**, 306—307
    secure electronic transactions (SET), 36
    Security Manager, flowcharts of, 301, **302—304**, 305, 305(*t*)
    Serial Port Profile (SPP), 191
    spoofing, 290
    storage of security-related information, 294
    trusted vs. untrusted devices, 25—26, 201—202, 292—293, 305, 305(*t*)
    unit key, 291
wireless access protocol (WAP), 36—37, **36**
Wireless Transport Layer Security (WTLS) protocol, 113
wireless wallets, 36
Security Manager, flowcharts of, 301, **302—304**, 305, 305(*t*)
segmentation and reassembly (SAR) packets, 150—155
segmentation, message, 96
semi-permanent link keys, 130
sequence number field, 80—81
sequencing, frames, 99
Serial Infrared (SIR) data link, 44
serial port emulation
    Fax Profile, 257
    Headset Profile, 243
Serial Port Profile (SPP), 182, 190—196, **190**, 286
    accept link/virtual connection, 192
    Acceptor devices, 191
    application-level procedures, 191—193
    bonding, 191
    Data Terminal Ready/Data Set Ready (DTR/DSR), 193
    Dialup Networking Profile, 248, 251
    error behavior, 195, 196
    establish link/virtual connection setup, 191—192
    Fax Profile, 256, 258
    File Transfer Profile, 268
    Flow Control On/Flow Control Off (FCON/FCOFF), 193
    General Inquiry Access Codes (GIAC), 195—196
    Headset Profile, 241, 246
    Initiator devices, 191
    inquiry scan, 196
    LAN Access Profile, 262
    Link Control (LC), 195
    link loss handling, 193
    Link Manager (LM), 195—196
    link policy, 195
    Logical Link Control and Adaptation Layer (L2CAP), 190, 192, 193, 194
    Modem Status Control (MSC), 193
    paging, 196
    power mode, 193
    Radio Frequency Communication (RFCOMM) protocol, 190, 192, 193
    register service record in local SDP database, 192—193
    Request to Send/Clear to Send (RTS/CTS), 193
    RS-232 control signals, 193—194
    RS-232 serial port emulation, 190
    security, 191
    Service Discovery Protocol (SDP), 192, 195
    signaling, 194
    Synchronization Profile, 280
    Transmitter On/Transmitter Off (XON/XOFF), 193
serial transmissions, 50—52, **52**, 83—86, **84**, 105
service discovery, 86—88, 122—123
    device discovery, 187
    Dialup Networking Profile, 254, 254(*t*)

# Index

discovery modes, 185
General Inquiry Access Codes (GIAC), 185, 186, 195—196, 209
Generic Access Profile (GAP), 183
idle mode, 186—187
Limited Inquiry Access Codes (LIAC), 185, 186
name discovery, 187
Salutation Architecture, 86—88
Serial Port Profile (SPP), 192, 195
service discovery application (SrvDscApp), 197, 200—202
Service Discovery Application Profile (SDAP), 196—210, **199**
Service Discovery Database, 278—279, 279(t), 284, 285(t), 286(t), 284
Service Discovery Protocol (SDP), 102(t), 105, 151, 152, 192, 195, 197, 202, 204—206, 211, 225, 236, 236(t), 241, 248, 257, 259, 260, 260(t), 262, 266—268, 267(t), 273—274, 273(t), 278—280, 279(t), 284, 285(t)
Service Location Protocol (SLP), 87
service records (SDP), 192—193
trusted vs. untrusted devices, 201—202
service discovery application (SrvDscApp), 197, 200—202
Service Discovery Application Profile (SDAP), 182, 196—210, **199, 203**, 286
client/server roles, 197—198
error behavior, 207, 210
General Inquiry Access Codes (GIAC), 209
inquiry/inquiry scan, 209
Link Control (LC), 208—210, 208—209(t)
Link Manager (LM), 206—207, 207(t)
link policy, 207
local device (LocDev), 197—198
Logical Link Control and Adaptation Layer (L2CAP), 196, 203—206
message sequence, 202
page scan, 210
paging, 210
pairing, 199—200
personal identification number (PIN), 196—197
protocol data units (PDUs), 205
remote device (RemDev), 197—198
service discovery application (SrvDscApp), 197, 200—202
Service Discovery Protocol (SDP), 197, 202, 204—206
service primitives, 205
signaling, 203
trusted vs. untrusted devices, 201—202
Service Discovery Database
Object Push Profile, 278—279, 279(t)
Synchronization Profile, 284, 285(t), 286(t)
Service Discovery Protocol (SDP), 85—86, 102(t), 105, 151, 152
Cordless Telephony Profile, 225, 236, 236(t)
Dialup Networking Profile, 248
Fax Profile, 257, 259, 260, 260(t)
File Transfer Profile, 268, 273, 273(t)
Generic Object Exchange Profile (GOEP), 211
Headset Profile, 241

LAN Access Profile, 262, 266—267, 267(t)
Object Push Profile, 274, 278—279, 279(t)
Serial Port Profile (SPP), 192, 195
Service Discovery Application Profile (SDAP), 197, 202, 204—206
service records, 192—193
Synchronization Profile, 280, 284, 285(t)
Service Location Protocol (SLP), 87
service primitives, 176—179
Configuration Response primitive, 177
Configure primitive, 177
Connect primitive, 176
Connect Response primitive, 177
Disable Connectionless Traffic primitive, 179
Disconnect primitive, 177
Enable Connectionless Traffic primitive, 179
Event Indication primitive, 176
Get Group Membership primitive, 179
Get Info primitive, 179
Group Add Member primitive, 178
Group Close primitive, 178
Group Create primitive, 178
Group Remove Member primitive, 178
Ping primitive, 179
Read primitive, 178
Service Discovery Application Profile (SDAP), 205
Write primitive, 177
service records (SDP), 192—193
Session layer, OSI model, 91(t), 95—96, **95**
sessions, 96
Shared Wireless Access Protocol (SWAP), 9—11
signaling, 170—173, **170**
Command Reject packet, 171
Configuration Request/Response packet, 171
Connection Request/Response packet, 171
Disconnection Request/Response packet, 172
Echo Request/Response packet, 172
Information Request/Response packet, 173
Logical Link Control and Adaptation Layer (L2CAP), 194, 203
Service Discovery Application Profile (SDAP), 203
signaling, 170—173, **170**, 170
simplex transmissions, Logical Link Control and Adaptation Layer (L2CAP), 152
slot offset, 133, 134(t)
smart cards, 34, **35**
sniff mode, 138—139, 139(t), 193
Socket Communications, 35
Space Division Multiple Access (SDMA), 327
specifications, 16
speed of transmissions, Bluetooth vs. infrared, 4
spoofing, 25, 290
spread-spectrum, 8, 9(t), 21, 60—66, **62**, 88
code division multiple access (CDMA), 61
direct sequencing, 63, 64
frequency-hopping, 61, 63, 65—66
interception of signals, 63
interference, 63, 66
jamming, 63

# Index

multipath conditions, 63
spreading technique, 62—63
time division multiple access (TDMA), 61
Standalone Dedicated Control Channel (SDCCH),
    Global System for Mobile (GSM), 321
start/stop bits, 53—55, **54**
state machine, Logical Link Control and Adaptation
    Layer (L2CAP), 155—157, **156**
station IDs, 76
subnets, 98, 110
subscriber identity modules (SIMs), 34
Sun Microsystems, 38
supervision of links, 146
supported features, 136, 136(*t*)
synchronizing data, 27—28
symmetrical links, 56
Sync Command Service, Synchronization Profile, 286
synchronization (sync) bits, 53—54
Synchronization Profile, 279—286, **280, 281**
    Connectable mode, 281
    content formats, 283
    features, 283—284, 284(*t*)
    General Discoverable mode, 281
    General Sync mode, 281—282
    Generic Object Exchange Profile (GOEP), 210, 214, 215, 280
    Initialization Sync mode, 281—282
    IrMC client/servers, 280—283
    IrOBEX protocol, 280
    Limited Discoverable mode, 281
    Link Management Protocol (LMP), 280
    Logical Link Control and Adaptation Layer (L2CAP), 280
    Object Exchange Protocol (OBEX), 280, 282—284, 285(*t*)
    Pairable mode, 281
    personal information management (PIM) applications, 279—280
    Radio Frequency Communication (RFCOMM) protocol, 280, 283
    Serial Port Profile, 280
    Service Discovery Database, 284, 285(*t*), 286(*t*)
    Service Discovery Protocol (SDP), 280, 284, 285(*t*)
    Sync Command Service, 286
    vCalendar, 283
    vCard, 283
    vMessage, 283
    vNote, 283
synchronous connection-oriented (SCO) links, 17, 57, 74—75, 77, 80, 82, 83, 104, 105, 123, 143—144, 143(*t*), 150, 220—222, 226, 229, 241, 242, 243, 248, 253, 257
synchronous transmissions, 50, 53, 57—60, 67—68, 74—75, 77, 80, 82, 83, 104, 105, 123, 143—144, 143(*t*), 150, 220—222, 226, 229, 241, 242, 243, 248, 253, 257
    continuously variable slope delta (CVSD) modulation, 58—60, **59**
    data communications, 57

pulse code modulation (PCM), 57—58, **58**
quantizing noise, 60
synchronous connection-oriented (SCO) links, 57
user synchonrous (US), 82, 83
voice communications, 57—58

## T

T-carrier, 57
3COM Corp., 15, 343
3G (*See* global 3G wireless)
TCP/IP, 20, 76, 87, 90, 96, 100, 102(*t*), 108—110
Technical Standard (TS) 07.10, 105
telephony applications, 10, 18, 50, 71—72, 105, 119
    Cordless Telephony Profile, 225—241, **226, 228**
    Intercom Profile, 218
    Telephony Control Specification Binary (TCS BIN), 101, 102(*t*), 107, 151, 152, 189, 219, 223, 224(*t*), 225, 226, 230
    wireless telephony applications (WTA), 115
    Telephony Control Specification Binary (TCS BIN), 101, 102(*t*), 107, 151, 152, 189, 219, 223, 224(*t*), 225, 226, 230
    Cordless Telephony Profile, 225, 226, 230
    Generic Access Profile (GAP), 189
    Intercom Profile, 219, 223, 224(*t*)
television, 19—21, 41—42
temporary link keys, 130—131
terminals, Cordless Telephony Profile, 227, 230—232
test modes, 147
Third Generation (*See* global 3G wireless)
Time Division Multiple Access (TDMA), 10, 11, 61, 71—73, **72**, 314—318, 327, 333, 338—339
Time Division Duplexing (TDD), 68—73, 333
    Global System for Mobile (GSM), 318, 321
Time Division Multiplexing (TDM), 23
timer events, Logical Link Control and Adaptation Layer (L2CAP), 160—161
timing accuracy, 134, 134(*t*)
token bucket size, 174
token rate, 174
Token Ring, 76
token ring networks, 85
topology, 23—25, **24**
Tornado, 87
Toshiba Corp., 14, 19, 20, 346
Total Access Communications System (TACS), 71, 311, 316
traffic control, 97, 98, 99
Transmission Control Protocol (TCP), 108, 109, 112—113
Transmitter On/Transmitter Off (XON/XOFF), 193
Transport layer, OSI model, 91(*t*), 96—97, **97**
travel using Bluetooth, 29, 31—32
TravelNote Connect, 29
trusted vs. untrusted devices, 25—26, 201—202, 292—293, 305, 305(*t*)

## U

Ultimate Headset (*See* Headset Profile)
UMTS Terrestrial Radio Access (UTRA), 329—330

# Index

unit key, 291
Universal Mobile Telecommunications System (UMTS), 329—334
Universal Serial Bus (USB), 51
Universal Wireless Communications Consortium (UWCC), 334, 338
usage models (*See also* profiles, general), 101, 118—120, 217
    Cordless Telephony Profile, 225—241, **226**, **228**
    Dialup Networking Profile, 248—256, **249**
    Fax Profile, 256—261, **261**, **262**
    File Transfer Profile, 189, 210, 214, 215, 268—273, **269**
    Generic Access Profile (GAP), 182—189, **183**, 211, 225, 239—241, 239—241(*t*), 243, 246—250, 247(*t*), 248(*t*), 254—256, 256(*t*), 258, 260—262, 265—266, 268, 286, 294
    Generic Object Exchange Profile (GOEP), 182, 188, 210—214, **212**, 268, 273, 274, 275, 280, 286
    Headset Profile, 241—248, **242**
    Intercom Profile, 119, 218—225, **218**, **219**
    Internet Bridge usage model, 110, 119
    IrDA Interoperability Specification, 188, 215—216
    LAN Access Profile, 261—268, **262**, **263**
    Object Push Profile, 210, 214, 215, 273—279, **274**, **275**
    Serial Port Profile (SPP), 182, 190—196, **190**, 241, 246, 248, 251, 256, 258, 262, 268, 280, 286
    Service Discovery Application Profile (SDAP), 182, 196—210, **199**, 286
    Synchronization Profile, 210, 214, 215, 279—286, **280**, **281**
user agents, WAE, 115
user asynchronous (UA), 82, 83
User Datagram Protocol (UDP), 96, 102, 102(*t*), 108, 109
user isochronous (UI), 82, 83
user synchonrous (US), 82, 83

## V

vCalendar, 103(*t*), 115, 116, **117**, 290
    Object Push Profile, 277
    Synchronization Profile, 283
vCard, 103(*t*), 115, 116—118, **118**, 290
    Object Push Profile, 277
    Synchronization Profile, 283
vending machines, 43
Versit Consortium, 116
Very Fast Infrared (VIFR) protocol, 4
video over Bluetooth, 19—21
virtual circuits/devices, 93, 312
Visa International, 35
vMessage
    Object Push Profile, 277
    Synchronization Profile, 283
vNote
    Object Push Profile, 278
    Synchronization Profile, 283

voice over Bluetooth (*See also* telephony applications), 18—19, 30, 46, 105
volume control of electronic devices (Q-Zone), 34
VXWorks, 87

## W

W-CDMA/W-CDMA/NA, 334—335, 337—338
WAP Applications Environment (WAE), 113—115, **114**
WAP portals, 111
web pages, 20(f)
wide area networks (WANs), 51
WIDEBAND, 335—337
Windows, 87
Wireless Application Environment (WAE) protocol, 103(*t*), 103
Wireless Application/Access Protocol (WAP), 36—37, 102—120, 103(*t*), 110—115, **112**
    content generators, 115
    encoding, standard content, 115
    hidden computing, 113
    JavaScript, 115
    user agents, WAE, 115
    vCalendar protocol, 115, 116, **117**
    vCard protocol, 115, 116—118, **118**
    WAP Applications Environment (WAE), 113—115, **114**
    wireless telephony applications (WTA), 115
    Wireless Transport Layer Security (WTLS) protocol, 113
    Wireless Markup Language (WML), 115
    Wireless Markup Language Script (WMLScript), 115
Wireless Domino Access, 46
wireless information devices (WIDs), 50
wireless LANs, 6—8, **7**, 46
    client/server architecture, 83—86, **84**
    interference, 21—22
    LAN adapters, 7
    personal area networks (PANs), 22
    spread-spectrum, 60—66, **62**
wireless markup language (WML), 37
Wireless Multimedia and Messaging Services (WIMS), 334
wireless telephony applications (WTA), 115
Wireless Transport Layer Security (WTLS) protocol, 113
Wireless User Group (WUG), Cordless Telephony Profile, 229, 234—235
wireless wallets, 36
wireline-to-wireless connections, 4—5, **5**, 106, **106**, 190
Wireless Markup Language (WML), 112, 115
Wireless Markup Language Script (WMLScript), 115
World Radio Conference (WRC), 326
Write primitive, 177

## X

X.25, 97
Xtraworx, 346

# About the Author

**Nathan Muller** is an independent consultant in Sterling, Virginia, specializing in advanced technology marketing, research and education. In his 30 years of industry experience, he has written extensively on many aspects of computers and communications, having published 19 books—including three encyclopedias—and over 2,000 articles about computers and communications in over 50 publications worldwide.

In addition, he is a regular contributor to the Gartner Group's Datapro Research Reports. He has participated in market research projects for Dataquest, Northern Business Information, and Faulkner Technical Reports. He also does custom projects for technology-oriented clients in the computer, telecommunications, and health care industries.

Muller has an M.A. in Social and Organizational Behavior from George Washington University. He has held numerous technical and marketing positions with such companies as Control Data Corporation, Planning Research Corporation, Cable & Wireless Communications, ITT Telecom and General DataComm Inc. His email address is nmuller@loudoun.com.